Offshore Wind Farms

Offshore Wind Farms

Special Issue Editors

María Dolores Esteban
José-Santos López-Gutiérrez
Vicente Negro Valdecantos

MDPI • Basel • Beijing • Wuhan • Barcelona • Belgrade • Manchester • Tokyo • Cluj • Tianjin

Special Issue Editors

María Dolores Esteban
Universidad Politécnica de
Madrid
Spain

José-Santos López-Gutiérrez
Universidad Politécnica de
Madrid
Spain

Vicente Negro Valdecantos
Universidad Politécnica de
Madrid
Spain

Editorial Office
MDPI
St. Alban-Anlage 66
4052 Basel, Switzerland

This is a reprint of articles from the Special Issue published online in the open access journal *Journal of Marine Science and Engineering* (ISSN 2077-1312) (available at: https://www.mdpi.com/journal/jmse/special_issues/offshore_wind_farms).

For citation purposes, cite each article independently as indicated on the article page online and as indicated below:

LastName, A.A.; LastName, B.B.; LastName, C.C. Article Title. *Journal Name* **Year**, *Article Number*, Page Range.

ISBN 978-3-03928-562-4 (Pbk)
ISBN 978-3-03928-563-1 (PDF)

Cover image courtesy of Kim Hansen.

Contents

About the Special Issue Editors

María Dolores Esteban (Ph.D.) is Associate Professor in Coastal Engineering at the Universidad Politécnica de Madrid and in Coastal Engineering and Renewable Energies at the Universidad Europea. She has approximately 15 years of experience in the fields of maritime engineering and renewable energies as an engineer in a private company and as a lecturer and researcher at the university. She has around 50 technical papers in SCI indexed journals and numerous oral presentations in international conferences. Her main fields of research are coastal and maritime engineering, harbor and ports, beaches, loads actions and effects in maritime structures, renewable energies like offshore wind, wave, and currents. She is very active in R&D projects, being an important asset to her research group the Environmental, Coast and Ocean Research Laboratory (ECOREL) at the Universidad Politécnica de Madrid.

José-Santos López-Gutiérrez (Ph.D.) is Associate Professor in Coastal Engineering at the Universidad Politécnica de Madrid. He has more than 25 years of experience in the fields of maritime engineering and renewable energies as a lecturer and researcher at the university. He has supervised 5 Ph.D. dissertations. He has around 60 technical papers published in SCI indexed journals and more than 60 oral presentations in national and international conferences. He is responsible of the research line on marine energies within the group of researchers of the Environmental, Coast and Ocean Research Laboratory (ECOREL) at Universidad Politécnica de Madrid. Apart from research fields related to marine energies, such as load and effect actions on marine structures for offshore wind installations, and wave and current exploitation, other research fields in which he is involved include coastal and maritime engineering, ports, nd beaches.

Vicente Negro Valdecantos (Ph.D.) is a Full Professor in Coastal Engineering He is the manager of the Environmental, Coast and Ocean Research Laboratory (ECOREL) at Universidad Politécnica de Madrid. He has supervised 13 Ph.D. dissertations. He has around more than 70 technical papers in SCI indexed journals and more than 40 scientific and dissemination articles at national level. He has more than 40 contributions in national and international conferences with peer arbitration. He is an epert in marine sciences in the Program of the Directorate General XII of Research in Brussels, Drafting Commission of the ROM (Puertos del Estado), being an Editorial Member of the ROM 0.3/91, Environmental Actions I: Waves; ROM 0.0/2002, General procedure and calculation bases in the maritime and port works project; ROM 1.0/09, Recommendations for the design of Breakwaters and ROM 2.0/11, Recommendations for the project and execution of Docking and Mooring Works.

Journal of
Marine Science and Engineering

Editorial

Offshore Wind Farms

María Dolores Esteban [1,2,*], José-Santos López-Gutiérrez [1,2] and Vicente Negro [1,2]

[1] Grupo de Investigación de Medio Marino, Costero y Portuario, y Otras Áreas Sensibles, Universidad Politécnica de Madrid, 28670 Madrid, Spain; josesantos.lopez@upm.es (J.-S.L.-G.); vicente.negro@upm.es (V.N.)

[2] Dept. Ingeniería Civil: Hidráulica, Energía y Medio Ambiente. C/Profesor Aranguren 3, Universidad Politécnica de Madrid (UPM), CP 28040, Madrid, Spain

* Correspondence: mariadolores.esteban@upm.es

Received: 9 February 2020; Accepted: 9 February 2020; Published: 14 February 2020

Keywords: offshore wind farm; marine energy; foundations; wind resource; wind turbine generators; electrical connection; numerical models; physical models; development; design and construction; operation and maintenance

---

In 2018, we were approached by the editorial team of the Journal of Marine Science and Engineering (MDPI editorial) to act as guest editors of a Special Issue related to offshore wind energy. This invitation was welcomed with great enthusiasm by the Guest Editorial Team. As soon as possible we started working on the project, with the great support of the Main Editor who has guided us to achieve a Special Issue that has exceeded our initial expectations. This Special Issue is entitled "Offshore Wind Farms". It is focused on the 7th Sustainable Development Goal (SDG), which is to ensure access to affordable, reliable, sustainable and modern energy for all.

Offshore wind energy is currently one of the most important sources of renewable energy around the world, and it is expected to increase very fast in installed power in the short term. In fact, offshore wind energy is part of the energy mix of some countries. The first offshore wind facility came into operation in the early 90s; the first commercial farms were built at the beginning of this century; currently only Europe exceeds 20 GW of offshore wind power installed. It can be stated that the offshore wind industry has lived a great technological evolution, where the challenges have been frequent, and the great professionals working in both the private and public sectors have allowed the great advances in innovation, making the sustainable development of this technology possible.

The aim of this is Special Issue was to put together papers that reflect the current state-of-the-art of the offshore wind industry, covering all the aspects that need to be taken into account for the planning, design, construction, operation and maintenance, and dismantling of the facilities, etc. The Special Issue invited contributions that deal with all the previously mentioned aspects (but is not limited to them), including the following topics: Legislation, environment, wind resources, foundations and support structures, wind turbine generators, electrical connection, etc.

This Special Issue achieved 13 published papers, with 4 feature papers. The papers are of very good quality, and deal with numerous topics related to offshore wind energy. They cover issues such as marine renewable energies on the Spanish coast [1], relevant factors for optimal locations for wind facilities [2], wind energy potential analysis on the Lebanese coast [3], the trailing-edge flap in large scale offshore wind farms [4], monopile foundations dimensional analysis [5], scour protection in monopile foundations [6], frequency response model tests in monopiles [7], gravity-based foundations [8], effect of the ice force in the support structures [9], dynamic response for submerged floating structures with different mooring configurations [10], tension leg platform wind turbines with non-rotating blades [11], the integration of offshore wind farms in future power systems [12], and transient overvoltage considering different electrical characteristics of vacuum circuit breaker [13].

It is a pleasure for the Guest Editorial Team to have put all these interesting manuscripts together. We thank the authors of the articles that have allowed this Special Issue to have such a high level.

Ass. Prof. Dr. M. Dolores Esteban

Ass. Prof. Dr. José-Santos López-Gutiérrez

Prof. Dr. Vicente Negro

Guest Editors of "Offshore Wind Farms"

Conflicts of Interest: The authors declare no conflict of interest.

References

1. Esteban, M.D.; Espada, J.M.; Ortega, J.M.; López-Gutiérrez, J.S.; Negro, V. What about Marine Renewable Energies in Spain? *J. Mar. Sci. Eng.* **2019**, *7*, 249. [CrossRef]
2. Gil-García, I.; García-Cascales, M.; Fernández-Guillamón, A.; Molina-García, A. Categorization and Analysis of Relevant Factors for Optimal Locations in Onshore and Offshore Wind Power Plants: A Taxonomic Review. *J. Mar. Sci. Eng.* **2019**, *7*, 391. [CrossRef]
3. Ibarra-Berastegi, G.; Ulazia, A.; Saénz, J.; González-Rojí, S. Evaluation of Lebanon's Offshore-Wind-Energy Potential. *J. Mar. Sci. Eng.* **2019**, *7*, 361. [CrossRef]
4. Cai, X.; Wang, Y.; Xu, B.; Feng, J. Performance and Effect of Load Mitigation of a Trailing-Edge Flap in a Large-Scale Offshore Wind Turbine. *J. Mar. Sci. Eng.* **2020**, *8*, 72. [CrossRef]
5. Sánchez, S.; López-Gutiérrez, J.S.; Negro, V.; Esteban, M.D. Foundations in Offshore Wind Farms: Evolution, Characteristics and Range of Use. Analysis of Main Dimensional Parameters in Monopile Foundations. *J. Mar. Sci. Eng.* **2019**, *7*, 441. [CrossRef]
6. Esteban, M.D.; López-Gutiérrez, J.S.; Negro, V.; Sanz, L. Riprap Scour Protection for Monopiles in Offshore Wind Farms. *J. Mar. Sci. Eng.* **2019**, *7*, 440. [CrossRef]
7. He, R.; Zhu, T. Model Tests on the Frequency Responses of Offshore Monopiles. *J. Mar. Sci. Eng.* **2019**, *7*, 430. [CrossRef]
8. Esteban, M.D.; López-Gutiérrez, J.S.; Negro, V. Gravity-Based Foundations in the Offshore Wind Sector. *J. Mar. Sci. Eng.* **2019**, *7*, 64. [CrossRef]
9. Liu, Y.; Li, X.; Zhang, Y. Analysis of Offshore Structures Based on Response Spectrum of Ice Force. *J. Mar. Sci. Eng.* **2019**, *7*, 417. [CrossRef]
10. Li, Y.; Le, C.; Ding, H.; Zhang, P.; Zhang, J. Dynamic Response for a Submerged Floating Offshore Wind Turbine with Different Mooring Configurations. *J. Mar. Sci. Eng.* **2019**, *7*, 115. [CrossRef]
11. Murfet, T.; Abdussamie, N. Loads and Response of a Tension Leg Platform Wind Turbine with Non-Rotating Blades: An Experimental Study. *J. Mar. Sci. Eng.* **2019**, *7*, 56. [CrossRef]
12. Fernández-Guillamón, A.; Das, K.; Cutululis, N.; Molina-García, Á. Offshore Wind Power Integration into Future Power Systems: Overview and Trends. *J. Mar. Sci. Eng.* **2019**, *7*, 399. [CrossRef]
13. Zhou, Z.; Guo, Y.; Jiang, X.; Liu, G.; Tang, W.; Deng, H.; Li, X.; Zheng, M. Study on Transient Overvoltage of Offshore Wind Farm Considering Different Electrical Characteristics of Vacuum Circuit Breaker. *J. Mar. Sci. Eng.* **2019**, *7*, 415. [CrossRef]

 Journal of
Marine Science and Engineering

Review

Gravity-Based Foundations in the Offshore Wind Sector

M. Dolores Esteban *, José-Santos López-Gutiérrez and Vicente Negro

Research Group on Marine, Coastal and Port Environment and other Sensitive Areas, Universidad Politécnica de Madrid, E28040 Madrid, Spain; josesantos.lopez@upm.es (J.-S.L.-G.); vicente.negro@upm.es (V.N.)
* Correspondence: mariadolores.esteban@upm.es

Received: 27 December 2018; Accepted: 24 January 2019; Published: 12 March 2019

Abstract: In recent years, the offshore wind industry has seen an important boost that is expected to continue in the coming years. In order for the offshore wind industry to achieve adequate development, it is essential to solve some existing uncertainties, some of which relate to foundations. These foundations are important for this type of project. As foundations represent approximately 35% of the total cost of an offshore wind project, it is essential that they receive special attention. There are different types of foundations that are used in the offshore wind industry. The most common types are steel monopiles, gravity-based structures (GBS), tripods, and jackets. However, there are some other types, such as suction caissons, triples, etc. For high water depths, the alternative to the previously mentioned foundations is the use of floating supports. Some offshore wind installations currently in operation have GBS-type foundations (also known as GBF: Gravity-based foundation). Although this typology has not been widely used until now, there is research that has highlighted its advantages over other types of foundation for both small and large water depth sites. There are no doubts over the importance of GBS. In fact, the offshore wind industry is trying to introduce improvements so as to turn GBF into a competitive foundation alternative, suitable for the widest ranges of water depth. The present article deals with GBS foundations. The article begins with the current state of the field, including not only the concepts of GBS constructed so far, but also other concepts that are in a less mature state of development. Furthermore, we also present a classification of this type of structure based on the GBS of offshore wind facilities that are currently in operation, as well as some reflections on future GBS alternatives.

Keywords: support structures; gravity-based structures; GBS; GBF

---

1. Introduction

The offshore wind sector can now be considered to be in a commercial stage of development [1]. According to a WindEurope report [2], in Europe during 2017, the total output of offshore wind turbines constructed during the year totaled 3148 megawatts (MW), which is the annual installed power record since the creation of this technology (in 2015, the annual installed capacity was very close to that of 2017, but the rest of the years do not exceed half of the installed capacity in 2017). At the end of 2017, there was a total of 15,780 MW of generation in Europe (double that at the end of 2014), installed in 92 offshore wind farms, located in 11 countries across Europe [2].

The total amount of installed power generation has been growing over time due to several factors, for instance: The installation of higher power wind turbines, larger numbers of wind turbines in each facility, the state of development of the offshore wind industry in different countries, the successes achieved in the sector, etc. This was accomplished by, among other reasons, moving to locations with greater depth, investigating new concepts for foundations and substructures, working to solve uncertainties in foundation design, implementing scour protection systems, and increasing the amount of investment to connect wind farms to energy grids [3,4].

The foundation is the key component in an offshore wind facility, representing the way through which the loads of the superstructure, in this case the wind turbine, are transmitted to the soil. Therefore, soil properties are essential in the selection of the type of foundation used in these facilities [5].

Different types of foundation have been developed for use in the offshore wind industry. The most common types are steel monopiles, gravity-based structures (GBS), tripods, and jackets [6]. Other types of foundations, such as suction caisson, concrete monopile, floating support, etc., are currently in a less advanced phase of development. This article focuses on GBS, also known as GBF (gravity-based foundations), being the type of foundation that is suitable in cases of soils with high bearing capacity. This is because of the way gravity structures support loads and transmit them to the soil [7,8].

Soils with a high load-bearing capacity allow the installation of shallow foundations [9], such as the GBS. On the other hand, in the case of low load-bearing capacity soils, the best solution is to use deep foundations or propose soil improvement techniques, such as piles, either as a monopile or as a foundation for jackets or tripods. Other key aspects when deciding on the type of foundation to use are the depth of the water, the loads associated with the characteristics of the wind turbine, and the climatic loads (wind, waves, marine currents, tidal range) means of manufacturing, installation, operation, dismantling, etc. In fact, there are many considerations to be measured when selecting the most appropriate foundation for a wind turbine facility. The cost of the foundation is around 35% of the overall cost of the project. This significant cost is an important factor when both designing and choosing the foundation to be used.

Most types of foundation used in offshore wind, both direct to the ground and floating supports, are concepts inherited from the offshore oil and gas industries [10,11]. In particular, it is clear that GBS foundation originated from the oil and gas industries, with a structure known as a Condeep (concrete deep-water structure) [12]. These are used for different water depths, with the maximum depth being 330 m (Troll Condeep, in the Norwegian sector of the North Sea, 1995). The first Condeep structure ever built was Ekofist I, in 1973, in the North Sea [13,14]. A Condeep is usually constructed in a fjord, given its good characteristics in terms of construction associated with sheltered waters and high-water depths [15].

On the other hand, concrete structures are also used at sea in breakwaters and quay walls [16], where many structures use the floating caisson technique [17]. In fact, some companies have tried to use a similar concept for offshore wind farms. It is important to understand the differences that exist between offshore wind structures and breakwaters. The function of offshore wind foundations is to support the wind turbine, and the function of the breakwater is to shelter the interior area, so that ships can securely carry out loading and unloading operations. It is for this reason that the breakwaters have to stop a large amount of wave energy. Furthermore, in the case of the supporting structures of the wind turbines, it is important that the wave loads be as low as possible and as transparent as possible to waves, with the objective of reducing the cost of the structure.

According to WindEurope's report, at the end of 2017, monopile foundation in Europe was at the top of the classification, with 3720 units (81.7%), followed by the jacket, with 315 units, and the gravity based foundation, with 283 units. There were only seven floating platforms constructed during this period, six of which were SPARs and one which was semi-submersible.

According to those statistics, there are some offshore wind facilities operating in Europe that have GBS-type foundations; however, the small percentage of such cases indicates that GBFs have not been widely used in the industry up until now. This is due to the ease of use of monopiles, which represent >80% of offshore wind foundations. This characteristic of monopiles, together with their reduced cost, has displaced other types of foundations from a strategic position in the sector. However, as water depth increases, some limitations appear around the use of monopiles, potentially causing other types of foundations to increase in use.

Alternatives to the monopile have to be considered in locations with a terrain where the driving-in of monopiles is difficult, for instance, rocky soils. In such cases, the GBS is expected to work well. The main advantages of the GBS are: The good behavior of similar structures in the oil and gas and port engineering industries; its suitability as a foundation in rocky or sandy soils, with its high bearing capacity, where pile driving can be complicated; and it being an alternative that can enrich market competitiveness and therefore reduce of costs in any industry. The main disadvantages of the GBS are: It has not had great acceptance in the wind industry up to now; it needs soil with specific geotechnical properties, such as high bearing capacity; in general, previous soil preparation is needed for correct support of the structure; the large occupation area in the seabed, with its associated environmental impact; and the necessary means of manufacture, transport, and installation.

Although this typology has not previously been widely used, there are certain opinions that have highlighted its advantages over other types of foundations for both shallow and deep water sites. In fact, the offshore wind industry is trying to introduce improvements to turn it into a competitive foundation alternative in the widest range of water depths. There has been recent open discussion in some of the most important conferences on offshore wind energy about possible competition between jacket, XXL monopile, and GBS foundation types in water depths of around 40 m.

The importance of the GBS foundation for the future of the offshore wind industry is not currently discussed. As a consequence, this article is about GBS foundations, since it is fundamental to achieve greater knowledge about this concept. The paper provides a review of the different existing GBS foundation concepts. For this, on one hand, the foundations of wind farms in Europe that are already in operation are analyzed, and on the other, the main existing concepts that are in a less mature development phase, either at the research or prototype level, are identified. In addition, this article includes a classification of GBS foundations, elaborated on by the authors, based on offshore wind farms examples in operation. The paper also includes some reflections on the future of GBS alternatives.

2. Objectives and Research Methodology

The main aim of this paper is to show the different existing alternatives of GBS foundations for offshore wind facilities, including the already constructed ones and others in an early stage of development.

For that, it was necessary to find all the offshore wind farms in operation that have GBS foundations, and to study each specific design. After that, the different GBS foundation concepts were identified and classified. Then, GBS alternatives in an early phase of study were analyzed. For all of this, an in-depth literature review was carried out.

Based on available information, a classification proposal for the already constructed GBS foundations is elaborated on by the authors, in order to clarify the different existing general concepts in operating wind farms. Furthermore, some reflections are given on other GBS concepts that have not yet been proven, as well as an analysis on the future on this type of concept.

3. State of the Art and Discussion

This section includes two parts. The first (3.1) identifies the different offshore wind farms in operation in Europe that have GBS foundations. The second (3.2) concerns the different concepts of GBS foundations. This second part includes not only the collected state of the art, but also a discussion on this information.

3.1. European Offshore Wind Farms with GBS Foundations

European offshore wind farms with GBS foundations were identified, even those already dismantled (Table 1); these were mainly culled from different reliable Internet sources [18,19].

Table 1. List of European offshore wind farms in operation with gravity-based structure (GBS) foundations.

Name of the Farm	Country	Year (COMISSIO-NING)	Total Power (MW)	Turbine Model	Depth (M)
Kårehamn	Sweden	2013	48	Vestas 3 MW	6–20
Vindpark Vänern	Sweden	2012	30	WWD 3 MW	-
Avedøre Holme	Denmark	2011	10.8	Siemens 3.6 MW	2
Rødsan II (Nysted II)	Denmark	2010	207	Siemens 2.3 MW	6–12
Sprogø	Denmark	2009	21	Vestas 3 MW	10–16
Thorntonbank Phase 1	Belgium	2009	30	Repower (Senvion) 5 MW	13–20
Lillgrund	Sweden	2007	110	Siemens 2.3 MW	4–13
Breitling	Germany	2006	2.5	Nordex 2.5 MW	0.5
Nysted I (Rødsan I)	Denmark	2003	166	Siemens 2.3 MW	6–10
Middelgrunden	Denmark	2001	40	Bonus, Siemens 2 MW	3–6
Tunø Knob	Denmark	1995	5	Vestas 500 kW	4–7
Vindeby (dismantled)	Denmark	1991	4.95	Bonus 450 kW	2–4

In analyzing Table 1, several conclusions can be drawn:

- Total wind farms: 12, one dismantled. Seven are in Denmark, which is the current leader in the use of GBS in offshore wind, three in Sweden, and one each in Belgium and Germany.
- Regarding year of commissioning: Two farms in 2009, and one each in 1991 (dismantled), 1995, 2001, 2003, 2006, 2007, 2010, 2011, 2012, and 2013. After 2013, there have been no facilities commissioned with GBS.
- Regarding the total power, the minimum is Breitling, with 2.5 MW and only one Nordex turbine, and the maximum is Nysted I (Rødsan I), with 166 MW and more than 70 turbine units.
- The nominal power of the wind turbines is between ~0.5 MW (Vindeby, with 0.45, dismantled, and Tunø Knob, with 0.5) and 5 MW (Thorntonbank Phase 1).
- The water depths of the sites are between 0.5 m (Breitling) and 20 m (Thorntonbank Phase 1).

3.2. GBS Foundation Types

A review of the different types of GBS foundations installed in offshore wind facilities was carried out. Section 3.2.1 includes the main information from this review. After that, Section 3.2.2, concerning new concepts for GBS foundations, includes some ideas at an early stage of development.

3.2.1. Proven Concepts of GBS Foundations

The first offshore wind farm with a GBS foundation was Vindeby, commissioned in 1991, dismantled in 2017, and located in between 2 and 4 m of water. Since then, different offshore wind facilities have been constructed with GBS foundations.

As a result of the analysis carried out here, a basic classification of the different GBS types is included. This classification includes first-, second-, and third-generation types of GBS foundations.

The first-generation GBS foundations correspond to the first offshore wind facilities with GBS foundations: Tunø Knob, commissioned in 1995 and located in between 4 and 7 m of water; and Middelgrunden (Figure 1), commissioned in 2001 and located in between 3 and 6 m of water. This first generation also corresponds to the first designs that were made of this type of foundation for the offshore wind industry.

Figure 1. Middelgrunden offshore wind farm foundations, reproduced from [20], with permission from Elsevier, 2019.

This type of GBS is a completely solid, reinforced concrete structure, without holes or cells. It is composed of a large-diameter slab; in the case of Middelgrunden, this is between 16.7 and 17.6 m and very thin. The said slab is attached to a small-diameter shaft that, in some cases, ends in the form of a cone as an icebreaker (Figure 1).

Since it is a solid structure, the weight to be taken into account for the transport and installation is very high compared to the weight of a structure with the same geometry, but with incorporated holes or cells. This typology was possible designed because Tunø Knob and Middelgrunden offshore wind farms were built in locations with very shallow water—between 3 and 7 m of water depth. While this type of concept can be considered to be a suitable solution for shallow depths, it is not viable for sites in deeper water. The main problem is that, in deeper water, the design of these structures, which work based on their own weight once they are in operation, leads to greater weights than if they are designed to incorporate holes or cells. These heavy structures are not easy to transport to the site or to install, as they need barges and cranes with special requirements. This makes this first generation of GBS unprofitable for greater depths.

Following this, designs corresponding to the second generation of GBS foundations were developed. Examples of this second generation are: Nysted I (or Rødsan I) (Figure 2), commissioned in 2003 and located in between 6 and 10 meters of water; Lillgrund (Figure 3), commissioned in 2007 and located in between 4 and 13 meters of water; Sprogø (Figure 4), commissioned in 2009 and located in between 10 and 16 meters of water; Rødsan II (or Nysted II) (Figure 5), commissioned in 2010 and located in between 6 and 12 meters of water; and Kårehamn (Figure 6), commissioned in 2013 and located in between 6 and 20 meters of water.

Figure 2. Nysted I (or Rødsan I) offshore wind farm foundations, with permission from Tech-Marine, 2019.

Figure 3. Lillgrund offshore wind farm foundations reproduced from [21], with permission from Tech-Marine, 2019.

Figure 4. Sprogφ offshore wind farm foundations [22], with permission from Aarsleff, 2019.

Figure 5. Rφdsan II (or Nysted II) offshore wind farm foundations, reproduced from [22], with permission from Aarsleff, 2019.

Figure 6. Kårehamn offshore wind farm foundations, reproduced from [20], with permission from Elsevier, 2019.

This type of GBS foundation is composed of a flat slab and a shaft, similar to that of the first generation but with the main difference being that the slab contains holes or cells. This means that if the weights of a first- and second-generation GBS structure were to be compared, both with the same geometry, the weight of the latter would be much lower. This lower weight allows more units to be transported in the same barge, from the port to the final location, in one trip. In addition, the cranes used in installation have less demanding requirements. Once the GBS foundation is installed, the holes or cells are filled with ballast, thus achieving the final design weight of the structure that allows it to be stable and resistant to loads.

The third generation of GBS foundations are the latest concepts that were built. An example of this generation is the first phase of the Thornton Bank offshore wind farm (Figure 7), commissioned in 2009 and located in between 13 and 20 meters of water. As can be seen in that figure, the structure has a conical shape in the lower part and a vertical shaft in the upper part. The structure is mostly hollow inside, not only the slab or lower part. This structure was designed to be transported using a semi-floating method, thus reducing the weight of the structure for that phase, which lowers the requirements of the transport vessels and cranes used in their installation. Once the structures are in place, the hollow area is filled with ballast, to provide the necessary weight to support the loads.

Figure 7. Thornton Bank offshore wind farm foundations, reproduced from [20], with permission from Elsevier, 2019.

The classification proposed for the GBS concepts used up to now in offshore wind farms already in operation is shown in Figure 8, which includes an example, a conceptual draft, and a photo of the different types of GBS foundations that correspond to the first, second, and third generations.

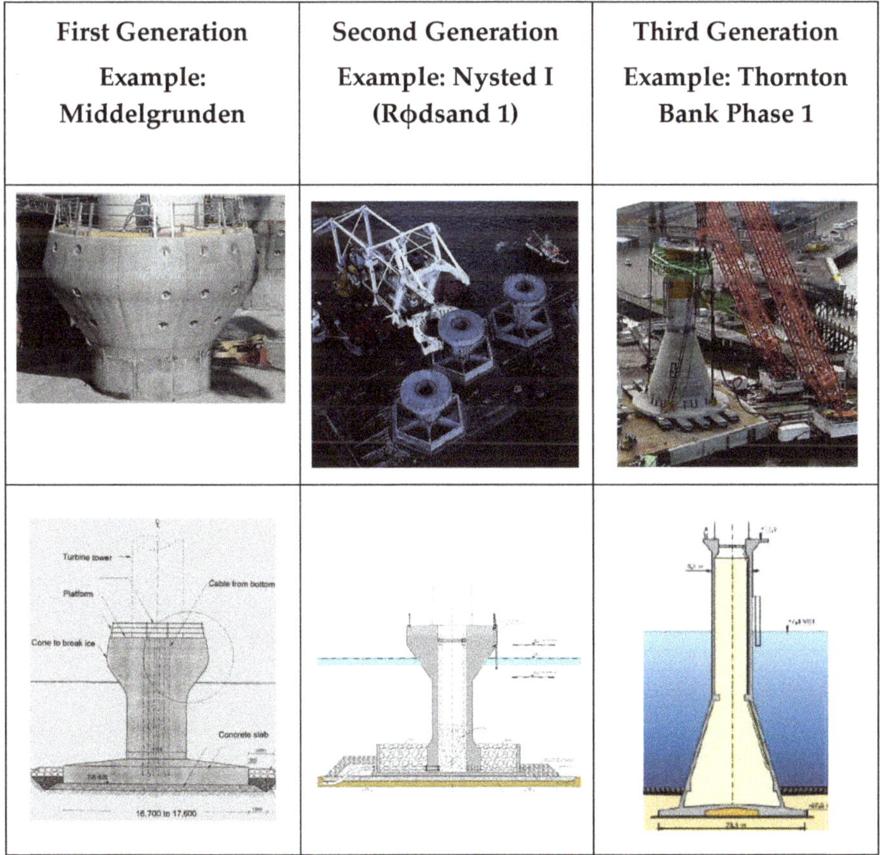

Figure 8. Classification proposal for the existing GBS concepts, reproduced from [20], with permission from Elsevier, 2019.

3.2.2. New Concepts of GBS Foundations

Now that the different concepts of operational offshore wind farms have been outlined, the following sections concern new types of GBS foundations that are in the research phase. Most of these new concepts are based on the F2F (floated to fixed) concept, which refers to a structure that behaves as a float during the transport phase from the port to its final location, and during the installation phase. During transport, it is necessary to have the support of small tugboats. By floating the structure, the need for vessels for transport is eliminated, except for the tugboats, thus reducing the cost of that phase. Some of these types of GBS foundations need special-purpose vessels, with the objective of transporting and installing the GBS structure and the wind turbine generator (WTG) together, with the WTG being pre-assembled in the port.

Crane-Free Gravity Base (Seatower)

The crane-free gravity base concept is a concrete structure with a relatively thin slab, an intermediate-length conical part, and a cylindrical shaft in the upper part. This concept was designed to be transported by its own flotation ability, so it is hollow inside, for which it needs the support of tugboats (Figure 9). It avoids the use of an expensive and weather-sensitive crane. According to Reference [23], this concept has been optimized for the logistics, from the manufacturing through to the decommissioning process.

Figure 9. Crane-free gravity base concept, reproduced from [23], with permission from Elsevier, 2019.

Gravitas Gravity Base (Arup/Costain/Hochtief)

The gravitas gravity base concept is shaped similarly to the crane-free gravity base type, also having a relatively thin slab, intermediate-length cone, and cylindrical shaft in the upper part. This structure is also self-floating and can be transported to its final location only with the assistance of small tugboats. According to www.arup.com, some characteristics of this concepts are: It is a reinforced concrete and ballasted gravity structure; it can be deployed in water depths up to 60 m; it can hold turbines generating up to 8 MW; it requires minimal seabed preparation because it can accommodate existing seabed slopes and surface sediments; its skirt has variants to suit specific seabed sediment conditions; the collar design for the turbine mast connection can accommodate an ~2° vertical alignment tolerance; there is the potential to repower it without replacing the foundation; the concrete base is configured for rapid construction using available construction skills; its construction is an onshore activity, which is tailored for ease of subsequent installation; it does not required deep water (10 m draft) for construction; the foundations are self-buoyant for ease of deployment to the wind farm location; it uses available and abundant standard tugs to install the foundations; installation is done by sinking, through a controlled influx of water, followed by sand/aggregate ballasting; and it includes scour protection, designed for minimum maintenance over the design life of the wind farm [24].

The key figures for a 35-m water depth, central North Sea environment conditions, and 6 MW output are:

- Air-gap concrete structure: 20 m.
- Hub height above LAT (Lowest Astronomical Tide): 90 m.
- Base outer diameter: 34 m.
- Outer diameter, caisson: 31 m.
- Outer diameter, top of shaft: 6 m.
- Concrete volume: 1919 m^3.
- Steel reinforcements: 720 tons.

Strabag Gravity Base (STRABAG)

STRABAG has two different gravity-base concepts. Both of them have a geometrical slab and a cylinder in the upper part. The concepts have in common joint transportation and installation of the foundation and the wind turbine generator, with preassembly being performed in port, thus reducing the number of operations carried out at sea during the installation phase. To be able to carry the floating foundation and wind turbine together, a specifically-purposed vessel is used, called STRABAG Carrier [25].

According to www.strabag-offshore.com, both these concepts use the pre-stressed concrete technique, they are suitable for water depths up to ~45 m, and they can be completely disassembled.

GBF Gravity Base (Ramboll/BMT Nigel Gee and Freyssinet)

The GBF gravity base concept is a concrete structure with a circular slab, a conical intermediate part, and a relatively small-diameter cylinder shaft in the upper part. This concept is not self-floating and requires a specific barge for the transport and installation of the GBS structure and the wind turbine generator together. This specific barge is called a transport and installation barge (TIB) [26].

The tower, nacelle, and rotor are assembled in the port quay before being lowered into the water. The TIB is ballasted down to the level of the base, then, upon connection, will refloat to the transportation depth. This concept was developed with the support of the Carbon Trust.

According to Reference [27], this type of foundation is suitable for water depths between 20 and 55 m, many seabed conditions, a distance offshore between 2 and 200 miles, and turbines with a unit power between 3 and 10 MW.

Other Gravity Base Structure Concepts

Other gravity base structures are described in Reference [28], both of them having similar shapes to the abovementioned concepts. These are, respectively, concepts from a collaboration between BAM Wind Energie and Van Oord, and the consortia of Skanska, Smit Marine Projects, and Grontmij. Both have two different parts—a slab and a shaft—and both are cylindrical, with a smooth transition between both with a conical shape.

Other Related Concepts

Other concepts than can be considered to be related to the GBS foundation concept are the Rockmat and the ocean brick system (OBS), both described below.

Rockmat (OFS: CETEAL/Cathie Associates/DVO) [29] is an innovative concept as a foundation for a wind turbine generator that can be used in rocky soils. It is a technology for the interface between the soil and different types of support structures, such as jackets, GBS, etc. It comprises a precast base to make the entire foundation self-floating and supportable by tugboats and is installed by ballasting with water and concrete. It is fixed in its final position through a combination of a grout injection system associated with a jack levelling system. Irregularities in the contact between the foundation and the seabed are filled with grout injections. After installation of the foundation, the wind turbine generator is installed.

According to www.rockmat.com, this concept has the following advantages: No need for previous soil preparation; no costly barge crane with weather restrictions, with only tugboats necessary for its installation; and reversible water ballasting. Installation is estimated to be 30 hours, with the support of one 100-ton bollard-pull tugboat, three 10-ton bollard-pull tugboats, and a barge to supply the concrete mixing unit and compressor.

The ocean brick system (OBS) (Technical University Braunschweig) is a modular system consisting of hollow precast blocs (10 m × 10 m × 10 m), piled like interconnected cubes, to create a stiff, light, strong structure. The structure can be constructed in a dry dock and floated to the site with the support of tugboats [30].

4. Conclusions

A review of the different operational European offshore wind farms with GBS foundations was carried out in this study. In total, there are only 13 of these, and they are located in Denmark, Germany, Sweden, Finland, and Belgium. The deepest water in all of those wind farms is 20 m, in the case of Thorntonbank Phase 1. Furthermore, it is important to note that since 2013, there have been no commissioned GBS facilities.

The current strength of the monopile in the offshore wind-power sector is evident, compared to other typologies. This is mainly due to the simplicity of the structure, which results in benefits during manufacturing, installation, and maintenance, as well as cost. With its 80% representativeness, the monopile has shown its clear dominance in the current market, where most sites have a water depth not exceeding 20–30 m. At greater depths, there is a battle to be more competitive between the XXL monopile, GBS, and jacket foundation types.

The different types of GBS foundations used in all the wind farms in operation in Europe were analyzed. Based on this analysis, a classification was proposed for the different types, which distinguishes between the first, second, and third generations.

Examples of first-generation GBS foundations are Tunϕ Knob and Middelgrunden. These are solid concrete structures, without cells or holes, corresponding to the first designs.

Examples of the second generation are Nysted I or Rϕdsan I, Lillgrund, Sprogϕ, Rϕdsan II or Nysted II, and Kårehamn. These foundations include holes or cells in the slab or lower part of the structure, which reduces their weight for transport and installation. Once the GBS structure is installed, the holes or cells are filled with ballast, achieving the final design weight that supports the design loads.

The only example of a third-generation structure is Thornton Bank. This concept has a conical shape with a hole or cell inside and not only in the slab or lower part, as in the second-generation models. This type of structure was planned to be semi-floating during the transport and installation phases, decreasing the weight of the foundation and reducing the lifting requirements. Once this foundation is placed on the seabed, the interior hole is filled with ballast to achieve the final design weight to support the design loads.

Other, nonproven concepts were analyzed in this study, some of them based on the F2F (floated to fixed) concept, which is a floating structure during the transport and installation phases, supported by small tugboats, which decreases the costs because of the self-buoyancy of the foundation and there being no need to use larger, more specific transport vessels. Another new concept needs special-purpose vessels to transport and install the GBS and WTG structures together, with the WTG being pre-assembled onshore. These transport vessels are designed specifically for each concept.

Some of these new concepts include the crane-free gravity base (Seatower), gravitas gravity base (Arup/Costain/Hochtief), Strabag gravity base (STRABAG), GBF gravity base (Ramboll/BMT Nigel Gee and Freyssinet), GBF gravity base (Ramboll/BMT Nigel Gee and Freyssinet), Rockmat (OFS: CETEAL/Cathie Associates/DVO), ocean brick system (Technical University Braunschweig), etc.

As mentioned above, new locations with greater water depths will begin a battle between the XXL monopile, GBS, and jacket types. The more real and well-analyzed options the market has, the more competitive this scenario could become. This is why it is expected that solutions similar to the third-generation concepts, characteristic of the first phase of the Thornton Bank offshore wind farm, will come online in the future, including more options to use F2F concepts where it is necessary to build special-purpose vessels.

The trend will likely be to have the entire wind turbine pre-assembled on the structure at the port, in order to reduce the number of operations carried out at sea, which involve greater costs and risks. The GBS concept is very interesting in that respect, and one example is the ELISA project in the Canary Islands, designed by Esteyco, which has a slab and a shaft, without a conical transition. The prototype for a 5 MW-output concept will use special-purpose vessels to help in the transportation of the entire GBS structure and wind turbine. As a nod to previous concepts, it has a telescopic tower and, using hydraulic jacks, manages to raise the different sections into their final positions, with the nacelle and rotor already installed on the last section of the tower.

Author Contributions: Conceptualization, M.D.E. and J.-S.L.-G.; Methodology, M.D.E. and V.N.; Investigation, M.D.E. and J.-S.L.-G.; Resources, V.N.; Writing—Original Draft Preparation, M.D.E.; Writing—Review and Editing, J.-S.L.-G. and V.N.

Funding: Authors give thanks the Agustín de Betancourt Foundation (FAB) for the support received over the past few years.

Conflicts of Interest: The authors declare no conflict of interest.

References

1. Esteban, M.D.; López-Gutiérrez, J.S.; Diez, J.J.; Negro, V. Methodology for the Design of Offshore Wind Farms. *J. Coast. Res. Coast. Educ. Res. Found.* **2011**, *64*, 496–500.
2. WindEurope. *Offshore Wind in Europe, 2017. Key Trends and Statistics 2017*; Technical Report; Wind Europe: Brussels, Belgium, 2017.
3. Esteban, M.D.; Diez, J.J.; López-Gutiérrez, J.S.; Negro, V. Why Offshore Wind Energy. *Renew. Energy* **2011**, *36*, 444–450. [CrossRef]
4. Luengo, J.; Negro, V.; García-Barba, J.; López-Gutiérrez, J.S.; Esteban, M.D. New Detected Uncertainties in the Design of Foundations for Offshore Wind Turbines. *Renew. Energy* **2019**, *131*, 667–677. [CrossRef]
5. Li, Y.; Cheng Ong, M.; Tang, T. Numerical Analysis of Wave-induced Poro-elastic Seabed Reponse Around a Hexagonal Gravity-based Offshore Foundation. *Coast. Eng.* **2018**, *136*, 81–95. [CrossRef]
6. Schallenberg-Rodríguez, J.; García-Montesdeoca, N. Spatial Planning to Estimate the Offshore Wind Energy Potential in Coastal Regions and Islands. Practical Case: The Canary Islands. *Energy* **2018**, *143*, 91–103. [CrossRef]
7. Negro, V.; López-Gutiérrez, J.S.; Esteban, M.D.; Matutano, C. Uncertainties in the Design of Support Structures and Foundations for Offshore Wind Turbines. *Renew. Energy* **2014**, *63*, 125–132. [CrossRef]
8. Esteban, M.D.; López-Gutiérrez, J.S.; Negro, V.; Matutano, C.; García-Flores, F.M.; Millán, M.A. Offshore Wind Foundation Design: Some Key Issues. *J. Energy Resour. Technol. ASME* **2015**, *136*. [CrossRef]
9. Das, B. *Shallow Foundations. Bearing Capacity and Settlement*; CRC Press Book: Boca Raton, FL, USA, 2017; Volume 400.
10. Mäkitie, T.; Andersen, A.D.; Hanson, J.; Normann, H.E.; Thune, T.M. Established sectors expediting clean technology industries? The Norwegian oil and gas sector's influence on offshore wind power. *J. Clean. Prod.* **2017**, *177*, 813–823. [CrossRef]
11. Zheng, X.Y.; Lei, Y. Stochastic Response Analysis for a Floating Offshore Wind Turbine Integrated with a Steel Fish Farming Cage. *Appl. Sci.* **2018**, *8*, 1228. [CrossRef]
12. Pérez-Fernández, R.; Lamas-Pardo, M. Offshore Concrete Structures. *Ocean Eng.* **2018**, *58*, 304–316. [CrossRef]
13. Eide, D.T.; Larsen, L.G. Installation of the Shell/Esso Brent B Condeep Production Platform. In Proceedings of the Offshore Technology Conference, Houston, TX, USA, 3–6 May 1976. [CrossRef]
14. Fernandes, J.F.; Bittencourt, T.; Helene, P. *A Review of the Application of Concrete to Offshore Structures*; American Concrete Institute, ACI Special Publication: Farmington Hills, MI, USA, 2008; pp. 377–392.
15. Kim, H.G.; Kim, B.J. Feasibility study of new hybrid piled concrete foundation for offshore wind turbine. *Appl. Ocean Res.* **2018**, *76*, 11–21. [CrossRef]

16. Smith, P.E. Types of Marine Concrete Structures. In *Design, Durability and Performance*; Elsevier: Amsterdam, The Netherlands, 2016; pp. 17–64.
17. Cejuela, E.; Negro, V.; Esteban, M.D.; López-Gutiérrez, J.S.; Ortega, J.M. From Julius Caesar to Sustainable Composite Materials: A Passage through Port Caisson Technology. *Sustainability* **2018**, *10*, 1225. [CrossRef]
18. Available online: www.4coffshore.com (accessed on 10 May 2017).
19. Ruiz de Temiño Alonso, I. Gravity Base Foundations for Offshore Wind Farms. Marine Operations and Installation Processes. Master in European Construction Engineering. Final Thesis. 2013. Available online: https://repositorio.unican.es/xmlui/handle/10902/3429 (accessed on 10 September 2018).
20. Esteban, M.D.; Couñago, B.; López-Gutiérrez, J.S.; Negro, V.; Vellisco, F. Gravity Based Support Structures for Offshore Wind Turbine Generators: Review of the Installation Process. *Ocean Eng.* **2015**, *110*, 281–291. [CrossRef]
21. Available online: www.tech-marine.dk (accessed on 22 February 2017).
22. Available online: www.aarsleff.com (accessed on 2 May 2017).
23. Available online: www.seatower.com (accessed on 12 March 2017).
24. Available online: www.arup.com (accessed on 1 May 2017).
25. Available online: www.strabag-offshore.com (accessed on 16 March 2017).
26. Available online: www.vinci-offshorewind.co-uk (accessed on 10 March 2017).
27. Available online: www.gbf.eu.com (accessed on 12 February 2017).
28. Available online: www.concretecentre.com/wind (accessed on 9 May 2017).
29. Available online: www.rockmat.com (accessed on 17 May 2017).
30. Oumeraci, H.; Pförtner, S.; Kudella, M.; Kortenhaus, A. Ocean Brick System (OBS) Used as a Foundation Structure for Offshore Wind Turbine. Coast, Marine Structures and Breakwater. 2009. Available online: https://www.icevirtuallibrary.com/doi/abs/10.1680/cmsb.41318.0008 (accessed on 10 May 2017).

Journal of
Marine Science and Engineering

Article

Dynamic Response for a Submerged Floating Offshore Wind Turbine with Different Mooring Configurations

Yane Li [1,2], Conghuan Le [1,2,*], Hongyan Ding [1,2,3], Puyang Zhang [1,2] and Jian Zhang [1,2]

[1] State Key Laboratory of Hydraulic Engineering Simulation and Safety, Tianjin University, Tianjin 300072, China; liyane0000@gmail.com (Y.L.); 966329@tju.edu.cn (H.D.); zpy@tju.edu.cn (P.Z.); jzjazz0420@gmail.com (J.Z.)
[2] School of Civil Engineering, Tianjin University, Tianjin 300072, China
[3] Key Laboratory of Coast Civil Structure Safety, Ministry of Education, Tianjin University, Tianjin 300072, China
* Correspondence: lch@tju.edu.cn

Received: 30 January 2019; Accepted: 15 April 2019; Published: 22 April 2019

Abstract: The paper discusses the effects of mooring configurations on the dynamic response of a submerged floating offshore wind turbine (SFOWT) for intermediate water depths. A coupled dynamic model of a wind turbine-tower-floating platform-mooring system is established, and the dynamic response of the platform, tensions in mooring lines, and bending moment at the tower base and blade root under four different mooring configurations are checked. A well-stabilized configuration (i.e., four vertical lines and 12 diagonal lines with an inclination angle of 30°) is selected to study the coupled dynamic responses of SFOWT with broken mooring lines, and in order to keep the safety of SFOWT under extreme sea-states, the pretension of the vertical mooring line has to increase from 1800–2780 kN. Results show that the optimized mooring system can provide larger restoring force, and the SFOWT has a smaller movement response under extreme sea-states; when the mooring lines in the upwind wave direction are broken, an increased motion response of the platform will be caused. However, there is no slack in the remaining mooring lines, and the SFOWT still has enough stability.

Keywords: floating offshore wind turbine (FOWT); mooring system; coupled dynamic response; broken mooring line; safety factor

---

1. Introduction

Over recent years, harnessing of offshore wind power usually has been concentrated in shallow water regions (<50 m) using fixed foundations, such as monopile, gravity, or jacket structures [1]. With the depletion of coastal resources, as well as the geographical and environmental constraints, the development of offshore wind power is bound to reach deep water to access abundant wind resources. Fixed foundations are limited at water depths of 30–50 m [2–5]. A series of floating wind turbine concepts has been proposed at various stages of development, which can be divided into three categories: spar, semi-submersible, and tension leg platform (TLP) [6–13]. In addition, the world's first commercial floating wind power project, Hywind Scotland pilot park, was put into operation in 2017 [14].

In recent years, several feasibility studies have been performed to investigate the economic viability of FOWT and optimize the design of both support structures and mooring systems [15–17]. The offshore code comparison collaboration continuation (OC4) DeepCWind semi-submersible floating offshore wind turbine (FOWT) model was simulated by Liu et al. [18]; a fully-coupled fluid-structure interaction system was analyzed in detail, and the impacts of wind turbine aerodynamics on the behavior of the floating platform and the mooring system responses were examined. Benassai [19] minimized the catenary mooring system weight for tri-floater floating offshore wind turbines and pointed out that the platform admissible offset and mooring line configuration significantly influence the weight of the mooring system. Yusuke [20] proposed a novel type of floating wind turbine with a single-point mooring system and examined two different configurations of the single point mooring system based on a tank test and a real sea test.

In order to ensure the FOWT safety, the aero-hydro-elastic-mooring coupled dynamic response of a floating wind turbine under extreme sea-states has to be checked. Utsunomiya et al. [21] evaluated the dynamic response of a spar-type FOWT under extreme sea-states using the dynamic analysis tool, which consists of a multibody dynamics solver, aerodynamic force evaluation library, hydrodynamic force evaluation library, and mooring force evaluation library. Mooring line damage is a key factor that influences the safety of the whole system. Several related studies have been conducted on the damaged mooring systems for floating offshore wind turbines. Benassai et al. [22] studied the performance changes of the OC4 DeepCWind semisubmersible with one broken mooring line and found that an accidental disconnection of one of the mooring lines changes the platform and turbine orientations, which might cause large nacelle yaw error. Li et al. [23] established a coupled aero-hydro-elastic numerical model to investigate the transient response of a spar-type FOWT in scenarios with fractured mooring lines, and they found that in terms of drift distance, it might be more dangerous to shut down the turbine when the wind load is in the opposite direction of drift. Ahmed et al. [24] established a simplified three-degrees-of-freedom (3-DOFs) model to analyze the transient motion of a truss-spar in the time domain after one or two mooring lines were broken. In the simulations, a quasi-static approach was applied to calculate the mooring loads.

A submerged FOWT (SFOWT) with a taut mooring system was proposed to support a 5-MW wind turbine for a water depth of 50–200 m [25–27]. This paper discusses the effects of mooring configurations on the dynamic response of SFOWT. Fully-coupled aero-hydro-servo-elastic time domain simulations were carried out using the code FAST [28] developed by NREL to simulate the dynamic response of SFOWT. TurbSim [29] is used to generate the 3D turbulent wind field, and Sesam/Wadam [30] is used to calculate the hydrodynamic coefficients of the SFOWT in the frequency domain. The coupled dynamic responses of SFOWT with different mooring configurations are investigated, and the dynamic responses of SFOWT with broken mooring lines are further analyzed by selecting the optimal mooring configurations.

2. Theoretical Calculations for SFOWT

2.1. Equation of Motion in the Time Domain

The equation of motion of SFOWT can be expressed as follows [31]:

$$[M + A_\infty]\ddot{\xi} + K\xi = F_1(t) + F_n\left(t, \dot{\xi}\right) + F_c\left(t, \dot{\xi}\right) + F_w(t) + F_m(t) \tag{1}$$

where M is the mass matrix of SFOWT, A_∞ is the added mass at infinity frequency, and $\xi, \dot{\xi},$ and $\ddot{\xi}$ are the 6-DOF displacements, velocities, and accelerations of the platform, respectively. K is the hydrostatic stiffness matrix. $F_1(t)$ is the first-order wave exciting forces. $F_n\left(t, \dot{\xi}\right)$ is the nonlinear drag force from Morison's equation. $F_w(t)$ is the wind-induced forces including aero-dynamic forces and tower drag forces. $F_m(t)$ is the restoring force resulting from the taut mooring lines. $F_c\left(t, \dot{\xi}\right)$ is the radiation damping force, which can be expressed as:

$$F_c\left(t, \dot{\xi}\right) = -\int_{-\infty}^{t} R(t - \tau)\dot{\xi}(\tau)d\tau \tag{2}$$

where $R(t)$ is the retardation function:

$$R(t) = \frac{2}{\pi} \int_0^\infty b(w) \cos(\omega t) d\omega \tag{3}$$

where b is the linear radiation damping matrix.

The hydrodynamic coefficients, such as the added mass, radiation damping, hydrostatic restoring stiffness, and first-order wave excitation force, are calculated in the frequency domain using the potential-based 3D diffraction/radiation code Wadam and then applied in the time domain.

2.2. Mooring Loads for SFOWT

The total pretension of the mooring system F^P is:

$$F^P = (\rho\nabla - m)g \tag{4}$$

where ρ is the density of sea water; ∇ is displacement; m is the total mass of the wind turbine system.

If mooring inertias and damping are ignored in a linear mooring system, the mooring load F_i^t can be calculated as follows [32]:

$$F_i^t = F_i^P - K_{ij}^t \xi_j \tag{5}$$

where F_i^P is the ith component of the total pretension; K_{ij}^t is the linearized restoring stiffness matrix of the mooring system; ξ_j is the ith degree of freedom displacement.

The restoring stiffness matrix of the mooring system can be obtained by the following formulas [27,33]:

$$K_{ij}^t = \begin{bmatrix} k_{11} & 0 & 0 & 0 & k_{15} & 0 \\ 0 & k_{22} & 0 & k_{24} & 0 & 0 \\ k_{31} & k_{32} & k_{33} & k_{34} & k_{35} & k_{36} \\ 0 & k_{42} & 0 & k_{44} & 0 & 0 \\ k_{51} & 0 & 0 & 0 & k_{55} & 0 \\ 0 & 0 & 0 & 0 & 0 & k_{66} \end{bmatrix} \begin{matrix} \text{Surge} \\ \text{Sway} \\ \text{Heave} \\ \text{Roll} \\ \text{Pitch} \\ \text{Yaw} \end{matrix} \tag{6}$$

$$k_{11} = k_{22} = \frac{F^P}{l_z} + \rho g A_w \frac{\delta_s}{l_z} \tag{7}$$

$$k_{33} = \frac{EA}{l} + \rho g A_w \tag{8}$$

$$k_{44} = \frac{\rho g I_x}{\cos^3 \varphi} + B(z_B - z_E) - G(z_G - z_E) + \frac{EI_{xx}}{l} \cos^2 \varphi + \rho g A_w (z_B - z_E)\delta_s + z_{GT}^2 k_{22} \tag{9}$$

$$k_{55} = \frac{\rho g I_y}{\cos^3 \theta} + B(z_B - z_E) - G(z_G - z_E) + \frac{EI_{yy}}{l} \cos^2 \theta + \rho g A_w (z_B - z_E)\delta_s + z_{GT}^2 k_{11} \tag{10}$$

$$k_{66} = r^2 \left(\frac{F^P}{l_z} + \rho g A_w \frac{\delta_s}{l_z} \right) \tag{11}$$

$$k_{15} = k_{51} = -z_{GT} k_{11} \tag{12}$$

$$k_{24} = k_{42} = z_{GT} k_{22} \tag{13}$$

$$k_{31} = k_{33} \xi_1 / (2l) \tag{14}$$

$$k_{32} = k_{33} \xi_2 / (2l) \tag{15}$$

$$k_{34} = z_{GT} k_{32} \tag{16}$$

$$k_{35} = -z_{GT} k_{31} \tag{17}$$

$$k_{36} = k_{33}\left(r^2 \xi_6\right)/(2l) \tag{18}$$

where B and G are the buoyancy and gravity of the SFOWT, respectively; l is the tether length; l_z is the mean vertical distance between the upper fairlead and seabed; E is the modulus of elasticity; A is the sectional area of a mooring line; ρ is the water density; g is the gravitational acceleration; A_w is the waterline area of SFOWT; r is the horizontal distance between the center of the column and the center of the cylinder-shaped pontoon; I_x and I_y are the inertia moment of the waterline about the x-axis and y-axis, respectively; I_{xx} and I_{yy} are the area moment of inertia of the SFOWT about the x-axis and y-axis, respectively; φ and θ are the rotational angles of roll and pitch, z_B, z_E, and z_G are the vertical coordinates of the buoyancy center, the upper fairlead, and the center of gravity, respectively; δ_s is the increment of the set-down motion, $\delta_s = l - l_z$; z_{GT} is vertical distance between the upper fairlead and the center of gravity.

3. Dynamic Response of SFOWT under Different Mooring Configurations

3.1. Structural Form of SFOWT

The SFOWT is shown in Figure 1, and its main parameters are listed in Tables 1 and 2. The submerged platform was composed of one column and four separated cylinder-shaped vertical pontoons connected by four rectangular horizontal pontoons. The column and pontoons were interconnected by four pipe members and cross braces. The wind turbine was mounted on the column. A NREL 5 MW baseline wind turbine was employed for the analysis, and its main parameters are listed in Table 3 [34].

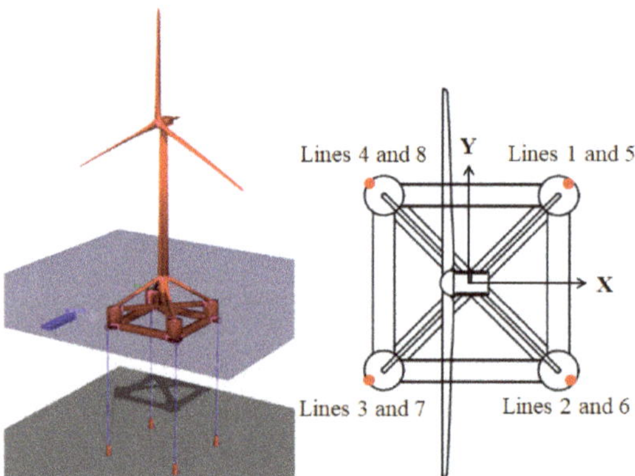

Figure 1. Overall model of the submerged floating offshore wind turbine (SFOWT).

Table 1. Main parameters of the floating platform.

Parameter	Value
Diameter of vertical pontoon	9 m
Height of vertical pontoon	12 m
Distance between vertical pontoons	40 m
Height of column	20 m
Column diameter	6 m
Mass of platform	2,734,200 kg
Mass moment of inertia in roll	7.818×10^8 kg m^2
Mass moment of inertia in pitch	7.818×10^8 kg m^2
Mass moment of inertia in yaw	1.359×10^9 kg m^2
COG of the platform during operation	$(0, 0, -16.75$ m$)$
Area of water plane during operation	51.45 m^2

Table 2. Main parameters of the mooring line.

Parameter	Value
Diameter of mooring line	0.127 m
Mass of per unit length	116.027 kg/m
Breaking strength	13,249 kN
Axial stiffness of mooring line (EA)	2.47×10^9 N

Table 3. Main parameters of the NREL 5 MW wind turbine.

Parameter	Value
Rated power	5 MW
Turbine control	Variable speed, collective pitch
Rotor diameter	126 m
Hub diameter	3 m
Hub height	90 m
Cut-in, rated, cut-out wind speeds	3, 11.4, and 25 m/s
Mass of impeller	110,000 kg
Mass of nacelle	240,000 kg
Mass of tower	347,460 kg
Centroid coordinates	(−0.2 m, 0, 74 m)

3.2. Dynamic Response

The safety factor of a mooring line can be expressed as follows:

$$F = \frac{P_B}{T_{max}} \tag{19}$$

where P_B and T_{max} are the breaking strength and maximum tension in the mooring line, respectively. The requirements of the minimum safety factor in different states are listed in Table 4.

Table 4. Requirements of the minimum safety factors [35].

State	Analysis Method	Safety Factor
Normal operation state (NOS)	Dynamic	1.67
Extreme sea-states	Dynamic	1.3
Broken mooring lines	Dynamic	1.0

The motion responses and internal force distributions of SFOWT under four different mooring configurations are shown in Table 5 and Figure 2, with a draft of 22 m, a water depth of 100 m, a wind speed of 11.4 m/s, a significant wave height of 3 m, and a peak period of 10 s generated by the JONSWAP spectrum with a peak enhancement factor of 3.3. The directions of wind and wave are along the X-axis. The sea-states are modeled using three-hour periods, and the statistical analysis was carried out using the time-series of 4000–5000 s. The pretension in mooring lines is shown in Table 6.

Table 5. Mooring configurations.

Mooring Configuration	No. of Vertical Mooring Lines	No. of Diagonal Mooring Lines	Inclination Angle
Configuration 1	8	-	-
Configuration 2	4	4	15°
Configuration 3	4	4	30°
Configuration 4	4	12	30°

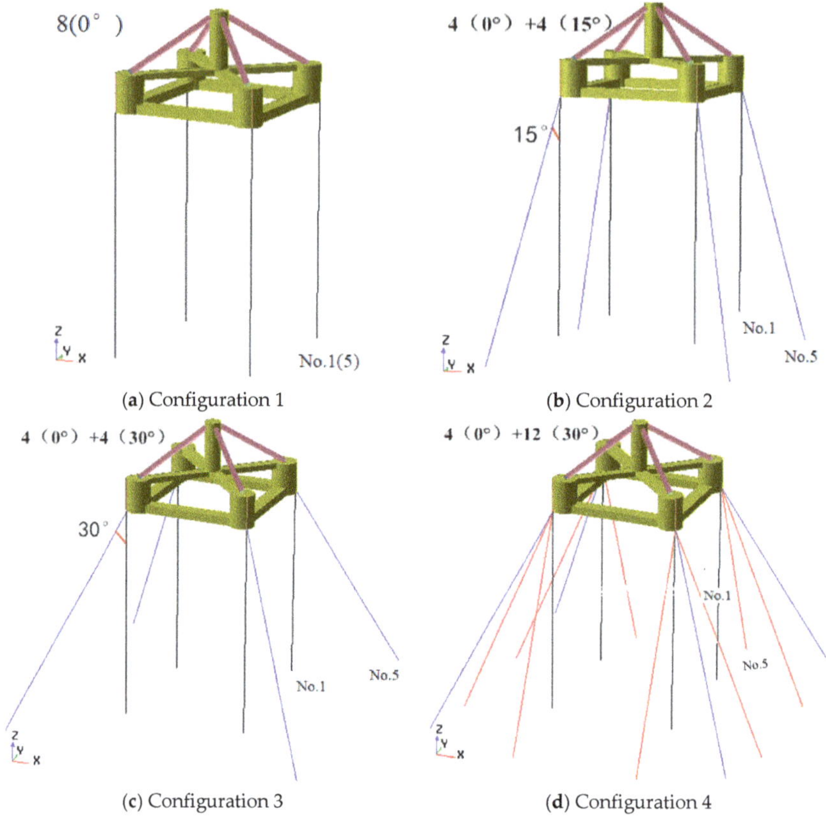

Figure 2. Four mooring configurations.

Table 6. Pretension in mooring lines.

Mooring Configuration	Pretension in Vertical Lines/kN	Pretension in Diagonal Lines/kN
Configuration 1	3600	-
Configuration 2	3600	3727
Configuration 3	3600	4157
Configuration 4	1800	2078

Figures 3 and 4 show the time series of platform motion under different mooring configurations and the corresponding motion statistics, respectively. The mooring lines provide a restoring force for the platform, affecting its motion response. As can be seen from Figures 3a and 4a, the surge equilibrium position under Configuration 1 had a larger displacement than the others; when there was a diagonal mooring line, the maximum value and standard deviation of surge were in the order of Configuration 2 > Configuration 3 > Configuration 4. In comparison to Configuration 2, the maximum value and standard deviation under Configuration 4 were reduced by 95.9% and 96.4%, respectively. The diagonal mooring lines provide a horizontal restoring force for the floating platform. As the tilt angle of the mooring line increased, the horizontal restoring force provided by the mooring lines increased. As shown in Figures 3b and 4b, the heave response of the platform under Configuration 2 was larger, because it had a smaller vertical restoring force than Configurations 1 and 4, and the large surge motion also induced the large heave motion due to the set-down motion. Figure 3c,d and Figure 4c,d show that the pitch and yaw responses were largest under Configurations 2, while the average of the pitch and yaw response was close to zero under the four configurations. As shown in Figures 3 and 4, the SFOWT had a smaller response and a better performance under Configuration 4.

Figures 5 and 6 show the time series of bending moment at the root of the blade and tower base under different mooring configurations and the corresponding statistics, respectively. There was little change in the bending moment at the root of the blade, which was due to the bending moment being dominated by wind load. The effect of the specific form of configurations was not obvious. The mean value of the tower base bending moment was mainly wind-induced, while its standard deviation was induced by both wind and wave loads; the standard deviation was also influenced by the motion response of the platform. From Figures 5b and 6b, it can be observed that the maximum value and standard deviation appeared under Configuration 2.

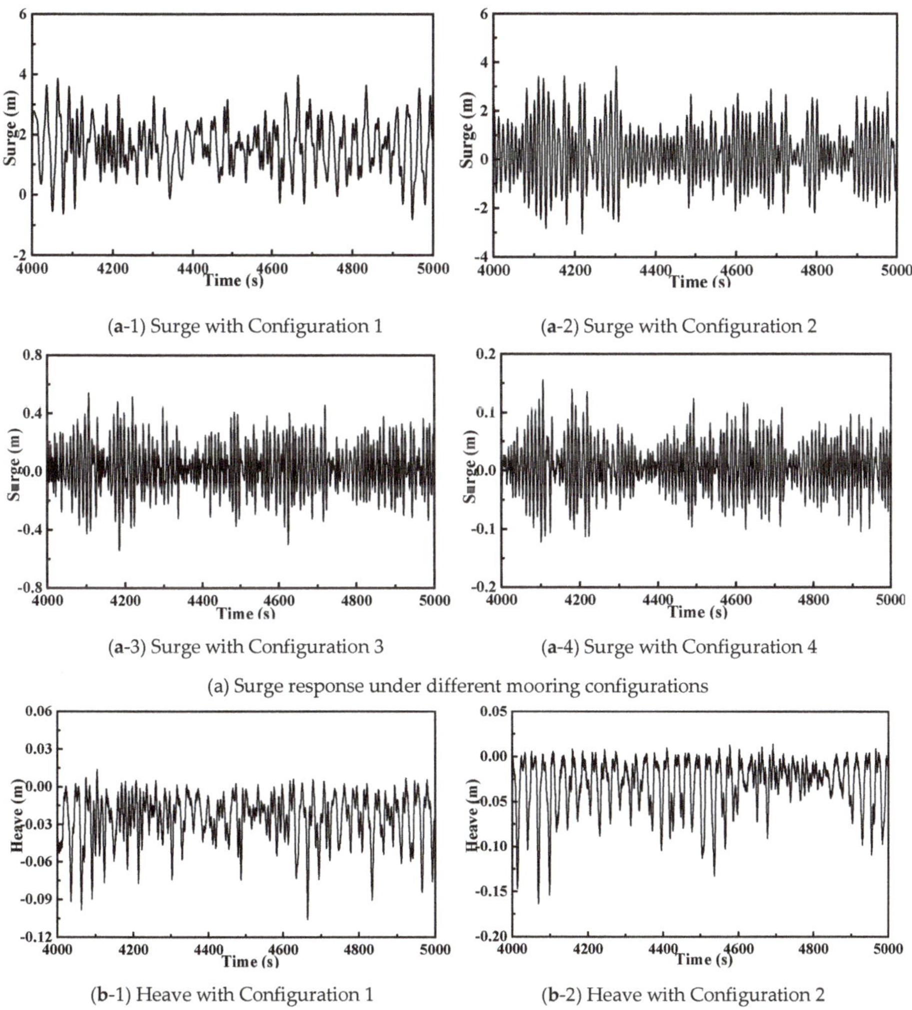

(**a**-1) Surge with Configuration 1   (**a**-2) Surge with Configuration 2

(**a**-3) Surge with Configuration 3   (**a**-4) Surge with Configuration 4

(a) Surge response under different mooring configurations

(**b**-1) Heave with Configuration 1   (**b**-2) Heave with Configuration 2

Figure 3. *Cont.*

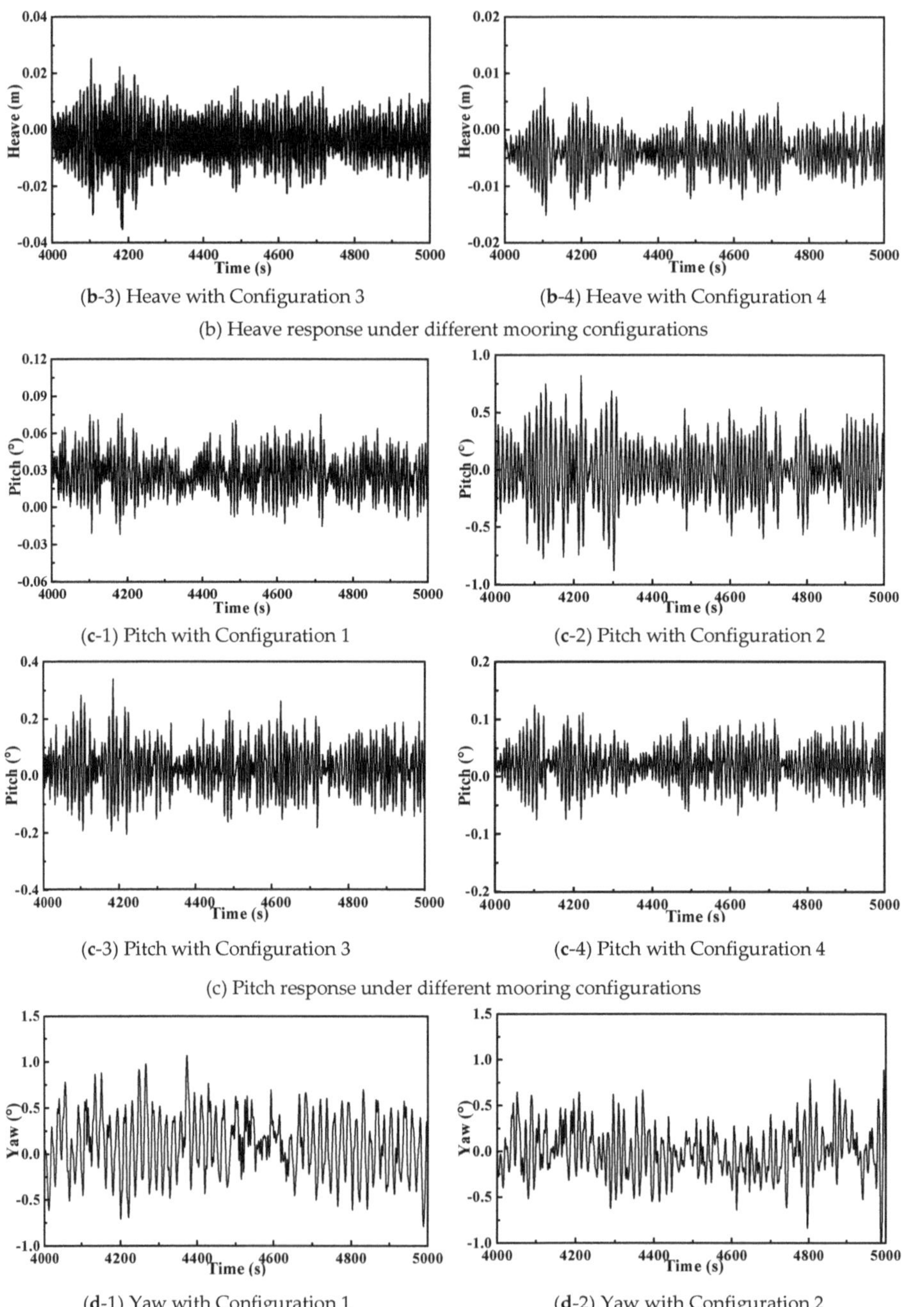

(**b**-3) Heave with Configuration 3   (**b**-4) Heave with Configuration 4

(b) Heave response under different mooring configurations

(**c**-1) Pitch with Configuration 1   (**c**-2) Pitch with Configuration 2

(**c**-3) Pitch with Configuration 3   (**c**-4) Pitch with Configuration 4

(c) Pitch response under different mooring configurations

(**d**-1) Yaw with Configuration 1   (**d**-2) Yaw with Configuration 2

Figure 3. *Cont.*

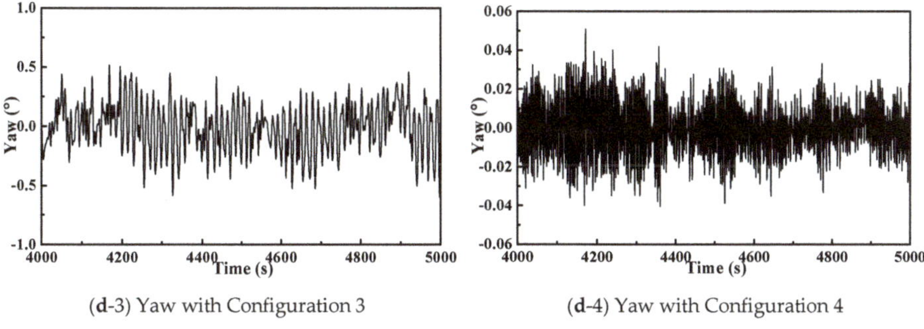

(**d**-3) Yaw with Configuration 3   (**d**-4) Yaw with Configuration 4

(d) Yaw response under different mooring configurations

Figure 3. Time series of the platform motion under different mooring configurations.

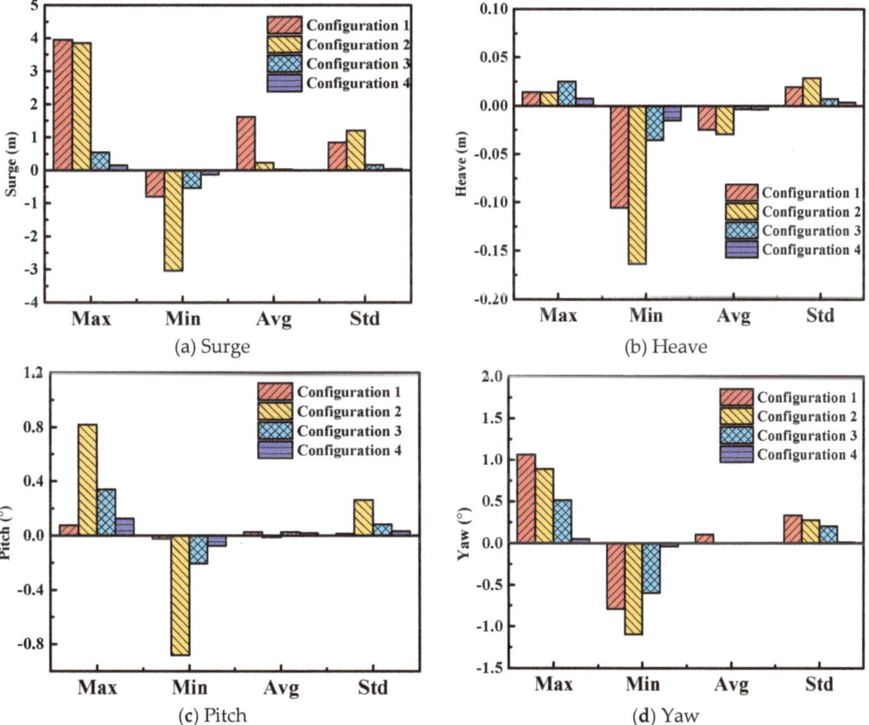

(a) Surge   (b) Heave

(c) Pitch   (d) Yaw

Figure 4. Motion statistics of the platform under different mooring configurations.

The time series of tension in Mooring Lines No. 1 (vertical) and No. 5 (diagonal) under different mooring configurations and the corresponding statistics are shown in Figures 7 and 8, respectively. The dynamic response of the platform drove the movement of the mooring lines, and the tension was influenced by wave loads. The smallest tensions appeared under Configuration 4.

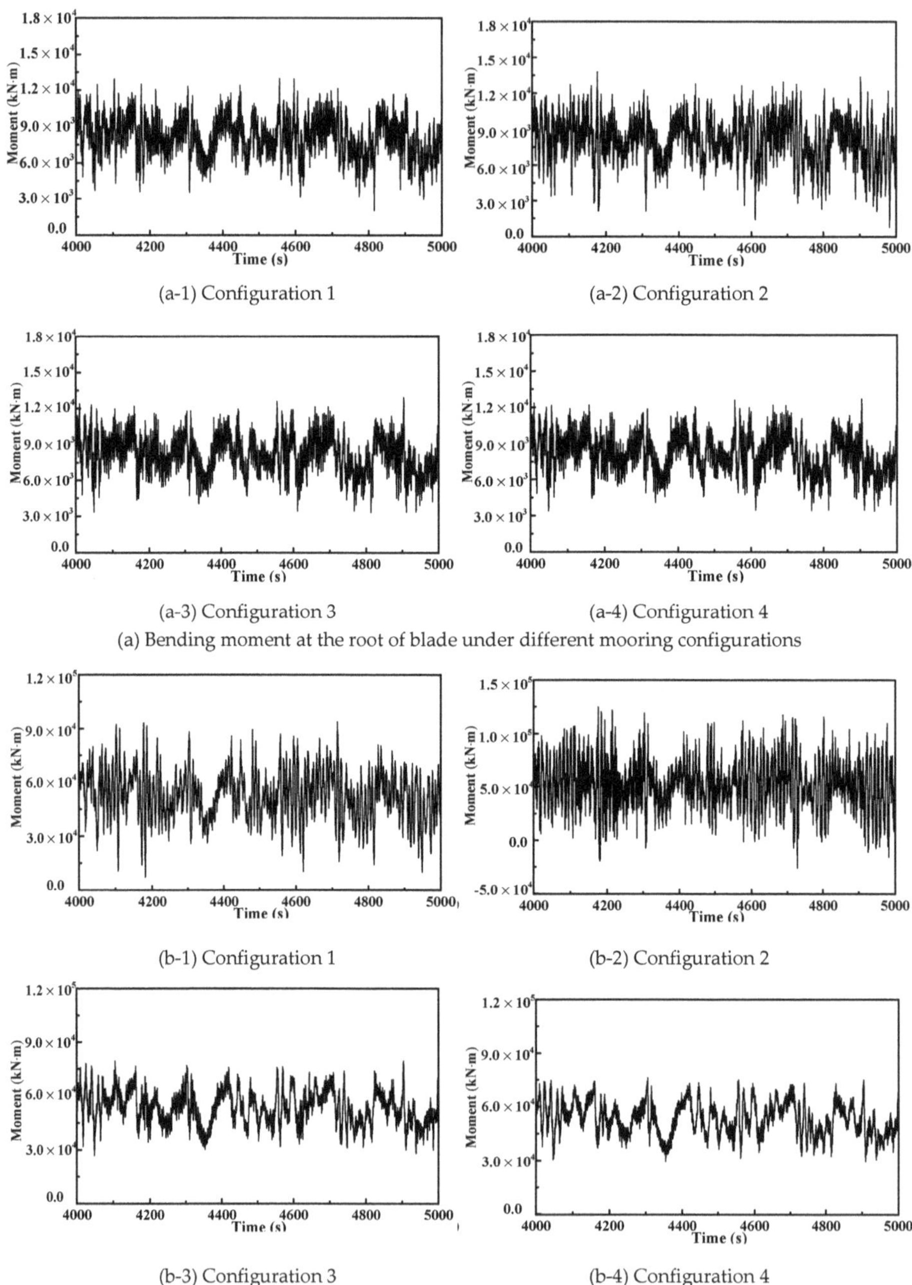

(a) Bending moment at the root of blade under different mooring configurations

(b) Bending moment at the tower base under different mooring configurations

Figure 5. Time series of the bending moment at the root of blade and tower base under different mooring configurations.

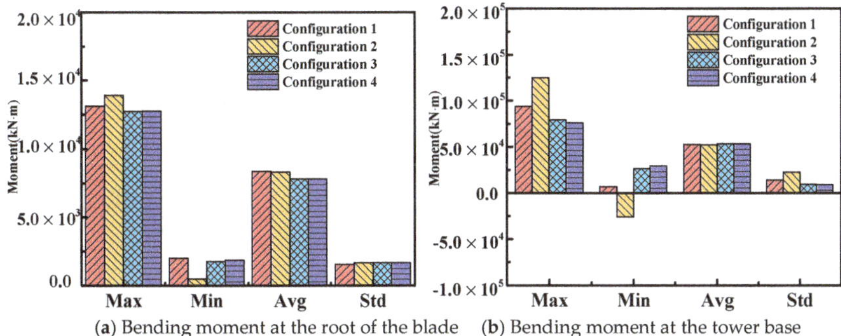

(**a**) Bending moment at the root of the blade (**b**) Bending moment at the tower base

Figure 6. Statistics of the bending moment at the root of blade and tower base under different mooring configurations.

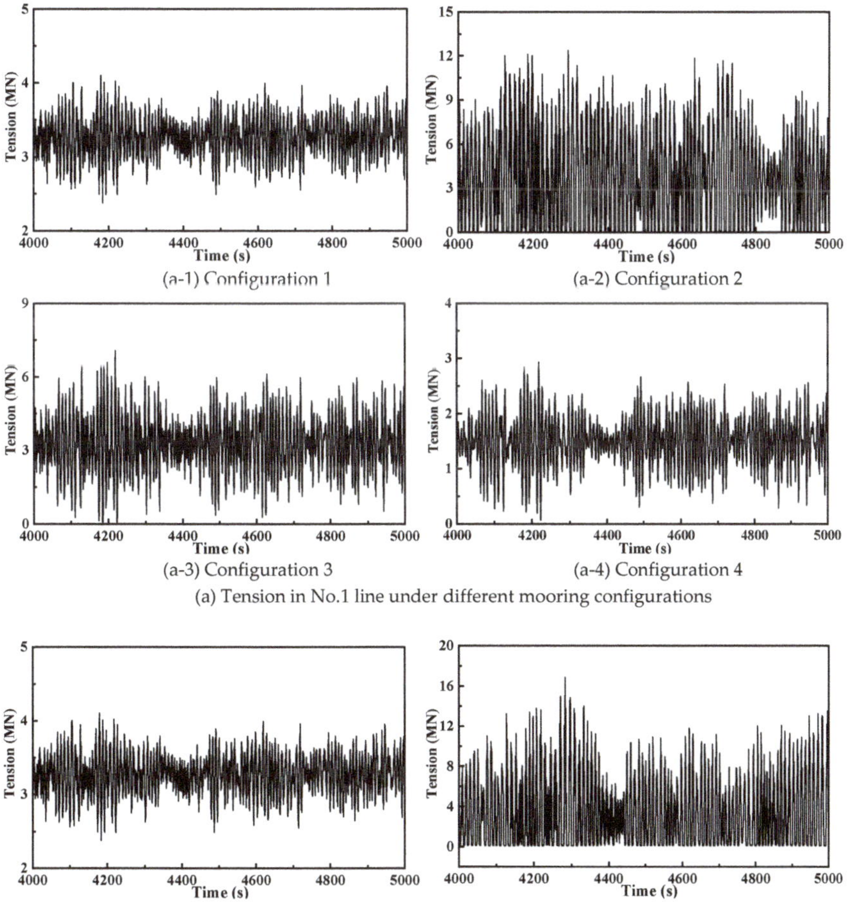

(a-1) Configuration 1

(a-2) Configuration 2

(a-3) Configuration 3

(a-4) Configuration 4

(a) Tension in No.1 line under different mooring configurations

(b-1) Configuration 1

(b-2) Configuration 2

Figure 7. *Cont.*

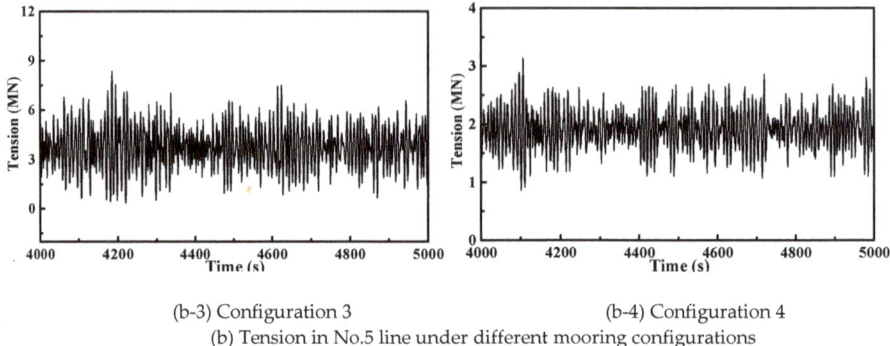

(b-3) Configuration 3   (b-4) Configuration 4

(b) Tension in No.5 line under different mooring configurations

Figure 7. Time series of tension on mooring lines in different mooring configurations.

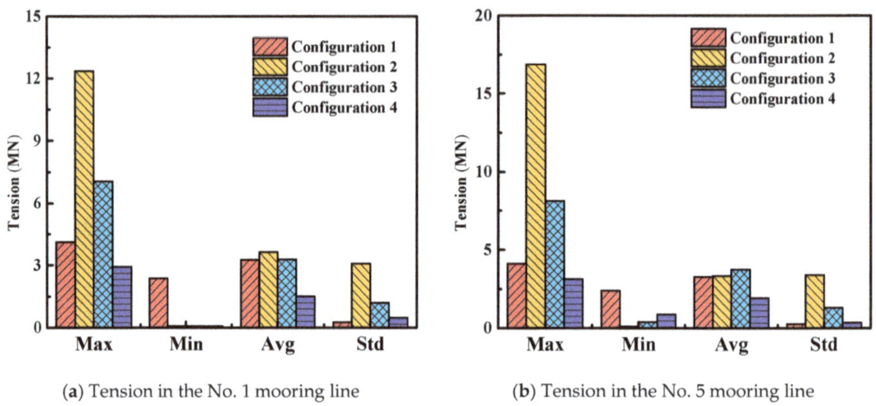

(**a**) Tension in the No. 1 mooring line   (**b**) Tension in the No. 5 mooring line

Figure 8. Statistics of tension in mooring lines under different mooring configurations.

The safety factor of mooring lines under different mooring configurations is shown in Figure 9. It can be seen that only Configuration 2 cannot meet the specification requirements.

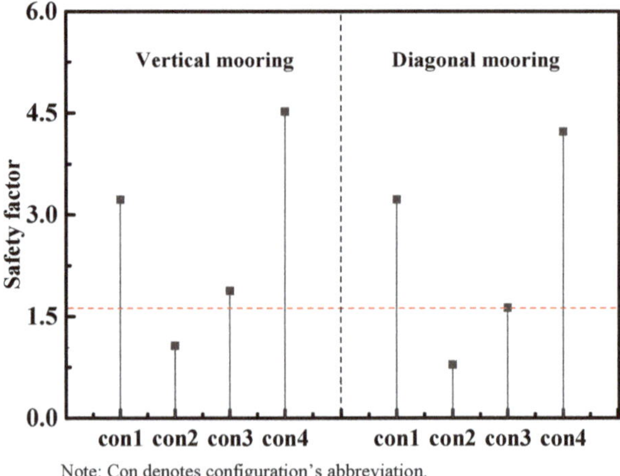

Note: Con denotes configuration's abbreviation.

Figure 9. Safety factor of mooring lines under different mooring configurations.

4. Dynamic Response of SFOWT under Extreme Sea-States

The SFOWT under Configuration 4 (i.e., four vertical lines and 12 diagonal lines with an inclination angle of 30°), as shown in Figure 10, were further studied in once-in-one-year and 50-years sea conditions of the East China Sea areas (see Table 7). Under extreme sea-states, the wind speed exceeded its cut-out value, and the wind turbine was under the condition of shutdown. The time histories of wind speed and wave elevation in the X-direction are shown in Figures 11 and 12, respectively; the power spectra of the wind speed and wave elevation are shown in Figures 13 and 14, respectively. We can see that the energy of the wind was mainly concentrated below 0.05 Hz, while the wave spectrum was mainly concentrated around 0.1 Hz.

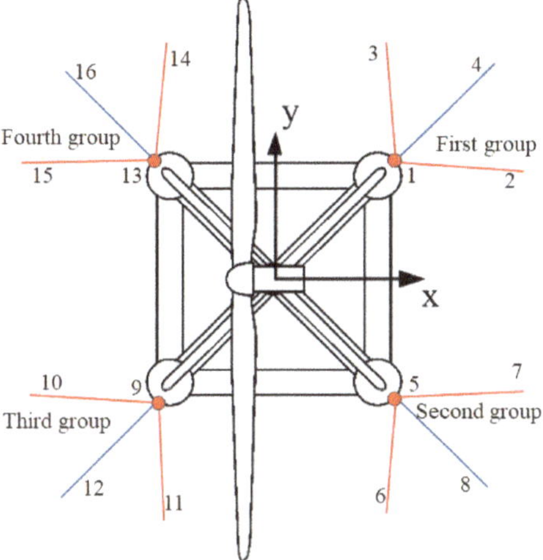

Figure 10. Mooring configuration of SFOWT.

Table 7. Environmental loads under extreme sea-states.

Extreme Sea-States	Wind Speed (m/s)	H_s (m)	T_p (s)
Once-in-1-year	30	7.1	11.9
Once-in-50-years	49	13	15.5

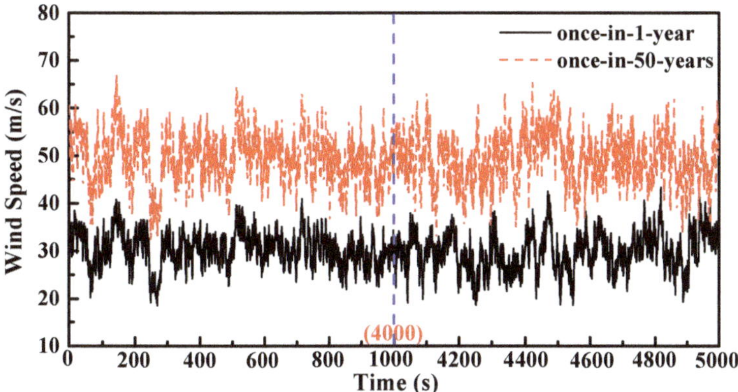

Figure 11. Time series of wind speed under extreme sea-states.

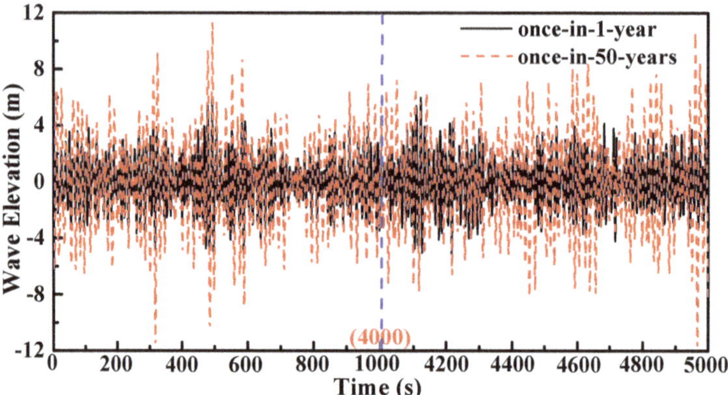

Figure 12. Time series of wave elevation under extreme sea-states.

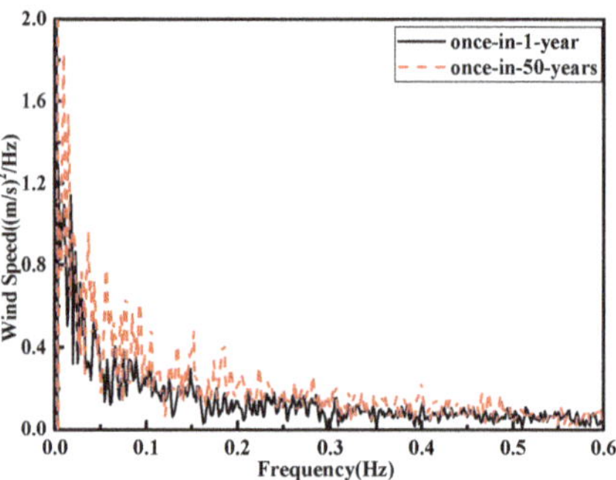

Figure 13. Power spectra of wind speed.

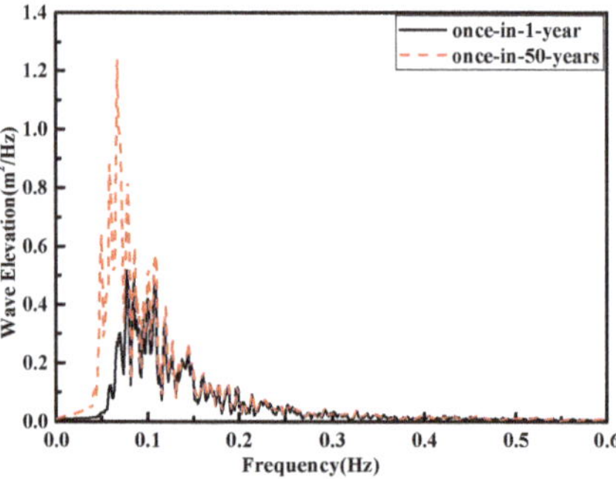

Figure 14. Power spectra of wave elevation.

The time series of platform surge motion and tension in the No. 1 mooring line under extreme sea-states are shown in Figures 15 and 16, respectively. It can be seen that the surge response suddenly increased around 300–340 s in the once-in-50-years sea condition, and the tension in the mooring line also quickly increased and then fell. This is because the wave elevation increased suddenly around 300–340 s (Figure 12), which led to a decrease of the restoring force on the platform and an increase of the motion response, and eventually caused a slack in mooring lines and instability of SFOWT in the once-in-50-years sea condition.

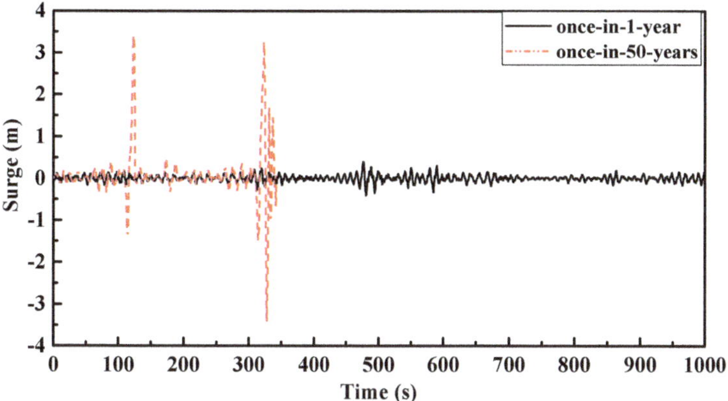

Figure 15. Time series of platform surge motion under extreme sea-states.

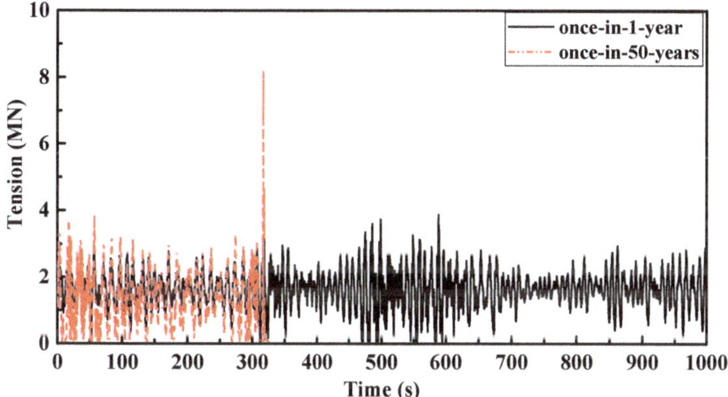

Figure 16. Time series of tension in the No. 1 mooring line under extreme sea-states.

The increase in the pretension in the mooring line can effectively avoid the slack in mooring lines and improve the movement performance of SFOWT. The vertical pretension was increased from 1800–2780 kN per mooring line by adjusting the ballast water of the platform. The corresponding motion statistics and power spectra of platform motion under extreme sea-states are shown in Figures 17 and 18, respectively. The platform had a small surge motion and a better heave performance due to the growing pretension under extreme sea-states. The yaw of SFOWT was small in the two sea conditions. In comparison to the once-in-one-year sea condition, the maximum value and standard deviation of yaw under the once-in-50-years sea condition increased by 176.8% and 319.8%, respectively.

As the wind loads are much smaller than wave loads in extreme sea-states because the turbine is parked, the spectra of motions were mainly dominated by the wave frequency response and resonant response. From Figure 18, it can be found that the surge motion was mainly dominated by the wave-frequency response; the heave motion was mainly induced by the wave-frequency response and set-down effect; and the pitch motion was mainly induced by wave-frequency and surge resonant response. The natural frequency on each DOF was higher than the wave frequency; the heave resonant response was the highest, i.e., 1.0 Hz.

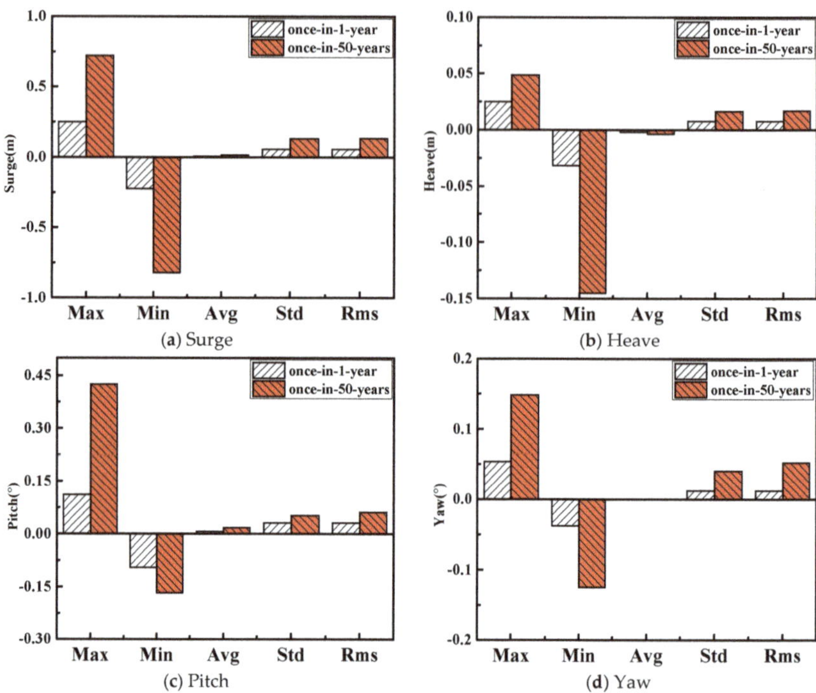

Figure 17. Motion statistics of the platform under extreme sea-states.

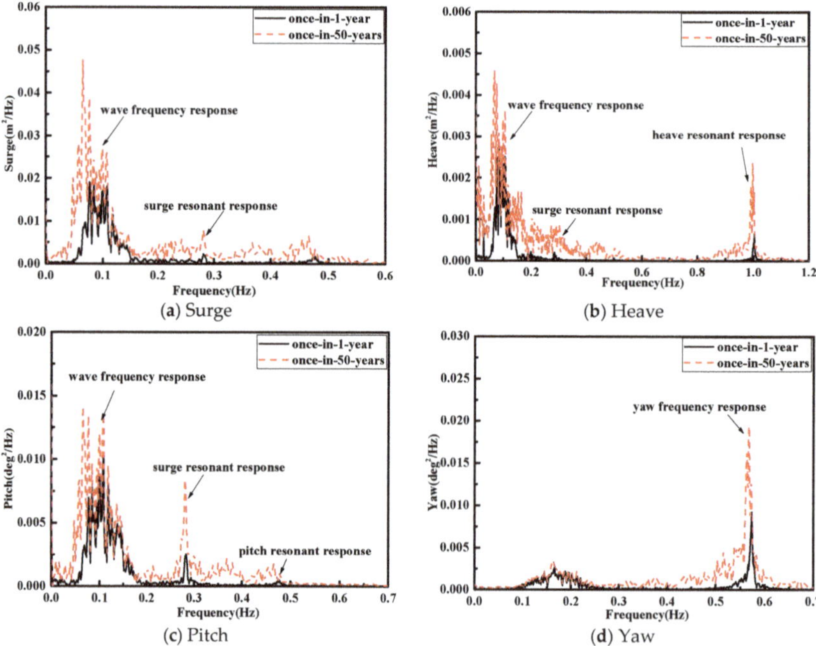

Figure 18. Power spectra of the platform motion under extreme sea-states.

The statistics of tension in vertical mooring lines (Nos. 1, 5, 9, and 13) and diagonal lines (Nos. 4, 6, 12, and 15) under once-in-one-year and 50-years sea conditions are shown in Figure 19. As shown in Figure 19, the No. 9 mooring line, which was located in the upwind direction of the SFOWT, had the largest average and standard deviation of tension among the vertical mooring lines, and the No. 15 mooring line, which was also located in the upwind direction of the SFOWT, had the largest maximum and standard deviation of tension among the diagonal mooring lines. In addition, the minimum tension in each line under extreme sea-states was larger than zero, indicating that they can provide enough stability.

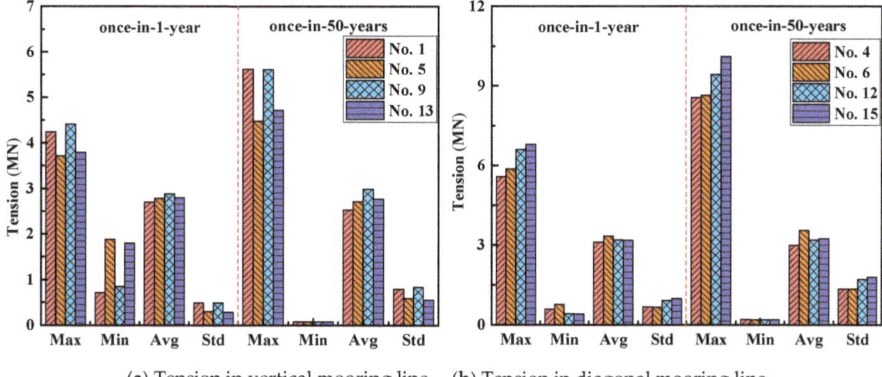

(a) Tension in vertical mooring line   (b) Tension in diagonal mooring line

Figure 19. Statistics of tension in the mooring lines under extreme sea-states.

Figure 20 shows the safety factor of mooring lines under extreme sea-states, indicating that the safety factor of each mooring line meets the design requirements in once-in-one-year and 50-years sea conditions. The safety factor of the No. 15 mooring line was the smallest, i.e., 1.31.

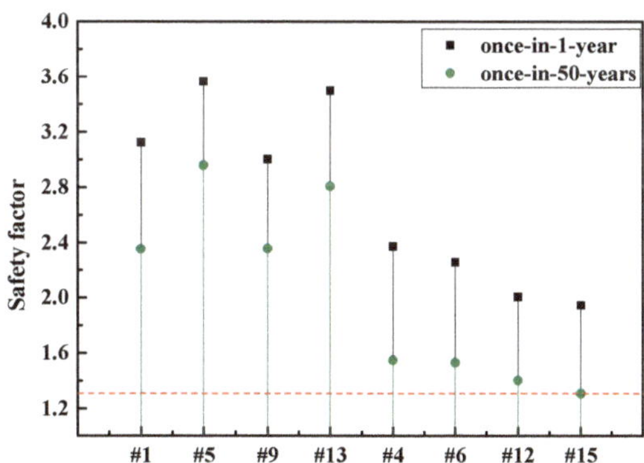

Figure 20. Safety factor of the mooring lines under extreme sea-states.

5. Dynamic Response of SFOWT with Broken Mooring Lines

There may be one or several broken mooring lines in a floating wind turbine under long-term environmental loads. The once-in-one-year sea condition was chosen to simulate an accident during operation [36]. Four cases were considered, as listed in Table 8.

The time series and motion statistics of platform motion with broken mooring lines are shown in Figures 21 and 22, respectively. It can be seen that in the course of 4100 s, the motion responses on all DOFs fluctuated and the amplitudes increased due to the wave frequency response. The motion response on each DOF of the platform in Case 1 was the largest. Because the No. 15 mooring line was in the upwind wave direction, as it consistently experienced a larger wave impact, an increasing motion response will be caused if it is broken, and the equilibrium position of yaw motion will be changed when the platform is at a new horizontal location. If the No. 1 mooring line is disconnected, the motion response will be smaller owing to its position in the downwind direction against the environmental loads. In comparison to Case 1, the maximum value and standard deviation of yaw in Case 2 were reduced by 83.5% and 60.2%, respectively.

Table 8. Conditions of broken mooring lines.

Case	Broken Mooring Line
Case 1	No. 15
Case 2	No. 1
Case 3	Nos. 1 and 12
Case 4	Nos. 1 and 13

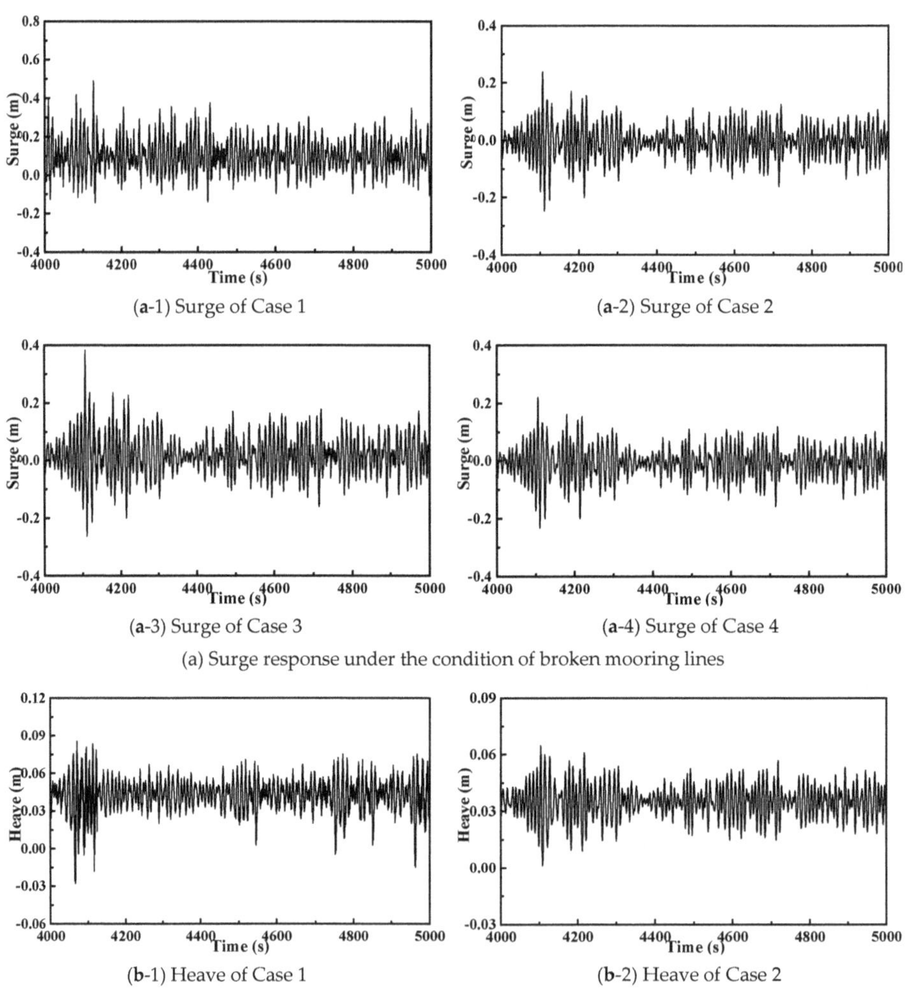

(a-1) Surge of Case 1   (a-2) Surge of Case 2

(a-3) Surge of Case 3   (a-4) Surge of Case 4

(a) Surge response under the condition of broken mooring lines

(b-1) Heave of Case 1   (b-2) Heave of Case 2

Figure 21. *Cont.*

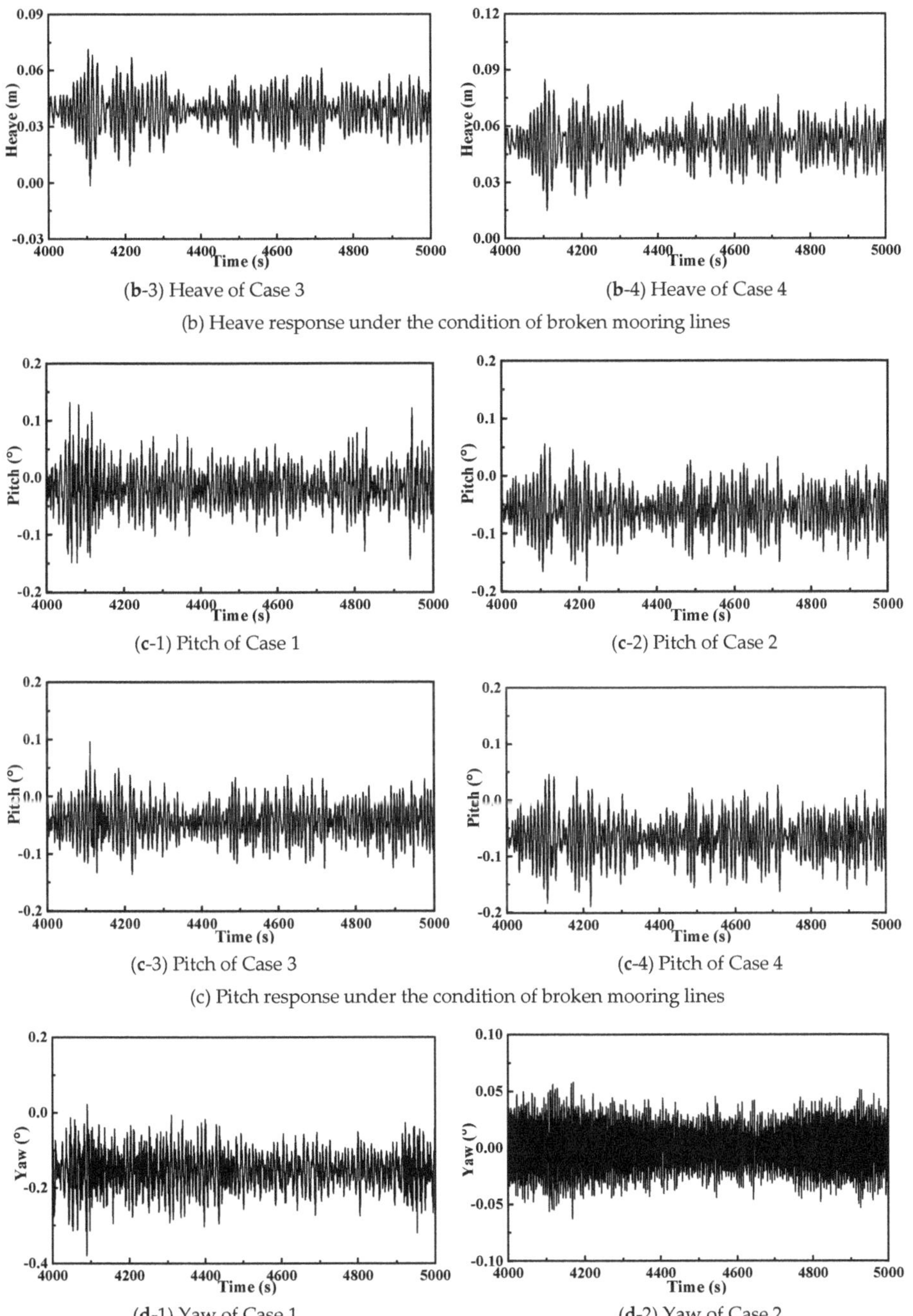

(**b**-3) Heave of Case 3   (**b**-4) Heave of Case 4

(b) Heave response under the condition of broken mooring lines

(**c**-1) Pitch of Case 1   (**c**-2) Pitch of Case 2

(**c**-3) Pitch of Case 3   (**c**-4) Pitch of Case 4

(c) Pitch response under the condition of broken mooring lines

(**d**-1) Yaw of Case 1   (**d**-2) Yaw of Case 2

Figure 21. *Cont.*

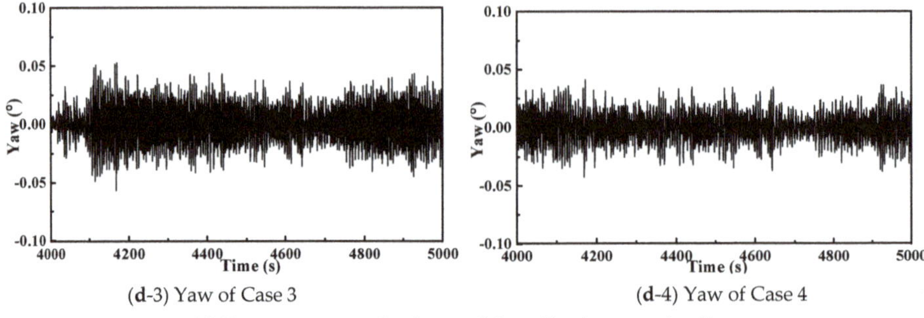

Figure 21. Time series of the platform motion under the condition of broken mooring lines.

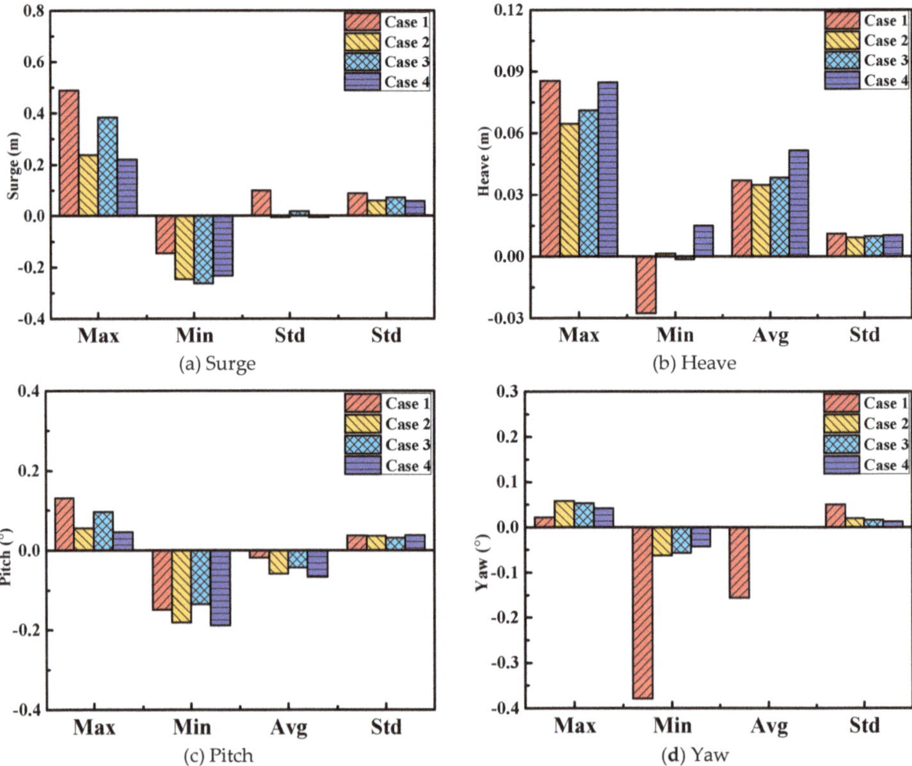

Figure 22. Motion statistics of the platform with broken mooring lines.

In all four cases, the motion responses of the platform were relatively smaller on each DOF, indicating that SFOWT had good stability in the condition with broken mooring lines.

The statistics of tension in vertical (Nos. 1, 5, 9, and 13) and diagonal (the one that has the maximum tension in each group) lines in the four different conditions with broken mooring lines are shown in Figure 23. When a certain line is broken, the tension in the mooring line with the same fairlead will increase. The tension of the mooring line in the upwind wave direction will become larger as it consistently experiences wave loads. The tension in the diagonal lines will be larger than those in vertical lines, which is due to the larger pretension in diagonal lines. In all four cases, there was no slack in mooring lines, which remained tight all the time.

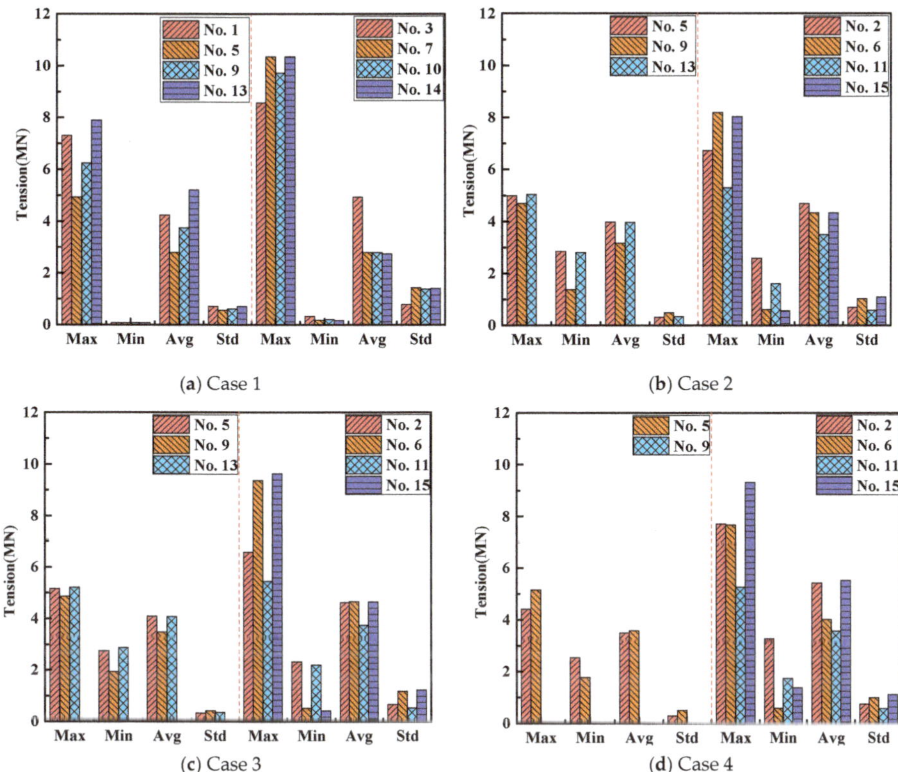

Figure 23. Statistics of tension in mooring lines in conditions with broken mooring lines.

6. Conclusions

A coupled dynamic model of SFOWT was established in this paper, and simulations of four different mooring configurations under normal sea conditions were performed. It was found that the motion response of the platform was smaller and the stability of SFOWT was better under Configuration 4 (i.e., four vertical lines and 12 diagonal lines with an inclination angle of 30°). In comparison to Configuration 2 (i.e., four vertical lines and four diagonal lines with an inclination angle of 15°), the maximum value and standard deviation under Configuration 4 were reduced by 95.9% and 96.4%, respectively. Since only the safety factor under Configurations 2 did not meet the requirement, the cross-sectional area of mooring line should be increased to reduce the tension. The safety factors of the No. 1 (vertical) and No. 5 (diagonal) mooring lines under Configuration 4 were 4.52 and 4.23, respectively.

The well-stabilized Configuration 4 was selected, and the pretension was increased to control the platform movement effectively by adjusting the ballast water of the platform. The coupled dynamic responses of SFOWT under extreme sea-states were checked. As the wind loads were much smaller than the wave loads in extreme sea-states because the turbine was parked, the spectra of the motions were mainly dominated by the wave frequency response and resonant response. It was found that this configuration can provide better surge and yaw performances, as well as better horizontal restraints; moreover, the motion responses of SFOWT on six degree of freedoms were smaller.

The influence of broken mooring lines on the SFOWT performance was investigated. Four failure cases of mooring line were considered, among which Case 1, which was the No. 15 mooring line in the upwind wave direction being broken, had the largest motion response. However, there was no slack in the remaining mooring lines, and the SFOWT still had enough stability.

Author Contributions: Conceptualization, H.D., C.L. and P.Z.; Methodology, C.L. and Y.L.; Software, Y.L. and C.L.; validation, C.L., H.D. and P.Z.; Formal Analysis, Y.L. and C.L.; Writing-Original Draft Preparation, Y.L. C.L. and J.Z.; Writing-Review & Editing, C.L. and Y.L.; Supervision, H.D., C.L. and P.Z; Project administration, H.D. and P.Z; Funding acquisition, H.D., P.Z. and C.L.

Funding: This research was funded by the National Natural Science Foundation of China (Grant Nos. 51679163 & 51779171), Innovation Method Fund of China (Grant No. 2016IM030100), Tianjin Municipal Natural Science Foundation (Grant No. 18JCYBJC22800) and State Key Laboratory of Coastal and Offshore Engineering (Grant No. LP1709).

Conflicts of Interest: The authors declare no conflict of interest.

References

1. Bachynski, E.E. *Design and Dynamic Analysis of Tension Leg Platform Wind Turbines*; Norwegian University of Science and Technology: Trondheim, Norway, 2014.
2. Wen, B.; Tian, X.; Dong, X.; Peng, Z.; Zhang, W. Influences of surge motion on the power and thrust characteristics of an offshore floating wind turbine. *Energy* **2017**, *141*, 2054–2068. [CrossRef]
3. Musial, W.; Butterfield, S.; Boone, A. Feasibility of floating platform systems for wind turbines. In Proceedings of the 42nd AIAA Aerospace Sciences Meeting and Exhibit, Reno, Nevada, 5 January 2004.
4. Roddier, D.; Cermelli, C.; Aubault, A.; Weinstein, A. WindFloat: A floating foundation for offshore wind turbines. *J. Renew. Sustain. Energy* **2010**, *2*, 033104. [CrossRef]
5. Hong, S.; Lee, I.; Park, S.H.; Lee, C.; Chun, H.-H.; Lim, H.C. An experimental study of the effect of mooring systems on the dynamics of a spar buoy-type floating offshore wind turbine. *Int. J. Nav. Archit. Ocean Eng.* **2015**, *7*, 559–579. [CrossRef]
6. Musial, W.; Butterfield, S.; Ram, B. Energy from offshore wind. In Proceedings of the Offshore Technology Conference, Houston, TX, USA, 1–4 May 2006.
7. Stewart, G.M.; Lackner, M.A. The effect of actuator dynamics on active structural control of offshore wind turbines. *Eng. Struct.* **2011**, *33*, 1807–1816. [CrossRef]
8. Arapogianni, A.; Genachte, A.; Ochagavia, R.M.; Vergara, J.P.; Castell, D.; Tsouroukdissian, A.R.; Korbijn, J.; Bolleman, N.C.F.; Huera-Huarte, F.J.; Schuon, F.; et al. *Deep Water: The next Step for Offshore Wind Energy*; European Wind Energy Association: Belgium, Brussels, 2013.
9. Bjoska, B.; Hanson, T.; Ruy, R.; Nielsen, F. Dynamic response and control of the hywind demo floating wind turbine. In Proceedings of the European Wind Energy Conference, Warsaw, Poland, 20–23 April 2010.
10. Weinzettel, J.; Reenaas, M.; Solli, C.; Hertwich, E.G. Life cycle assess of a floating offshore wind turbine. *Renew. Energy* **2009**, *34*, 742–747. [CrossRef]
11. Aubault, A.; Cermelli, C.; Roddier, D. WindFloat: A floating foundation for offshore wind turbines-part III: Structural analysis. In Proceedings of the ASME 2009 28th International Conference on Ocean, Offshore and Arctic Engineering, Honolulu, HI, USA, 31 May–5 June 2009; American Society of Mechanical Engineers: New York, NY, USA, 2009; pp. 213–220.
12. Chengxi, L.; Jun, Z. Nonlinear Coupled Dynamics Analysis of A Truss Spar Platform. *China Ocean Eng.* **2016**, *30*, 835–850.
13. Lefebvre, S.; Collu, M. Preliminary design of a floating support structure for 5 MW offshore wind turbine. *Ocean Eng.* **2012**, *40*, 15–26. [CrossRef]
14. Polaris Wind Power Network. The Success Assembly of the First Floating Offshore Wind Power Base in the World. Polaris Wind Power. 2017. Available online: http://news.bjx.com.cn/html/20170207/806667.shtml (accessed on 7 February 2017).
15. Coulling, A.J.; Goupee, A.J.; Robertson, A.N.; Jonkman, J.M.; Dagher, H.J. Validation of a FAST semi-submersible floating wind turbine numerical model with DeepCwind test data. *J. Renew. Sustain. Energy* **2013**, *5*, 023116. [CrossRef]

16. Duan, F.; Hu, Z.; Niedzwecki, J.M. Model test investigation of a spar floating wind turbine. *Mar. Struct.* **2016**, *49*, 76–96. [CrossRef]

17. Benassai, G.; Campanile, A.; Piscopo, V.; Scamardella, A. Optimization of Mooring Systems for Floating Offshore Wind Turbines. *Coast. Eng. J.* **2015**, *57*, 1550021-1–1550021-9. [CrossRef]

18. Liu, Y.; Xiao, Q.; Incecik, A.; Peyrard, C.; Wan, D. Establishing a fully coupled CFD analysis tool for floating offshore wind turbines. *Renew. Energy* **2017**, *112*, 280–301. [CrossRef]

19. Benassai, G.; Campanile, A.; Piscopo, V.; Scamardella, A. Ultimate and accidental limit state design for mooring systems of floating offshore wind turbines. *Ocean Eng.* **2014**, *92*, 64–74. [CrossRef]

20. Yasunori, N.; Yusuke, M. Research and development about the mechanisms of a single point mooring system for offshore wind turbines. *Ocean Eng.* **2018**, *147*, 431–446.

21. Utsunomiya, T.; Yoshida, S.; Ookubo, H.; Sato, I.; Ishida, S. Dynamic analysis of a floating offshore wind turbine under extreme environmental conditions. *J. Offshore Mech. Arct. Eng.* **2014**, *136*, 02094. [CrossRef]

22. Bae, Y.H.; Kim, M.H.; Kim, H.C. Performance changes of a floating offshore wind turbine with broken mooring line. *Renew. Energy* **2017**, *101*, 364–375. [CrossRef]

23. Li, Y.; Zhu, Q.; Liu, L.; Tang, Y. Transient response of a SPAR-type floating offshore wind turbine with fractured mooring lines. *Renew. Energy* **2018**, *122*, 576–588. [CrossRef]

24. Ahmed, M.O.; Yenduri, A.; Kurian, V.J. Evaluation of the dynamic responses of truss SPAR platforms for various mooring configurations with damaged lines. *Ocean Eng.* **2016**, *123*, 411–421. [CrossRef]

25. Ding, H.; Han, Y.; Zhang, P. Dynamic analysis of a new type of floating platform for offshore wind turbine. In Proceedings of the 26th International Ocean and Polar Engineering Conference, Rhodes, Greece, 26 June–2 July 2016.

26. Han, Y.; Le, C.; Ding, H.; Cheng, Z.; Zhang, P. Stability and dynamic response analysis of a submerged tension leg platform for offshore wind turbines. *Ocean Eng.* **2017**, *129*, 68–82. [CrossRef]

27. Le, C.; Li, Y.; Ding, H. Study on the coupled dynamic responses of a submerged floating wind turbine under different mooring conditions. *Energies* **2019**, *12*, 418. [CrossRef]

28. Jonkman, J.M. Dynamics of Offshore Floating Wind Turbines-Model Development and Verification. *Wind Energy* **2009**, *12*, 459–492. [CrossRef]

29. Jonkman, B.J. TurbSim User's Guide: Version 1.50. Available online: https://www.nrel.gov/docs/fy09osti/46198.pdf (accessed on 26 January 2019).

30. Det Norske Veritas. *SESAM User Manual HydroD*; Det Norske Veritas: Oslo, Norway, 2013.

31. Bae, Y.H.; Kim, M.H. Rotor-floater-tether coupled dynamic analysis including second-order sum-frequency wave loads for a mono-column-TLP-type FOWT (floating offshore wind turbine). *Ocean Eng.* **2013**, *61*, 109–122. [CrossRef]

32. Zhao, Y.S.; Yang, J.M.; He, Y.P. Coupled dynamic response analysis of a multi-column tension-leg-type floating wind turbine. *China Ocean Eng.* **2016**, *30*, 505–520. [CrossRef]

33. Li, Y.; Tang, Y.-G.; Zhu, Q.; Qu, X.-Q.; Wang, B.; Zhang, R.-Y. Effects of second-order wave forces and aerodynamic forces on dynamic responses of a TLP-type floating offshore wind turbine considering the set-down motion. *J. Renew. Sustain. Energy* **2017**, *9*, 063302. [CrossRef]

34. Jonkman, J.; Butterfield, S.; Musial, W.; Scott, G. *Definition of a 5MW Reference Wind Turbine for Offshore System Development*; National Renewable Energy Laboratory (NREL): Golden, CO, USA, 2009.

35. DNV. *Offshore Standard DNV-OS-J103. Design of Floating Wind Turbine Structures*; Det Norske Veritas: Oslo, Norway, 2013.

36. DNV. *DNV-RP-C205 Environmental Conditions and Environmental Loads*; Det Norske Veritas: Oslo, Norway, 2010.

Journal of
**Marine Science
and Engineering**

Review

What about Marine Renewable Energies in Spain?

María Dolores Esteban [1,2,*], Juan Manuel Espada [1], José Marcos Ortega [3],
José-Santos López-Gutiérrez [2] and Vicente Negro [2]

[1]   Departamento de Ingeniería Civil, Universidad Europea, 28040 Madrid, Spain
[2]   Grupo de Investigación de Medio Marino, Costero y Portuario, y Otras Áreas Sensibles, Universidad
   Politécnica de Madrid, 28670 Madrid, Spain
[3]   Departamento de Ingeniería Civil, Universidad de Alicante, Ap. Correos 99, 03080 Alacant/Alicante, Spain
*   Correspondence: mariadolores.esteban@upm.es; Tel.: +34-917407272

Received: 4 July 2019; Accepted: 26 July 2019; Published: 30 July 2019

Abstract: Renewable energies play a fundamental role within the current political and social
framework for minimizing the impacts of climate change. The ocean has a vast potential for
generating energy and therefore, the marine renewable energies are included in the Sustainable
Development Goals (SDGs). These energies include wave, tidal, marine currents, ocean thermal, and
osmotic. Moreover, it can also be included wind, solar, geothermal and biomass powers, which their
main use is onshore, but in the near future their use at sea may be considered. The manuscript starts
with a state-of-the-art review of the abovementioned marine renewable energy resources worldwide.
The paper continues with a case study focused on the Spanish coast, divided into six regions: (I)
Cantabrian, (II) Galician, (III) South Atlantic, (IV) Canary Islands, (V) Southern Mediterranean, and
(VI) Northern Mediterranean. The results show that: (1) areas I and II are suitable for offshore wind,
wave and biomass; (2) areas III and V are suitable for offshore wind, marine current and offshore
solar; area IV is suitable for offshore wind, ocean wave and offshore solar; (3) and area VI is suitable
for offshore wind, osmotic and offshore solar. This analysis can help politicians and technicians to
plan the use of these resources in Spain.

Keywords: renewable energies; ocean energy; offshore wind; wave; tidal; marine currents;
ocean thermal

---

1. Introduction

In recent years, there have been some social and political concerns about climate change and
the high dependence on fossil fuels. In order to prevent a big problem, different climate and energy
policies are being created to achieve environmental sustainability. Renewable energies are called to
play an essential role in this process. The ocean has been an integral part of human civilization and its
development for a long time. Although its potential use for power generation has been the object of
different patents, only some technologies able of taking advantage of ocean energy resources are in a
mature stage of development. This is a key point taking into account that oceans and seas have the
potential to play an important role in the supply of clean and endless energy. Oceans and seas contain
large quantities of energy potential. In fact, ocean energy resources are estimated with a potential of
around 120,000 TWh/year, enough to satisfy more that 400% of the current global demand for electricity.

The great potential shown by ocean energies is due in part to the huge ranges of possibilities:
wave, tidal, marine currents, ocean thermal, and osmotic (salinity gradient) energies. In addition to
these ones, it is essential to take into account those created in a first moment for onshore locations, and
later thought for sites in the sea: wind and solar (onshore on land and offshore on sea), geothermal and
biomass. Nowadays, there are some barriers to achieve an adequate development of that sector [1].
These barriers are mainly: the state of the technology with a lot of important challenges both in

the short and medium term, the high capital cost in the first project of each technology, the lack of experience and environmental aspects, etc. These barriers are smaller or larger depending on each of the oceanic technologies and its current development stage.

In recent years, this industry has seen encouraging signs, with some of the technologies showing significant progress. This is the case for offshore wind and tidal energy, which can be considered mature enough to be ready for their commercial development. Regardless, other technologies follow a slower learning curve. Based on this, the forecasts in the short, medium, and long term are not ambitious at all [2]. With the objective of promoting the sector, various initiatives have emerged to take advantage of its energetic potential. One of them is the Implementation Agreement of Oceanic Energy Systems (IEA-OES), which aims to have installed 337 GW of capacity worldwide in 2050, objective difficult to be achieved with current figures.

Another key aspect in the development of this type of project is the Levelized Cost of Electricity (LCOE) [3]. LCOE value is very high in the case of prototype projects, having a lot of room for improvement. It is important to take into account that the energy generated must have a competitive cost [4]. On the other hand, the use of green energy has other benefits. In fact, governments must support this type of project with some incentives to help the companies involved, above all in the early stages of the development of those technologies. On the other hand, a better perception of the citizens about marine renewable energy is essential, because nowadays there is a great ignorance of the possibilities of the ocean to offer clean and endless energy. A high percentage of citizens only know their disadvantages.

In Spain, the case is not different, and the commitments on renewable energy must be achieved. Although the development of onshore facilities is not negligible at all, offshore ones have not yet been developed except with R&D projects, patents and prototypes. So, the road ahead is huge in this area. For that, it is essential to know the available resource. A lot of studies have been carried out to estimate it, but mainly focused on one category, for instance, offshore wind or wave power. There are some examples of these studies focused on different areas in the world: one of themis about the analysis of methodologies for the assessment of wind resource in European Seas [5]; another one is only focused in offshore wind around Korean peninsula [6]; another one is about wave energy along the Cornish coast (UK) [7]; another one analyzes the renewable mix to satisfy the needs of energy of Pantelleria, a real island in the Mediterranean Sea, focusing on starting to study minor island, including the study of different renewable sources: wind, solar and wave [8]; another one studies the possible energy independence in Malta based on the use of Wave Energy Converters [9]; another one is about wave energy resource variation in the coast of Ireland [10]; and another one analyzes the influence of air density changes in the offshore wind potential over Northeastern Scotland [11].

Furthermore, there are some resource analyses in Spain: one of them determines the wave energy resource in the Estaca de Bares area (Spain) [12]; another one includes the evaluation of wave energy potential in the Spanish coast [13]; another one is about a review of combined wave and offshore wind energy [14]; another one is focused in wave energy in Menorca (Spain) [15]; another one is related to wave energy trends and its variation, and is very important to be taken into account in the Bay of Biscay [16]; and another one is about seasonal corrections due to the use of real values of air density for the assessment of offshore wind energy potential in the Iberian Peninsula [17].

Regarding wave energy, it is can be stated that UK and Spain are pioneering countries. In Spain, some Wave Energy Converters (WEC) were installed in Mutriku breakwater, and this facility is in operation [18–20].

Some investigations used different algorithms that included an importance sampling-based expectation-maximization algorithm for sequence detection in a single-photon avalanche diode underwater optical wireless communication [21]. Other studies are focused on water desalination applied to Sicilia with the objective of solving the problem in water supply for areas with a chronic debt of water [22]. In other cases, the researchers analyze energy saving in public transport using hydrogen produced by renewable sources instead of fossil fuels [23].

The main objective of this review paper is to analyze the marine renewable energy resource in the Spanish coast, based on a preliminary and prospective analysis of the possibilities of applying the different types of energies that can be exploited at the sea in the different areas of the Spanish coast. This review gives approximate numbers. If an accuracy number is looked in a specific area, it will be necessary to carry out a detailed analysis in that area. Marine renewable energies are included in the Sustainable Development Goals (SDGs), especially in the SDG-7 (Affordable and Clean Energy, with the objective of ensuring access to affordable, reliable, sustainable and modern energy for all), and in the SDG-14 (Life below Water, with the aim of conserving and sustainable using the oceans, seas and marine resources for sustainable development).

To achieve the abovementioned objective, the first step of the work is based on the need of using renewable energies, in this case marine energies, to achieve a reduction on the polluting gases in the atmosphere. For this, it is essential to carry out an exhaustive search of information about the state of the art of marine energies, the potential and the forecast at short, medium and long term, etc. This is done for all the types of renewable energies to be harnessed in seas and oceans: offshore wind, wave, tidal, marine currents, ocean thermal, osmotic (salinity gradient), solar, geothermal and biomass.

The state-of-the-art is necessary to address the next parts of the work, consisting of carrying out a detailed study and analysis of the potential of marine energies firstly in the world based on a review study, and later along the Spanish coast. For this, the Spanish coast is divided into six different areas: (I) Cantabrian coast, (II) Galician coast, (III) South Atlantic coast, (IV) Canary Islands coast, (V) Southern Mediterranean coast, and (VI) Northern Mediterranean coast. The energy potential associated to each area is indicated. So, optimal areas for each type of marine energy are identified.

This paper gives the basis to achieve the established objectives for the installation of renewable energies in Spain, in the sense of incorporating marine energies, which have been left aside until now in the country. In Spain, it has mainly only been carried out theoretical studies, tests in laboratory and some prototypes.

2. Marine Energy Resource in the World: State-Of-The-Art

A review of energy resource to be exploited in the sea around the world is summarized in this section. The order followed related to the different type of marine energies is: (1) offshore wind, (2) wave, (3) tidal, (4) marine currents, (5) ocean thermal, (6) osmotic, (7) biomass, (8) geothermal and (9) solar. To end this section, LCOE figures are discussed.

There are some specific studies that focus on one or more types of marine renewable energies. Regardless, there is no research that includes all the information of all the above-mentioned types together. On the other hand, there are some specific studies for some countries and some of the types of the abovementioned energies, but not including all the countries and all the types. This can be seen with some examples: wave [24], tidal, ocean thermal, ocean current and salinity gradient in Iran, wave energy in Asia [25], wave energy in the Mediterranean Sea [26], wave energy in Baltic Sea [27], offshore wind energy in UK [28], among others.

2.1. Offshore Wind Energy

Offshore wind is similar to onshore wind with some important changes in the protection of all the components against the aggressive marine environment, and with higher costs due to the difficulties related to the construction and operation in the sea [29]. The main parameter to analyze the wind resource is the average wind speed at the hub height [30]. This is the first value to be analyzed when identifying potential sites for wind facilities. For more detailed studies of wind resource assessment, other aspects must be considered, as the variability in the time and the horizontal and vertical space, wind direction, turbulence intensity, etc. The wind turbine generator model selected influences clearly when calculating the gross and net production of the facility.

World wind potential, including onshore and offshore, is estimated about 20,000–50,000 TWh/year, representing offshore wind more than 70% of those figures [31]. Figure 1 shows the annual average wind speed of offshore wind at 90 m high, obtained using NOAA's Blended Sea Winds global offshore wind dataset. That dataset included wind stresses and vector winds on the ocean surface, with a grid of 0.25°, and with different time resolutions: monthly, daily and 6-h. It includes data from 1986 until 2006 [32]. Wind speeds represented in Figure 1 were generated from via satellite observations and in-situ sensors. The direction was a combination of the products of National Centers for Environmental Prediction (NCEP) and European Center for Medium-Range Weather Forecasts (ECMWF). The use of both products together has gaps for obtaining wind speed values. Some complex algorithms were created to fill those gaps, and there are some numerous research groups trying to continue improving those results [33,34].

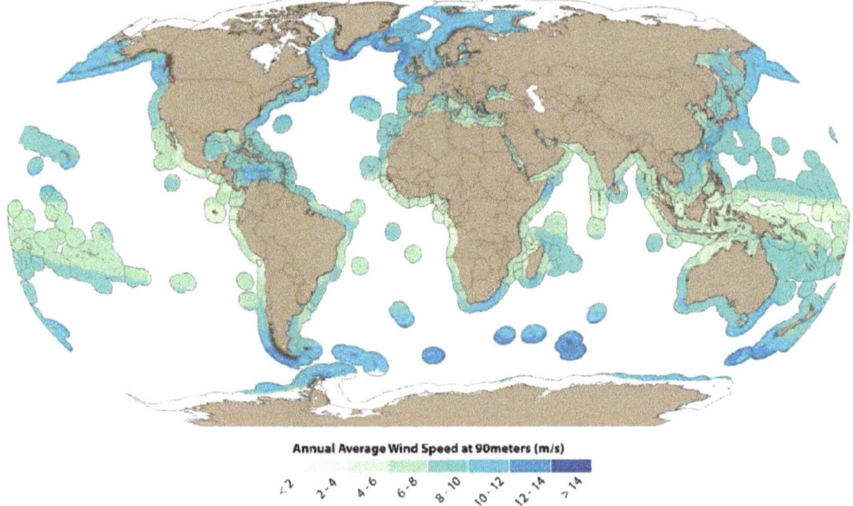

Figure 1. Annual average wind speed, in m/s, at 90 m high (Reproduced with permission from [34]. European Centre for Medium-Range Weather Forecasts, 2019).

2.2. Wave Energy

Wave energy is a great potential marine renewable energy. Different studies trying to determine the wave energy potential have been carried out in the world. There are some of those studies focused on specific countries, and even in specific areas within a country [35]. It is important to differentiate between wave energy potential, expressed in kW/m or kWh/year units, and wave energy production, only in kWh/year unit. The wave energy potential is the existing energy to be harnessed in a specific location. The wave energy production is the produced energy in the site using a specific wave energy converter [36].

For wave energy potential estimation, the only parameters to be taken into account are the average wave climate, including the combination of significant wave heights and peak wave periods. Knowing the wave spectrum type to be considered (JONSWAP, TMA, Pierson-Moskowitz, etc. [37]), it is easy to obtain the wave potential in a location. Characteristics of wave energy converters must be considered: minimum and maximum depth of the site, way of energy extraction, matrix power, etc. With all those data, the wave energy production (kWh/year) can be calculated [38].

According to International Energy Agency (IEA) estimations, the wave energy potential can oscillate between 8000 and 80,000 TWh/year [39]. Figure 2 shows the wave power distribution in the world. It is not homogeneous, with some areas characterized by more than 100 kW/m, and some of them below 10 kW/m [38].

Figure 2 is based on Mork et al. [40]. The global wave power dataset used was the default calibrated wave data set included in WorldWaves. WorldWaves included model data for the period between 1997 and 2006, a 10-year period, with 6-h frequencies and 0.5° grid, calibrated and validated with global TOPEX and JASON altimeter wind speed and wave height data. The calibration includes the consideration of some wave buoy records.

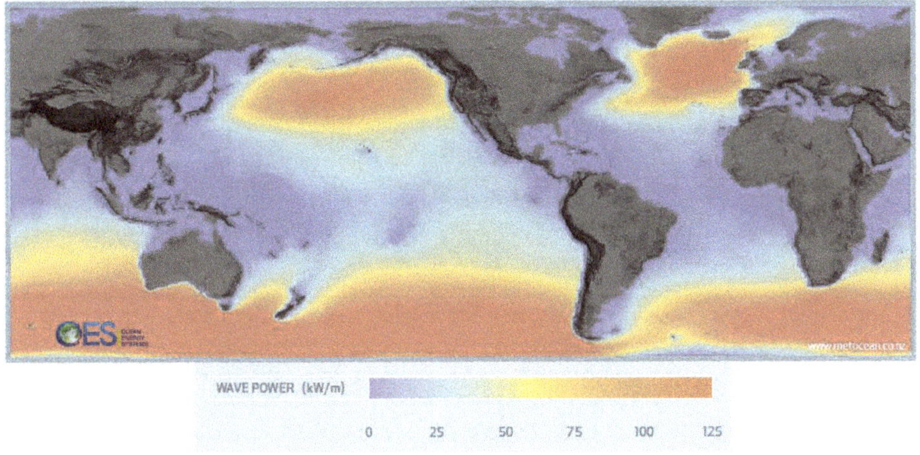

Figure 2. Wave power distribution, in kW/m (Reproduced with permission from [41,42]. International Energy Agency—Ocean Energy Systems, 2014, 2011).

2.3. Tidal Energy

Tidal energy is caused by the gravitational attraction generated by the Moon and the Sun. Those gravitational forces are responsible for causing the ascent and descent movement of the seal level. Those movements are described by a wave with a period between 12 and 24 h. It is known as astronomical tides [43]. The tidal range associated with the astronomical tide is different throughout the planet, with places where the tidal range can be neglected and with locations where it exceeds 10 m [44]. Although astronomical tides are subject to general conditions, they may be affected by local conditions, such as wave reflection, the depths of the seabed, the shape of the coast, the existence of river mouths, resonance effects, etc.

The energy generated by the tides has been estimated around 26,000 TWh/year [39]. Ocean tidal ranges can be determined based on harmonic analysis using the formula included in Figure 3, where O1 and K1 are diurnal tidal constituents, M2 and S2 are semidiurnal tidal constituents, MHWS is mean high-water springs and MLWS is mean low water springs. The tidal range is expressed in meters. Tides can be exploited in two different ways. The first of these ways consists of the construction of dams, which serve as a barrier to maintain water at different levels on both sides of the dam, so that through turbines they can take advantage of the kinetic energy derived from the abovementioned difference in levels. The second way takes advantage of the speed of currents due to differences in sea level without the need of building dams. This process is similar to the use of the energy of ocean currents, and in fact some ocean currents harnessed are tidal currents.

To be able to produce electricity based on tidal dams, the location has to provide a tidal range greater than 5 m [43,45,46]. This strict condition reduces the number of viable sites for the use of tidal energy, as it can be observed in Figure 3.

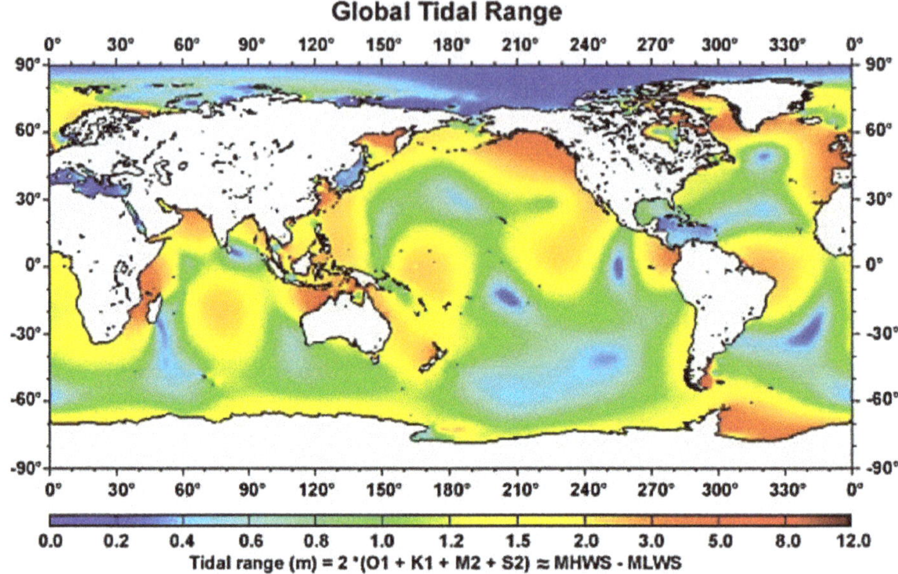

Figure 3. Tidal range, in m, around the world (Reproduced with permission from [47]. University of Graz Wegener Center for Climate and Global Change, 2019).

2.4. Marine Current Energy

Ocean currents are due to a mix of periodic and aperiodic water movements. There are seasonal and short duration changes, and a lot of oscillatory and sporadic movements superimposed to the general oceanic circulation. Current observation is a very complex challenge [48]. The origin cause of marine currents is mainly the difference of temperature, salinity, etc. adding to them the influence of the wind, tides, Earth rotation, etc. [49–51].

Generally, the most intense marine currents are caused by the effect of the wind and the tides. In case of being due to tides, it is usual to name it a tidal stream. Tidal stream is the most harnessed marine current energy type at the moment. Therefore, this paper is going to be focused on tidal stream when marine current energy is mentioned.

Focusing on tidal stream, the bathymetry can also help to produce a remarkable increase in its characteristics, mainly in narrowing areas with a consequent velocity increase [52,53]. The energy capacity of this renewable energy source is high, estimated around 800 TWh/year [39]. This potential can be increased if it could be taken advantage of not only tidal currents. The speed of currents has to be greater than 2 m/s for the harnessing of this type of energy [54].

2.5. Ocean Thermal Energy

Ocean thermal energy is generated as a consequence of a process that begins with the incidence of solar rays on the sea, increasing its temperature [55]. Sea water allows the rays to penetrate through its surface, being able to increase the temperature of lower layers of sea water. The penetration will depend, among others, on the water turbidity state. In lower layers the radiation decreases and, therefore, the temperature decreases. This has as a consequence a vertical distribution of temperatures. In fact, this can also be explained as solar energy which is stored as heat in the surface layer of the ocean, and that heat is distributed to water depths around 100 m due to the waves and surface currents. Cold water is found deeper due to its higher density, and this different mass of water is moved by ocean currents. In the world, the temperature difference is between 10 and 25 °C, with the maximum values close to the Equator [56].

The vertical distribution is marked by two areas with a big change in the temperature between them. The temperature difference between the marine surface and one thousand meters depth in a specific location can allow energy production. This is known as ocean thermal energy. The harvesting of this type of energy is based on the difference between deep cold water and warm surface water; this allows running a heat engine to produce electricity. The mechanism consists of the shallow ocean water heat a liquid in the engine characterized by having a low boiling point. The liquid turns into vapor and it moves a turbine. This vapor cools with the deep water, restarting the generation cycle.

To have a profitable project, the difference in temperature must be at least 20 °C. This condition limits the possibilities of finding a suitable location for ocean thermal energy facilities. The most interesting ones are those located in equatorial and subtropical areas, where the minimum temperature on the surface is around 24 °C and temperatures at depths around one thousand meters are usually around 5 °C. The estimate for ocean thermal energy resource around the world is between 30,000 and 90,000 TWh/year, being usable around 10,000 TWh/year [39]. Figure 4 shows the temperature difference in degrees along the Earth.

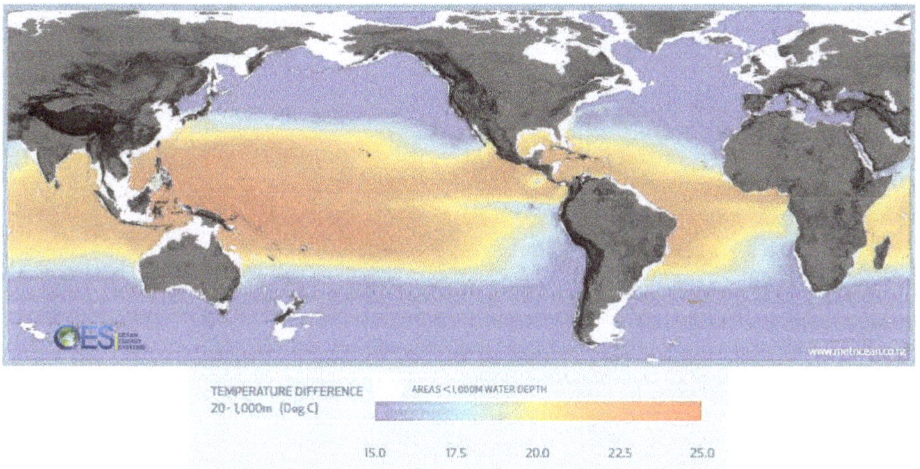

Figure 4. Ocean thermal energy: difference of temperatures, in degrees (Reproduced with permission from [42]. Ocean Energy Systems, 2011).

2.6. Osmotic Energy

Osmotic energy is based on the difference that occurs in the osmotic pressure due to the difference in salinity between fresh water versus saltwater [57]. This process occurs when the two fluids come into contact causing a balance in the concentration of salt between them. It is essential that the facilities are located in river mouths looking for the salinity difference [58]. However, the mouths of rivers can present limitations for the installation of this type of facility due to possible conflicts with other uses or activities usually developed in the rivers' mouth.

The energy potential is estimated about 2000 TWh/year [31]. Figure 5 shows salinity concentration in seas and oceans around the world, with greater salinity concentration in places like the Mediterranean Sea, and less salinity concentration in the Arctic Ocean.

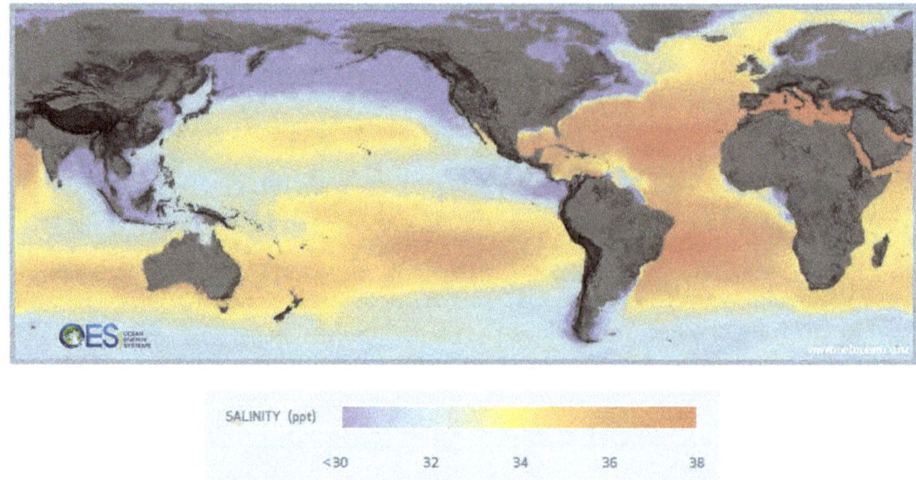

Figure 5. Salinity concentration in the world (Reproduced with permission from [42]. Ocean Energy Systems, 2011).

2.7. Biomass Energy

Biomass can be defined as the entire living mass of both animal and vegetal origin existing on the Earth. Most of the biomass can be burned for energy production, in the form of heat, electricity or fuel through different types of treatments [59]. With regard to the biomass in the sea, algae are used to obtain biofuels, based on their function of using the sun energy to transform carbon dioxide, water and organic nutrients into oxygen and vegetable biomass [60]. These have microbes constituting 90% of the biomass, which gives them a lot of energy. It favors its conversion into various biofuels thanks to its low energy density.

Two types of algae can be differentiated: the microalgae from the aquatic systems and the larger algae known as macro algae that grow attached to stable substrates coming mainly from the seabed. Although microalgae have been studied in recent years for biodiesel production, marine macroalgae have awakened recently a great interest for obtaining different biofuels due to its chemical composition and ability to produce large biomass. Macroalgae have important advantages as a source of biofuels [61], if it is compared to other raw materials. They have a higher growth than other plants used until now, and its large-scale cultivation is feasible, profitable and it does not occupy land or require fresh water [62].

The exploitation of this type of cultivation areas is currently very limited. There are around 300,000 hectares of cultivation in the world, which represent more than 90% of global production, estimated at 24.9 million tons. The International Federation of Aquaculture (FAO) indicated that the production of macroalgae increases by 6% every year [63]. Figure 6 shows the natural distribution of macroalgae in shallow waters, and the capacity of coastal areas for biogas and the percentage of local Net Primary Productivity (NPP).

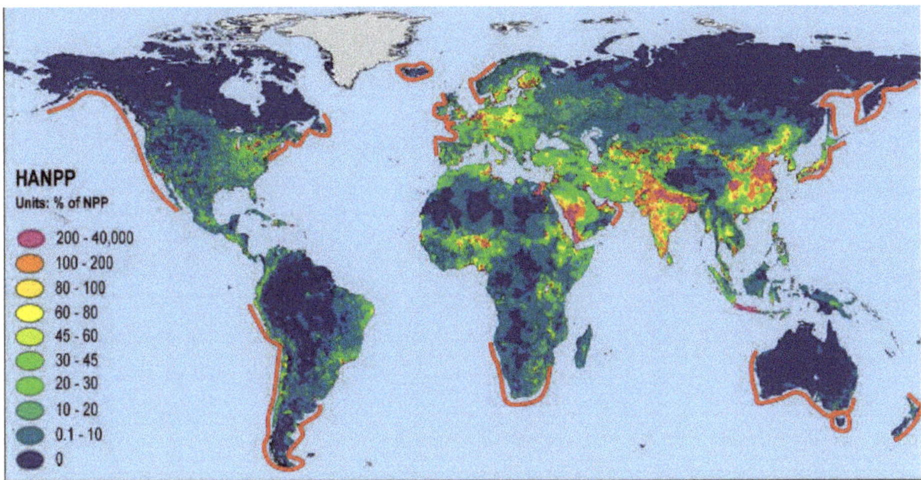

Figure 6. Natural distribution of macroalgae in shallow waters and the capacity of coastal areas for biogas and the percentage of local Net Primary Productivity (NPP). HANPP is Human Appropriation of Net Primary Production (Reproduced with permission from [64]. BioMed Central Ltd., 2012).

2.8. Geothermal Energy

Underwater geothermal energy is based on the existence of deep cracks in the seabed generated by divergent movements of the plates. These movements produce cracks, allowing vertical transfer of the magmatic heat from the mantle to the floor of the ocean. The sea cold water, when coming into contact with these cracks, warms up and chemically changes, producing hot water enriched with hydrogen sulfide expelled through the cracks produced by the movements of the plates, and known as hydrothermal vents. The hot water can be expelled up to 400 °C, getting a high heat flow for its transformation into electrical energy.

Different studies were carried out to analyze the scope of hydrothermal ventilations along the Earth's crust. There are around 65,000 km of oceanic ridges, representing 30% of all the heat released in the Earth, of which a significant number of ventilations have been obtained hydrothermal vents located at depths greater than 2000 m (Figure 7). They are usually found in the Pacific and Atlantic Oceans, although there are some cases in the Mediterranean Sea and in the Indian Ocean. Only 13,000 km of the existing ocean ridges have been explored up to the moment, and only around 3900 km (about 30% of the surface explored) have good skills to be exploited.

Marine geothermal energy has a very high potential, but hardly quantifiable, since at present there is no defined system that allows knowing the real and usable potential of this source. This is because the state of technological development is not very mature. There are some estimates, considering only the explored areas, indicating that more than 1000 GW can be extracted [65,66].

2.9. Solar Energy

Solar energy can be used for electricity production by direct conversion of solar radiation into electric current passing through solar or photovoltaic cells. Its implementation to date has been done mainly onshore, but some companies in the sector have developed some designs with the objective of being able to take advantage of this energy in the surface of seas and oceans [67]. Maritime surface represents more than 70% of the global surface of the earth. Figure 8 shows the solar radiation marine great potential, with around 18 TWe only in the black points.

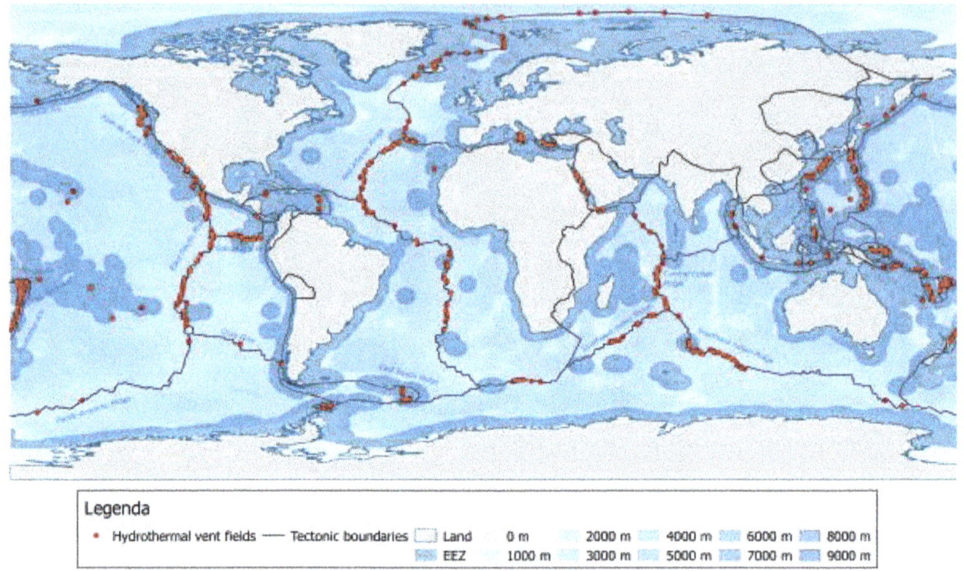

Figure 7. Submarine ventilation around the Earth [65].

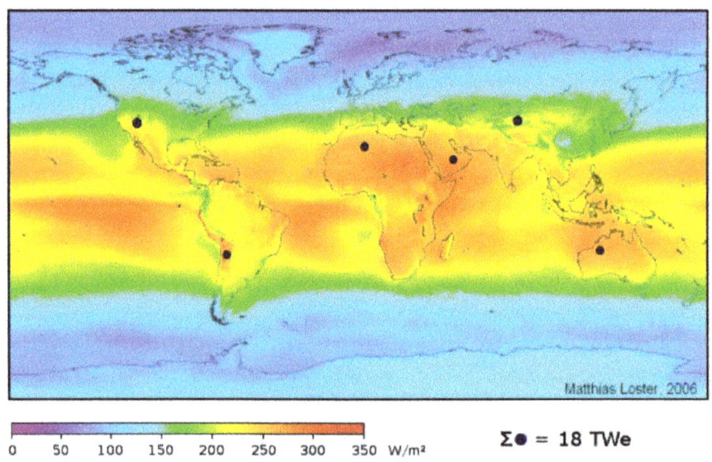

Figure 8. Distribution of solar radiation in the world (Reproduced with permission from [68]. Matthias Loster, 2010).

As an outline of the state of this technology, Figure 9 shows a summary of the power installed in 2015 and the forecasts for 2020 and 2030, focused on Europe considering: (a) waves, currents and tides, (b) offshore wind, (c) onshore solar, and (d) offshore wind.

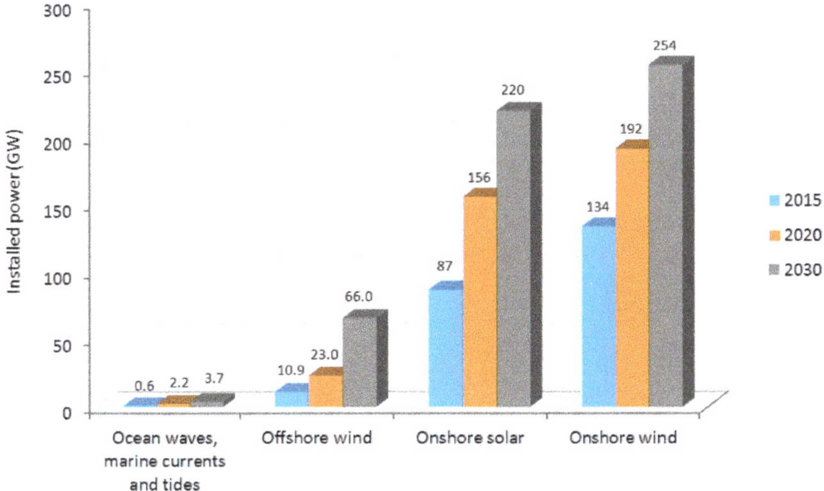

Figure 9. Global view of onshore and offshore renewable facilities in Europe: installed power, in GW, in 2015, and forecasts for 2020 and 2030.

2.10. Levelized Cost of Energy (LCOE) Data

For investments decision-making, energy industry uses economical models to determine and compare the energy cost of different technologies. The most known one is the levelized cost of power (LCOE). It depends on capital, operating and fuel costs. The value of LCOE is obtained by dividing the sum of costs over lifetime by the sum of electrical energy produced over lifetime. It is important to take into account that most of renewable technologies are capital intensive but have low operational costs, while fossil fuel-based technologies may be cheaper to construct but much more expensive to operate.

Anyway, it is important to clarify that LCOE values are more reliable when there are a lot of installed facilities in a specific technology. So, the existing and reliable current LCOE data (median values) are given for offshore wind: 0.1434 $/kWh, for wave: 0.3263 $/kWh, for tidal and marine currents together: 0.3263 $/kWh. The rest of technologies included in the paper have not a reliable LCOE value. On the other hand, it is recommended to know LCOE values of other onshore renewable technologies: 0.0777 $/kWh in onshore wind, 0.1320 $/kWh in solar photovoltaic, 0.2466 $/kWh in solar thermal, 0.0479 $/kWh in hydropower, 0.0951 $/kWh in biomass, 0.0635 $/kWh in geothermal [69]. Although there are many uncertainties regarding the LCOE values in the medium term, it is expected that they will stabilize and even decrease as the learning curve progresses.

3. Practical Application: Marine Energy Resource in the Spanish Coast

Spain has a larger potential of marine energies as has been indicated by Montoya et al. [70], mainly focusing that work on onshore renewable energies, including a brief section about marine energies. The Spanish coast has more than 7800 km; the culture and the economy are very linked to the ocean in coastal areas. The study and exploitation of marine renewable energies are destined to play an essential role in the medium term. Many investigations have been carried out in the field of marine renewable energies in last years; specialized technology centers and demonstration plants have been installed focused on advancing in this field, which will be essential to achieve progress and development. Based on the investigation and different tests and errors, it will be possible to reach the sufficient maturity of the installations of the use of marine renewable energies so that its commercial development is feasible.

Spain has a very heterogeneous coastline in terms of energy potential. It belongs to a peninsula (the Iberian Peninsula), connected in the Northeast with France. In addition, the Spanish territory also has two archipelagos of islands: The Canary and Balearic Islands. Canary Islands are located in the Atlantic Ocean, southwest of the Iberian Peninsula and close to the African coast. Balearic Islands are located in the Mediterranean Sea, east of the Iberian Peninsula. For this reason, the Spanish coast cannot be studied as a single area since a great mistake would be made. The best alternative to this situation is to zone this coast in areas with homogeneous energy characteristics. Therefore, this paper divides the Spanish coast into six different areas: (I) Cantabrian coast, (II) Galician coast, (III) South Atlantic coast, (IV) Canary Islands coast, (V) Southern Mediterranean coast, and (VI) Northern Mediterranean coast (Figure 10). Areas I, II, III, V and VI correspond to the Iberian Peninsula, with the Balearic Islands included in Area VI. Area IV refers to the Canary Islands.

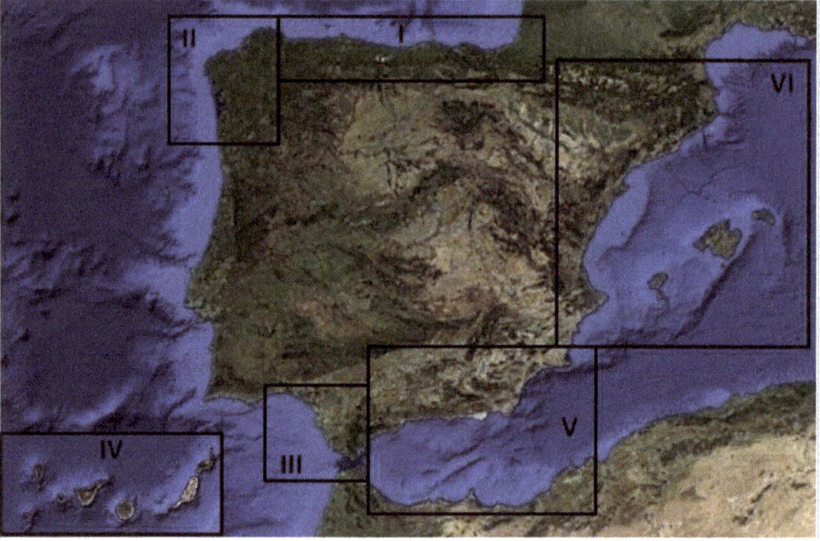

Figure 10. Zoning created for the study of marine renewable resource in the Spanish coast.

As it is done in the previous section, the order followed for the analysis in this section is: (1) offshore wind, (2) wave, (3) tidal, (4) marine currents, (5) ocean thermal, (6) osmotic, (7) biomass, (8) geothermal and (9) solar. So, the different types of energy are evaluated in the six areas in which the Spanish coast is divided.

3.1. Offshore Wind Energy

Wind resource study is based on wind speed map from Spanish IDAE (*"Instituto para la Diversificación y Ahorro de la Energía"*, in Spanish language) covering a coastal strip of 24 nautical miles, and with the data of the annual average wind speed at 80 m high (Figure 11) [71].

This analysis considers next classification depending on the annual average wind speed:

- Low: annual average wind speed lower than 6 m/s, in white.
- Medium: Low: annual average wind speed between 6 and 7 m/s, in greenish.
- Medium: High: annual average wind speed between 7 and 8 m/s, in yellow and orange.
- High: annual average wind speed higher than 8 m/s, in pink and reddish tones.

The values of the annual average wind speed (m/s) are shown in Table 1 for all the areas previously defined, at different water depths (20, 50 and 100 m). The values were obtained after making the average of the annual speeds in different points of each specific zone that make up the different areas. In Cantabrian coast (Area I), specific zones defined are País Vasco, Asturias and Cantabria. In Galician coast (Area II), there is only one specific zone, Galicia. In South Atlantic coast (Area III), the specific zones are Huelva, Cádiz and Estrecho-Ceuta. In the Canary Islands coast (Area IV), there is only one specific zone, Canary Islands. In Southern Mediterranean coast (Area V), the specific zones are Málaga, Almería, Melilla and Murcia. In Northern Mediterranean coast (Area VI), the specific zones are Valencia, Cataluña and Islas Baleares (Balearic Islands).

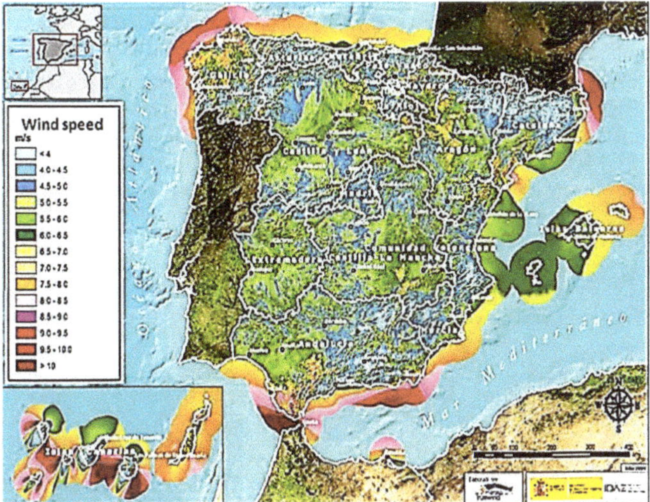

Figure 11. Map with the annual average wind speed, in m/s, at 80 m high (Reproduced with permission from [71]. IDEA, 2019).

Table 1. Annual average wind speed, in m/s, in all the areas in the Spanish coast.

Area	Specific Zone	20 m Depth	50 m Depth	100 m Depth
(I) Cantabrian Coast	País Vasco	6.0–6.5	6.5–7.0	6.5–7.0
	Asturias	6.0–6.5	7.0–7.5	7.5–8.0
	Cantabria	6.0–6.5	6.5–7.0	7.0–7.5
(II) Galician Coast	Galicia	6.5–7.0	7.0–7.5	8.0–8.5
(III) South Atlantic Coast	Huelva	6.5–7.0	7.0–7.5	7.5–8.0
	Cádiz	8.0–8.5	8.5–9.0	>10.0
	Estrecho-Ceuta	8.5–9.0	>10.0	>10.0
(IV) Canary Islands Coast	Canary Island	6.0–6.5	6.5–7.0	7.0–7.5
(V) Southern Mediterranean Coast	Málaga	6.5–7.0	7.0–7.5	7.0–7.5
	Almería	7.5–8.0	8.0–8.5	8.5–9.0
	Melilla	6.5–7.0	7.0–7.5	7.5–8.0
	Murcia	6.5–7.0	7.0–7.5	7.0–7.5
(VI) Northern Mediterranean Coast	Valencia	5.5–6.0	5.5–6.0	5.5–6.0
	Cataluña	5.5–6.0	6.0–6.5	6.5–7.0
	Islas Baleares	6.0–6.5	6.5–7.0	6.5–7.0

As it is observed in Table 1, the most interesting zones for offshore wind facilities are Cádiz at 100 m depth and Estrecho-Ceuta at 50 and 100 m depth, both in Area III (South Atlantic coast) with more than 10 m/s of average wind speed at 80 m high.

Zones and areas with an average wind speed above 7 m/s are: Asturias at 50 and 100 m depth and Cantabria at 100 m depth, both in Area I (Cantabrian coast), Galicia at 50 and 100 m depth in Area II (Galician coast), Huelva at 50 and 100 m depth, Cádiz at 20, 50 and 100 m depth, and Estrecho-Ceuta at

50 and 100 m depth, the three ones in Area III (South Atlantic coast), Canary Island at 100 m depth in Area IV (Canary Island coast), and Málaga at 50 and 100 m depth, Almería at 20, 50 and 100 m depth, Melilla at 50 and 100 m depth and Murcia at 50 and 100 m depth the four ones in Area V (Southern Mediterranean coast). Area VI (Northern Mediterranean coast) has values below 7 m/s.

3.2. Wave Energy

To analyze the potential of wave energy on the Spanish coast, IHCantabria wave atlas is used [72,73]. It allows knowing the average wave potential in different locations of the Spanish coast (Figure 12). For the study of each area, IDAE annual mesh maps are used, allowing obtaining the detailed characteristics of the resource, both considering the power and direction of waves (Figure 13). Wave energy resource is studied for depths of 20, 50 and 100 m, and also for deep waters.

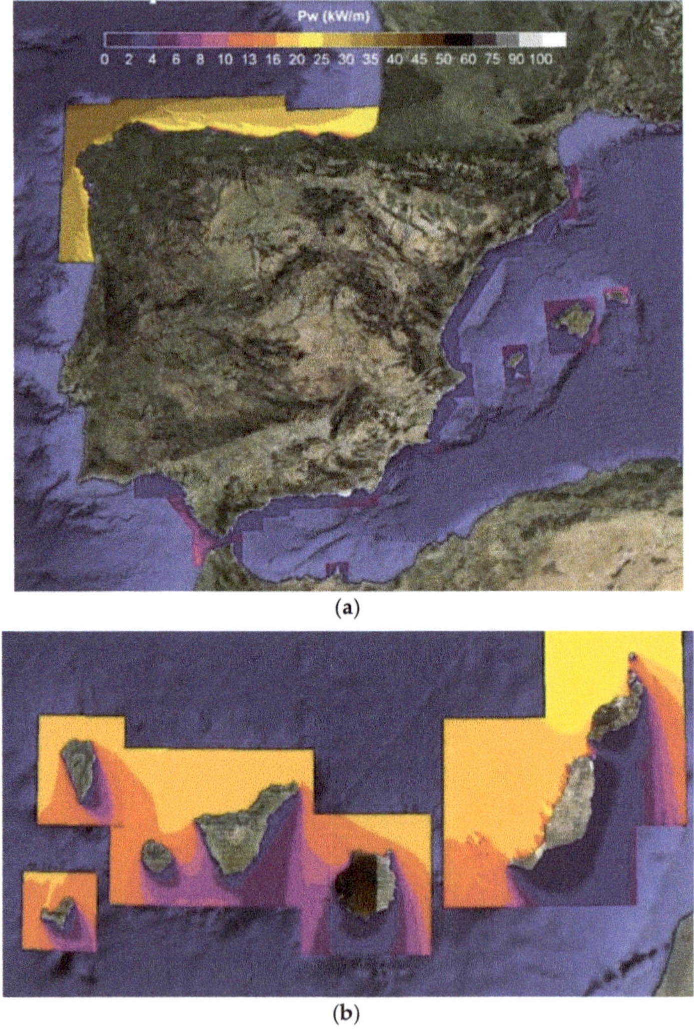

Figure 12. Annual average wave power, in kW/m: (**a**) in the Iberian Peninsula and (**b**) Canary Island coast (Reproduced with permission from [72]. IHCantabria, 2019).

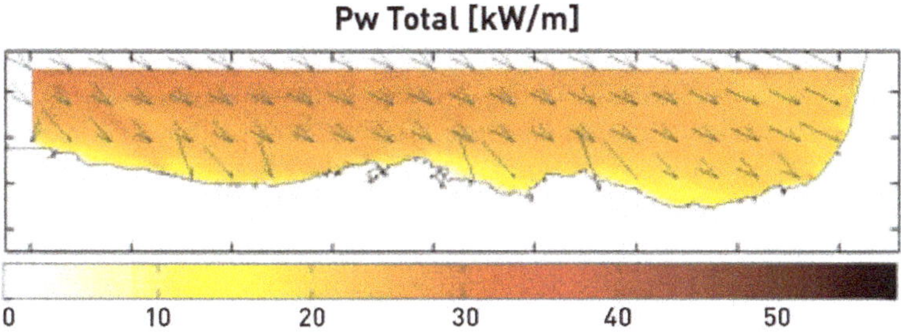

Figure 13. Annual average wave power, in kW/m, in the west part of the Area I (Cantabrian coast) (Reproduced with permission from [73]. IDEA, 2019).

Based on Atlas data, Table 2 is prepared. It includes the values of the annual average wave power for different depths in kW/m for each area and specific zone inside each area. Input data are from IDAE through the GOW (Global Ocean Waves) database. The specific zones defined for each area are the same defined for the offshore wind energy in Section 3.1.

Table 2. Annual average wave power, in kW/m, in all the areas in the Spanish coast.

Area	Specific Zone	20 m Depth	50 m Depth	100 m Depth	Deep Waters
	País Vasco	12.20	12.18	19.89	26.68
(I) Cantabrian coast	Cantabria	14.94	18.98	22.94	30.97
	Asturias	13.55	17.48	21.51	35.14
(II) Galician coast	Galicia	17.08	25.21	30.26	37.97
	Huelva	2.53	3.06	3.42	4.50
(III) South Atlantic coast	Cádiz	2.21	2.87	3.02	3.85
	Estrecho-Ceuta	2.37	2.94	3.24	4.31
(IV) Canary Islands coast	Canary Island	7.53	8.94	9.70	13.59
	Málaga	1.63	2.04	2.52	3.11
(V) Southern Mediterranean coast	Almería	1.34	1.82	2.24	2.96
	Melilla	3.20	3.50	4.65	5.40
	Murcia	1.07	1.43	2.25	3.08
	Valencia	1.70	2.30	2.86	2.96
(VI) Northern Mediterranean coast	Cataluña	1.62	2.26	3.13	4.17
	Islas Baleares	2.01	3.63	4.56	5.11

Analyzing previous results, the most interesting areas for harnessing the wave resource power, are:

- Galician coast, the area with the greatest potential in Spain, with values maximum above 35 kW/m.
- Cantabrian coast, with an average of maximum values around 30 kW/m.
- Canary Island coast, with a potential higher than 10 kW/m, specifically on the northern facades if the islands.

The worst locations for wave energy facilities in the Spanish coast are: Murcia, Almería, Cataluña, Málaga, Valencia, Islas Baleares, Cádiz, Estrecho-Ceuta, Huelva and Melilla.

3.3. Tidal Energy

The resource of tidal energy is studied based on the data recorded in "Puertos del Estado" tidal gauges network [74], located in different coastal areas, mainly ports location.

Table 3 includes different aspects to be considered in the analysis: HAT (High astronomical tide), MHWS (Mean high water springs), MHWN (Mean high water neaps), MSL (Mean sea level), MLWN (Mean low water neaps), MLWS (Mean low water springs) and LAT (Lowest astronomical tide). LAT in all the cases is taken as the zero level of the harbor where the tidal gauge is located.

Table 3. Tidal data, in cm, in all the areas in the Spanish coast.

Area	Zone	HAT	MHWS	MHWN	MSL	MLWN	MLWS	LAT
(I) Cantabrian coast	Bilbao	476	432	314	237	156	40	0
	Cantabria	526	480	362	286	205	90	0
	Gijón	509	468	350	274	195	84	0
(II) Galician coast	Ferrol	439	499	285	217	147	40	0
	Coruña	493	453	342	273	202	98	0
	Vilagarcía	431	394	284	220	153	50	0
	Vigo	416	379	272	207	141	41	0
(III) South Atlantic coast	Huelva	404	363	264	203	145	47	0
	Bonanza	359	323	227	173	114	46	0
(IV) Canary Islands coast	Lanzarote	347	311	226	179	132	54	0
	Fuerteventura	314	278	196	152	109	33	0
	Las Palmas	311	278	199	158	116	43	0
	Tenerife	295	262	188	150	111	44	0
	Gomera	248	221	157	127	98	37	0
	La Palma	278	245	175	142	107	45	0
	El Hierro	285	253	186	158	129	67	0
(V) Southern Mediterranean coast	Málaga	105	-	-	61	-	-	0
	Motril	95	-	-	57	-	-	0
	Melilla	64	-	-	32	-	-	
	Almería	66	-	-	40	-	-	0
(VI) Northern Mediterranean coast	Gandía	19	-	-	2	-	-	0
	Valencia	30	-	-	12	-	-	0
	Barcelona	50	-	-	29	-	-	0
	Ibiza	52	-	-	36	-	-	0

In Cantabrian coast (Area I), the zones defined are Bilbao, Cantabria and Gijón. In Galician coast (Area II), the zones are Ferrol, Coruña, Vilagarcía and Vigo. In South Atlantic coast (Area III), the zones are Huelva and Bonanza. In Canary Islands coast (Area IV), the zones are Lanzarote, Fuerteventura, Las Palmas, Tenerife, Gomera, La Palma and Hierro. In Southern Mediterranean coast (Area V), the zones are Málaga, Motril, Melilla and Almería. In Northern Mediterranean coast (Area VI), the zones are Gandía, Valencia, Barcelona and Ibiza.

Based on buoys data and current technical minimum criteria before mentioned, tidal range greater than 5 m, the construction of tidal range facilities is not viable in the Spanish coast.

3.4. Marine Currents Energy

To analyze the resource of the energy of marine currents along the Spanish coasts, a study is made with the values obtained in different buoys of "Puertos del Estado" [74], installed along the Spanish coast. Table 4 shows marine current speeds in each buoy, including maximum and average speeds between 2007 and 2016. This is the only information available about marine currents in Spain, so the analysis is based on it. Regardless, it will be interested to develop a measurement campaign to determine marine current speed along the water column.

Table 4. Marine current speed, in cm/s, in all the areas in the Spanish coast between 2007 and 2016.

Area	Buoy		2007	2008	2009	2010	2011	2012	2013	2014	2015	2016
(I) Cantabrian coast	Bilbao-Vizcaya	Max	61	54	53	83	55	81	83	89	63	81
		Min	22	15	24	25	21	23	31	27	28	25
	Cabo de Peñas	Max	70	61	53	76	93	75	70	60	64	83
		Min	22	13	19	24	26	27	20	21	23	31
(II) Galician coast	Estaca de Bares	Max	53	56	55	57	58	55	50	62	77	128
		Min	20	18	17	22	19	16	18	19	19	47
	Villano-Sisargas	Max	60	63	69	68	81	63	95	83	101	82
		Min	22	20	23	27	28	22	27	30	29	32
	Cabo Villano	Max	62	42	56	68	64	56	96	62	69	99
		Min	19	16	19	25	17	24	35	25	21	44
(III) South Atlantic coast	Golfo de Cádiz	Max	79	53	64	60	48	74	97	61	73	70
		Min	30	21	26	27	22	26	37	30	24	29
(IV) Canary Islands coast	Gran Canaria	Max	69	56	66	54	49	74	96	84	84	66
		Min	27	20	24	21	21	26	40	31	24	20
	Tenerife	Max	71	59	64	61	63	87	80	110	73	60
		Min	23	18	21	15	18	23	25	26	14	16
(V) Southern Mediterranean coast	Cabo de Gata	Max	89	93	83	86	86	101	79	80	127	116
		Min	33	35	29	42	40	51	35	28	28	35
(VI) Northern Mediterranean coast	Cabo de Palos	Max	55	67	64	74	59	120	96	81	125	81
		Min	21	25	20	27	22	52	46	45	26	44
	Valencia	Max	41	56	47	108	63	59	47	62	57	63
		Min	15	18	15	19	15	17	13	17	25	24
	Tarragona	Max	70	56	77	71	69	73	69	63	97	78
		Min	27	21	28	34	40	32	30	26	39	42
	Dragonera	Max	-	-	76	69	79	92	70	58	89	73
		Min	-	-	24	22	35	33	27	25	34	32

In Cantabrian coast (Area I) the buoys are Bilbao-Vizcaya and Cabo de Peñas. In Galician coast (Area II), the buoys are Estaca de Bares, Villano-Sisargas and Cabo Villano. In South Atlantic coast (Area III), the buoy is Golfo de Cádiz. In Canary Islands coast (Area IV), the buoys are Gran Canaria and Tenerife. In Southern Mediterranean coast (Area V), the buoy is Cabo de Gata. In Northern Mediterranean coast (Area VI), the buoys are Cabo de Palos, Valencia, Tarragona and Dragonera.

The most interesting area for harnessing of marine current power in Spain is Gibraltar Straight, with average values above 1.5 m/s and maximum figures higher than 2.5 m/s. Table 4 show that there are only 9 maximum values above 100 cm/s: 128 cm/s (Estaca de Bares, year 2016), 127 cm/s (Cabo de Gata, year 2015), 125 cm/s (Cabo de Palos, 2015), 120 cm/s (Cabo de Palos, year 2012), 116 cm/s (Cabo de Gata, year 2016), 110 cm/s (Tenerife, year 2014), 108 cm/s (Valencia, year 2010), 101 cm/s (Villano-Sisargas, year 2015), 101 cm/s (Cabo de Gata, year 2012).

3.5. Ocean Thermal Energy

To analyze the resource of ocean thermal energy along the Spanish coast, a study of the values in "Puertos del Estado" buoys was performed [74]. The values of maximum, minimum and average temperatures are established for each buoy, considering data in the last 10 years. Table 5 includes this data for all the areas in the Spanish coast.

In Cantabrian coast (Area I), the buoys are Costera Bilbao, Pasajes, Gijón, Bilbao-Vizcaya and Cabo de Peñas. In Galician coast (Area II), the buoys are Estaca de Bares, Villano-Sisargas and Cabo Silleiro. In South Atlantic coast (Area III), the buoys are Golfo de Cádiz, Cádiz and Tarifa. In Canary Islands coast (Area IV), the buoys are Gran Canaria, Tenerife Sur and Tenerife. In Southern Mediterranean coast (Area V), the buoys are Málaga, Cabo de Gata and Melilla. In Northern Mediterranean coast (Area VI), the buoys are Cabo de Palos, Alicante, Valencia, Costera Valencia, Tarragona, Barcelona, Palamós, Dragonera and Capdevera.

Knowing that the minimum value of temperature in the surface it around 24 degrees, it can be stated that there are not sites in the Spanish coast suitable for taking advantage of this type of energy. This is also indicated in Figure 4, with temperature differences in deep and surface water below 20 degrees. Therefore, it is not viable to install a device for capturing ocean thermal energy in the Spanish coast.

Table 5. Temperature data, in degrees, in "Puertos del Estado" buoys, in all the areas in the Spanish coast.

Area	Buoy	Maximum Temperature	Minimum Temperature	Average Temperature
(I) Cantabrian coast	Costera Bilbao	15.81	13.91	14.86
	Pasajes	16.35	14.42	15.38
	Gijón	16.21	14.03	15.12
	Bilbao-Vizcaya	16.67	14,37	15.52
	Cabo de Peñas	15.89	13.87	14.88
(II) Galician coast	Estaca de Bares	16.52	14.79	15.65
	Villano-Sisargas	16.94	14.09	15.51
	Cabo Silleiro	15.50	14.43	14.96
(III) South Atlantic coast	Golfo de Cádiz	19.76	16.55	18.15
	Cádiz	18.92	15.71	17.31
	Tarifa	18.41	15.25	16.83
(IV) Canary Islands coast	Gran Canaria	21.95	20.22	21.08
	Tenerife Sur	22.67	20.59	21.63
	Tenerife	22.78	20.87	21.82
(V) Southern Mediterranean coast	Málaga	19.60	15.31	17.45
	Cabo de Gata	19.02	16.01	17.51
	Melilla	20.03	16.94	18.48
(VI) Northern Mediterranean coast	Cabo de Palos	19.16	16.35	17.75
	Alicante	18.86	16.47	17.67
	Valencia	17.75	15.45	16.60
	Costera Valencia	17.34	14.90	16.03
	Tarragona	17.25	14.67	15.96
	Barcelona	17.61	15.23	16.42
	Palamós	17.04	14.63	15.83
	Dragonera	18.97	16.31	17.64
	Capdepera	18.63	16.56	17.59

3.6. Osmotic Energy

For osmotic energy study, the input data were several "Puertos del Estado" buoys [74], establishing maximum and average salinity concentration in the last 10 years (Table 6). In order for the use of this energy, it is necessary that the facilities are located close to river mouths so that this difference in salinity occurs.

In Cantabrian coast (Area I) the buoys are Bilbao-Vizcaya and Cabo de Peñas. In Galician coast (Area II), the buoys are Estaca de Bares, Villano-Sisargas and Cabo Silleiro. In South Atlantic coast (Area III), the buoy is Cádiz. In Canary Islands coast (Area IV), the buoys are Gran Canaria and Tenerife. In Southern Mediterranean coast (Area V), the buoy is Cabo de Gata. In Northern Mediterranean coast (Area VI), the buoys are Cabo de Palos, Valencia and Tarragona.

The most interesting areas for this type of energy are Catalonian and Valencia coast, with salinity concentration about 27 parts per thousand. Rivers with mouth in these areas are: Almanzora, Segura, Júcar, Turia or Guadalaviar, Mijares, Ebro, Llobregat, Ter and Fluviá.

3.7. Biomass Energy

In Spain the natural resources of the luminaria, a specific type of algae, are very limited. They are mainly on the Atlantic coast, specifically the northern area corresponding to Cantabrian and Galician coast. In the case of the Cantabrian coast, this resource has diminished its abundance in the last decade, and in some places it has disappeared. At present, it is abundant in some locations in Galicia.

Table 6. Salinity concentration, in parts per thousand, in "Puertos del Estado" buoys, in all the areas in the Spanish coast.

Area	Buoy	Maximum Salinity Concentration	Average Salinity Concentration
(I) Cantabrian coast	Bilbao-Vizcaya	35.53	35.32
	Cabo de Peñas	35.56	35.38
(II) Galician coast	Estaca de Bares	35.71	35.55
	Villano-Sisargas	35.61	35.43
	Cabo Silleiro	35.87	35.71
(III) South Atlantic coast	Cádiz	36.60	36.45
(IV) Canary Islands coast	Gran Canaria	36.87	36.68
	Tenerife	37.46	37.23
(V) Southern Mediterranean coast	Cabo de Gata	37.51	37.13
(VI) Northern Mediterranean coast	Cabo de Palos	37.50	37.18
	Valencia	37.99	37.70
	Tarragona	38.25	38.02

3.8. Geothermal Energy

It can be concluded that Spain does not have hydrothermal vents along its coasts based on the analysis of the world map of submarine ventilation locations (Figure 7). Therefore, it is not possible to exploit this marine energy resource in the country.

3.9. Solar Energy

To analyze the resource of solar energy along the Spanish coast, the values of solar irradiation annual averages are established in each area, considering for that values between 1985 and 2005 (Figure 14).

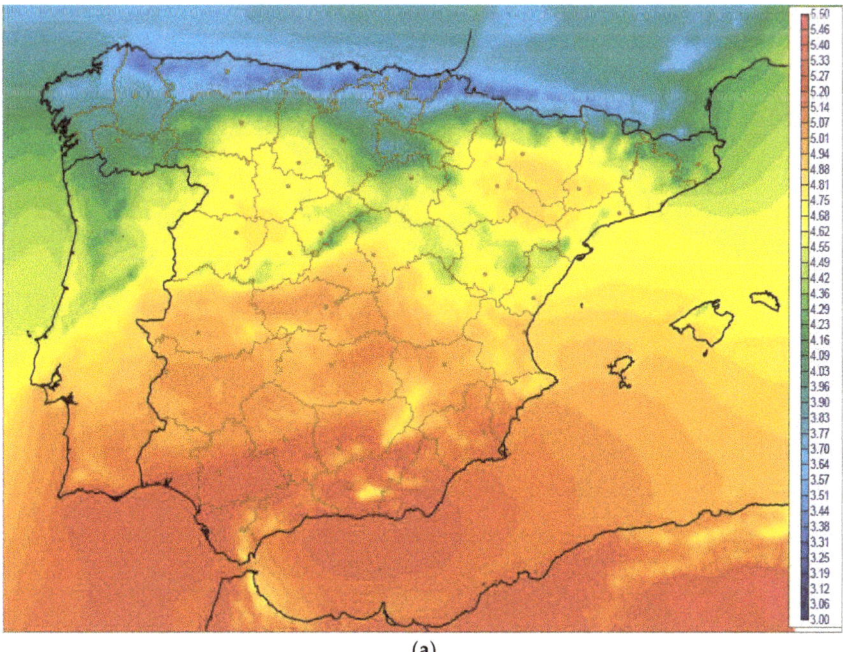

(a)

Figure 14. *Cont.*

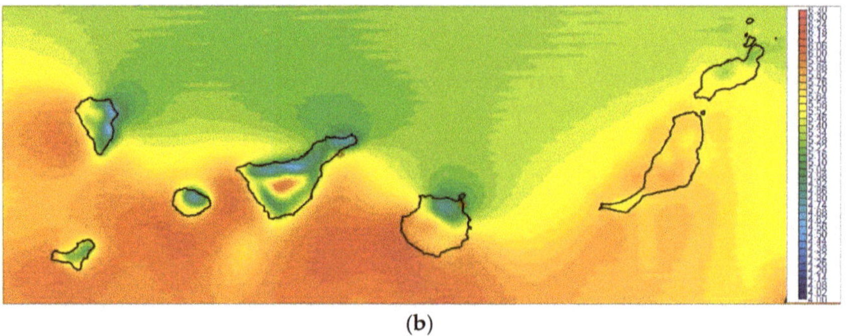

(b)

Figure 14. Radiation solar map: (**a**) Iberian Peninsula and (**b**) Canary Island. The values in the color bars are the average annual radiation solar, in kWh/m^{2}day (Reproduced with permission from [75]. Aemet, 2019).

Table 7 shows the average annual radiation in each area, and specific zones. In Cantabrian coast (Area I), the Zones are País Vasco, Cantabria and Asturias. In Galician coast (Area II), the only zone is Galicia. In South Atlantic coast (Area III), the zones are Huelva, Cádiz and Ceuta. In Canary Islands coast (Area IV), the only zone is Canary Island. In Southern Mediterranean coast (Area V), the zones are Málaga, Almería, Murcia and Melilla. In Northern Mediterranean coast (Area VI), the zones are Valencia, Cataluña and Baleares (Balearic Islands).

Table 7. Radiation solar values, in kWh/m^{2}day in all the areas in the Spanish coast.

Area	Zone	Average Annual Radiation
(I) Cantabrian coast	País Vasco	3.53
	Cantabria	3.66
	Asturias	3.57
(II) Galician coast	Galicia	4.16
(III) South Atlantic coast	Huelva	5.21
	Cádiz	5.28
	Ceuta	4.90
(IV) Canary Islands coast	Canary Island	4.72
(V) Southern Mediterranean coast	Málaga	5.19
	Almería	5.28
	Murcia	5.13
	Melilla	5.08
(VI) Northern Mediterranean coast	Valencia	4.89
	Cataluña	4.51
	Baleares	4.77

The most interesting areas for solar exploitation are:

1. The southern Atlantic coast and the southern Mediterranean coast with values above 5 kWh/m^{2}day, representing the area with the highest radiation in the Spanish coast.
2. Canary Islands, specifically the southern facade with values above 4.50 kWh/m^{2}day.
3. The northern Mediterranean coast, specifically in Valencia, Tarragona and the Balearic Islands, with values higher than 4.50 kWh/m^{2}day.

4. Conclusions

The paper consists of a preliminary and prospective analysis of the possibilities of applying the different types of energies that can be exploited at sea in the Spanish coast. In case of looking for specifying the final numbers of a certain area of lesser extent a little better, it is clear that it will be necessary to carry out specific studies in the study area.

This section, which shows the main conclusions of this work, has been divided into the general ones and those related to the Spanish case of study.

Firstly, the general conclusions related to renewable energies to be exploited in the sea are the following:

- Ocean energy resource, including other types of energies to be exploited in the sea, is huge, and there are different types of marine energies with an estimated theoretical potential, being able to satisfy the current electrical demand of the whole world.
- However, given its state of evolution and its technological development, the use of this type of energies is very scarce. Offshore wind and tidal facilities present a higher degree of development. On the contrary, wave, marine current, ocean thermal, osmotic, biomass, geothermal and offshore solar energies are in an early phase of development.
- The current installed power of ocean energy represents a very small percentage, in comparison with other sources of renewable energies. It will be necessary an important consolidation of the sector that allows increasing the installed capacity competing with sources of energy. This is not expected to happen in the short term, except in the case of offshore wind energy.

Finally, the main conclusions related to the specific case of study focused on the Spanish coast are shown next:

- Currently in Spain, the marine energy sector has a very poor role in the energy mix. This may be due to the great economic crisis suffered by the country in recent years, with the consequent reduction of investments in the renewable energy sector and especially in the marine sector.
- In order to ensure that energies to be exploited in the sea can play a fundamental role in Spain, a great support will be needed from local and national administrations, with initiatives that allow for the correct development and growth of these types of energies.
- The Spanish coast has good characteristics for the use of marine renewable energies, but not all of them. Feasible marine renewable energies on the Spanish coast are: offshore wind, ocean wave, osmotic, ocean currents, biomass and offshore solar.
- The Spanish coast is divided in six different areas: (I) Cantabrian coast, (II) Galician coast, (III) South Atlantic coast, (IV) Canary Islands coast, (V) Southern Mediterranean coast, and (VI) Northern Mediterranean coast (Figure 10).
- As a result of the analysis conducted in the present study, Areas I and II are suitable for offshore wind, wave and biomass. Areas III and V are suitable for offshore wind, marine current and offshore solar. Area IV is suitable for offshore wind, ocean wave an offshore solar, and Area VI is suitable for offshore wind, osmotic and offshore solar. This has been summarized in Figure 15.

J. Mar. Sci. Eng. **2019**, 7, 249

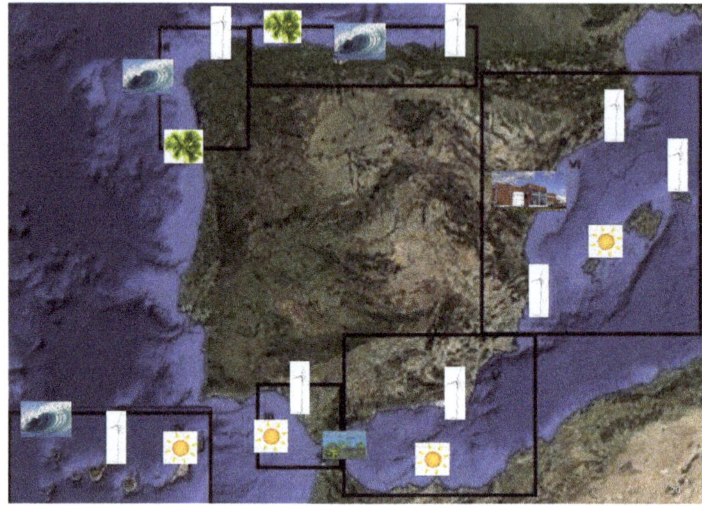

Figure 15. Summary of the feasibility study of the different types of marine renewable energies in the Spanish coast.

Author Contributions: Conceptualization, M.D.E. and J.-S.L.-G.: Methodology, V.N.; Investigation, J.M.E.; Resources, J.M.O.; Writing—Original Draft Preparation, M.D.E.; Writing—Review & Editing, J.M.O.

Funding: This research was funded by the Agustín de Betancourt Foundation (FAB).

Conflicts of Interest: The authors declare no conflict of interest. The funders had no role in the design of the study; in the collection, analyses, or interpretation of data; in the writing of the manuscript, or in the decision to publish the results.

Abbreviations

ECMWF	European Center for Medium-Range Weather Forecasts.
FAO	International Federation of Aquaculture.
GOW	Global Ocean Waves.
HANPP	Human Apropriation of Net Primary Production.
HAT	High Astronomical Tide.
IDAE	Instituto para la Diversificación y Ahorro de la Energía.
IEA	International Energy Agency.
IEA-OES	Implementation Agreement of Oceanic Energy Systems.
IHCantabria	Instituto de Hidráulica Ambiental de la Universidad de Cantabria.
JASON	Joint Altimetry Satellite Oceanography Network.
JONSWAP	JOint North Sea WAve Project.
K1	diurnal tidal constituent.
LAT	Lowest Astronomical Tide.
LCOE	Levelized Cost of Electricity.
M2	semidiurnal tidal constituent.
MHWN	Mean High Water Neaps.
MHWS	Mean High Water Springs.
MLWN	Mean Low Water Neaps.
MLWS	Mean Low Water Springs.
MSL	Mean Sea Level.
NCEP	National Centers for Environmental Prediction.
NPP	Net Primary Productivity.
NOAA	National Oceanic and Atmospheric Administration.
O1	diurnal tidal constituent.

S2	semidiurnal tidal constituent.
SDG	Sustainable Development Goal.
TMA	Texel-Mardsen-Arsloe.
TOPEX	The Ocean Topography Experiment.
WEC	Wave Energy Converter.

References

1. Esteban, M.D.; Diez, J.J.; López-Gutiérrez, J.S.; Negro, V. Integral Management Applied to Offshore Wind Farms. *J. Coast. Res.* **2009**, *56*, 1204–1208.

2. Ocean Energy Europe. *European Commission Issue Paper on Ocean Energy Industry Response*; Technical Report; Ocean Energy Europe: Brussels, Belgium, 2016.

3. Esteban, M.D.; Diez, J.J.; López-Gutiérrez, J.S.; Negro, V. Why Offshore Wind Energy? *Renew. Energy* **2011**, *36*, 444–450. [CrossRef]

4. IRENA. *Ocean Energy Technology: Innovation, Patents, Market Status and Trends*; Technical Report; IRENA: Abu Dhabi, UAE, 2014.

5. Sempreviva, A.M.; Barthelmie, R.J.; Pryor, S.C. Review of Methodologies for Offshore Wind Resource Assessment in European Seas. *Surv. Geophys.* **2008**, *29*, 471–497. [CrossRef]

6. Oh, K.; Kim, J.; Lee, J.; Ryu, K. Wind Resource Assessment Around Korean Peninsula for Feasibility Study on 100 MW Class Offshore Wind Farm. *Renew. Energy* **2012**, *42*, 217–226. [CrossRef]

7. Van Nieuwkoop, J.C.C.; Smith, H.C.M.; Smith, G.H.; Johanning, L. Wave Resource Assessment Along the Cornish Coast (UK) from a 23-year Hindcast Dataset Validated Against Buoy Measurements. *Renew. Energy* **2013**, *58*, 1–14. [CrossRef]

8. Franzitta, V.; Rao, D.; Curto, D.; Viola, A. Greening Island: Renewable Energies Mix to Satisfy Electrical Needs of Pantelleria in Mediterranean Sea. In Proceedings of the Oceans 2016 MTS/IEEE, Monterey, CA, USA, 19–23 September 2016.

9. Franzitta, V.; Curto, D.; Rao, D. Energetic Sustainability Using Renewable Energies in the Mediterranean Sea. *Sustainability* **2016**, *8*, 1164. [CrossRef]

10. Penalba, M.; Ulazia, A.; Ibarra-Berastegi, G.; Ringwood, J.; Sáenz, J. Wave Energy Resource Variation off the West Coast of Ireland and its Impact on Realistic Wave Energy Converters' Power Absortion. *Appl. Energy* **2018**, *224*, 205–219. [CrossRef]

11. Ulazia, A.; Nafarrate, A.; Ibarra-Berastegi, G.; Sáenz, J.; Carreno-Madinabeitia, S. The Consequences of Air Density Variations over Northeastern Scotland for Offshore Wind Energy Potential. *Energies* **2019**, *12*, 2635. [CrossRef]

12. Iglesias, G.; Carballo, R. Wave Energy Resource in the Estaca de Bares Area (Spain). *Renew. Energy* **2010**, *35*, 1574–1584. [CrossRef]

13. López-Gutiérrez, J.S.; Esteban, M.D.; Negro, V. Wave Energy Potential Assessment and Feasibility Analysis of Wave Energy Converters. Case Study: Spanish Coast. *J. Coast. Res.* **2018**, *85*, 1291–1295. [CrossRef]

14. Pérez-Collazo, C.; Greaves, D.; Iglesias, G. A Review of Combined Wave and Offshore Wind Energy. *Renew. Sustain. Energy Rev.* **2015**, *42*, 141–153. [CrossRef]

15. Sierra, J.P.; Mösso, C.; González-Marco, D. Wave Energy Resource Assessment in Menorca (Spain). *Renew. Energy* **2014**, *71*, 51–60. [CrossRef]

16. Ulazia, A.; Penalba, M.; Ibarra-Berastegi, G.; Ringwood, J.; Sáenz, J. Wave Energy Trends over the Bay of Biscay and the Consequences for Wave Energy Converters. *Energy* **2017**, *141*, 624–634. [CrossRef]

17. Ulazia, A.; Ibarra-Berastegi, G.; Sáenz, J.; Carreno-Madinabeitia, S.; González-Rojí, S.J. Seasonal Correction Owing to Use of Actual Air Density in Estimation of Offshore Wind Energy Potential: Case of the Iberian Peninsula. *Sustainability* **2019**, *11*, 3648. [CrossRef]

18. Ibarra-Berastegi, G.; Sáenz, J.; Ulazia, A.; Serras, P.; Esnaola, G.; García-Soto, C. Electricity Production, Capacity Factor, and Plant Efficiency Index at the Mutriku Wave Farm (2014–2016). *Ocean Eng.* **2018**, *147*, 20–29. [CrossRef]

19. Mustapa, M.A.; Yaakob, O.B.; Ahmed, Y.M.; Rheem, C.K.; Koh, K.K.; Adnan, F.A. Wave Energy Device and Breakwater Integration: A Review. *Renew. Sustain. Energy Rev.* **2017**, *77*, 43–58. [CrossRef]

20. López, I.; Andreu, J.; Ceballos, S.; Martínez de Alegría, I.; Kortabarria, I. Review of Wave Energy Technologies and the Necessary Power Equipment. *Renew. Sustain. Energy Rev.* **2013**, *27*, 413–434. [CrossRef]

21. Huang, N.; Chen, M.; Xu, W.; Wang, J.W. Incorporating Importance Sampling in EM Learning for Sequence Detection in SPAD Underwater OWC. *IEEE Access* **2019**, *7*, 4529–4537. [CrossRef]

22. Viola, A.; Franzittia, V.; Trapanese, M.; Curto, D. Nexus Water & Energy: A case Study of Wave Energy Converters (WECs) to Desalination Applications in Sicily. *Int. J. Heat Technol.* **2016**, *34*, 1–8.

23. Franzitta, V.; Curto, D.; Milone, D.; Trapanese, M. Energy Saving in Public Transport Using Renewable Energy. *Sustainability* **2017**, *9*, 106. [CrossRef]

24. Zabihian, F.; Fung, A.S. Review of Marine Renewable Energies: Case Study of Iran. *Renew. Sustain. Energy Rev.* **2011**, *15*, 2461–2474. [CrossRef]

25. Khojasteh, D.; Mahmood, S.; Glamore, W.; Iglesias, G. Wave Energy Status in Asia. *Ocean Eng.* **2018**, *169*, 344–358. [CrossRef]

26. Besio, G.; Mentaschi, L.; Mazzino, A. Wave Energy Resource Assessment in the Mediterranean Sea on the Basis of a 35-year Hindcast. *Energy* **2016**, *94*, 50–63. [CrossRef]

27. Jakimavicius, D.; Kriauciuniene, J.; Sarauskiene, D. Assessment of Wave Climate and Energy Resources in the Baltic Sea Nearshore (Lithuanian Territorial Water). *Oceanologia* **2018**, *60*, 207–218. [CrossRef]

28. Cavazzi, S.; Dutton, A.G. An Offshore Wind Energy Geographic Information System (OWE-GIS) for Assessment of the UK's Offshore Wind Energy Potential. *Renew. Energy* **2016**, *87*, 212–228. [CrossRef]

29. Kaldellis, J.K.; Apostolou, D.; Kapsali, M.; Kondili, E. Environmental and Social Footprint of Offshore Wind Energy. Comparison with Onshore Counterpart. *Renew. Energy* **2016**, *92*, 543–556. [CrossRef]

30. Mohammadzadeh, S.; Jalilinasrabady, S.; Fujii, H.; Farabi-Asl, H. A Comprehensive Approach for Wind Power Plant Potential Assessment, Application to Northwestern Iran. *Energy* **2018**, *164*, 344–358. [CrossRef]

31. European Wind Energy Association (EWEA). *The European Offshore Wind Industry—Key Trends and Statistics 2014*; Technical Report; European Wind Energy Association: Bruxelles, Belgium, 2015.

32. Arent, D.; Sullivan, P.; Heimiller, D.; Lopez, A.; Eurek, K.; Badger, J.; Jorgensen, H.E.; Kelly, M. *Improved Offshore Wind Resource Assessment in Global Climate Stabilization Scenarios*; NREL Technical Report; NREL: Golden, CO, USA, 2012.

33. Zhang, H.M.; Reynolds, R.W.; Bates, J.J. Blended and gridded high resolution global sea surface wind speed and climatology from multiple satellites: 1987–present. In Proceedings of the American Meteorological Society, Annual Meeting, Atlanta, GA, USA, 28 January–3 February 2006.

34. European Centre for Medium-Range Weather Forecasts Web Page. Available online: www.ecmwf.int (accessed on 10 January 2019).

35. Iglesias, G.; Carballo, R. Offshore and Inshore Wave Energy Assessment: Asturias (N Spain). *Energy* **2010**, *35*, 1964–1972. [CrossRef]

36. Peter, J.; Frigaard, P.; Friis-Madsen, E.; Sorensen, H. Prototype Testing of the Wave Energy Converter Wave Dragon. *Renew. Energy* **2006**, *31*, 181–189.

37. Kim, G.; Mu, W.; Soo, K.; Jun, K.; Eun, M. Offshore and Nearshore Wave Energy Assessment Around the Korean Peninsula. *Energy* **2011**, *36*, 1460–1469. [CrossRef]

38. Esteban, M.D.; López-Gutiérrez, J.S.; Negro, V.; Laviña, M.; Muñoz-Sánchez, P. A New Classification of Wave Energy Converters Used for Selection of Devices. *J. Coast. Res.* **2018**, *85*, 1.286–1.290. [CrossRef]

39. Ocean Energy Systems Web Page. Available online: www.ocean-energy-systems.org (accessed on 12 January 2019).

40. Mork, G.; Barstow Stephen, B.; Kabuth, A.; Pontes, M.T. Assesingthe Global Wave Enery Potential. In Proceedings of the OMAE 2010, 29th International Conference on Ocean, Offshore Mechanics and Arctic Engineering, Shanghai, China, 6–11 June 2010.

41. IEA-OES (International Energy Agency—Ocean Energy Systems). *Annual Report 2013*; Technical Report; IEA-OES: Paris, France, 2014.

42. OES (Ocean Energy Systems). *An International Vision for Ocean Energy*; Technical Report; OES: Porto, Portugal, 2011.

43. Rourke, F.O.; Boyle, F.; Reynolds, A. Tidal Energy Update 2009. *Appl. Energy* **2010**, *87*, 398–409. [CrossRef]

44. Water, S.; Aggidis, G. Tidal Range Technologies and State of the Art in Review. *Renew. Sustain. Energy Rev.* **2016**, *59*, 514–529. [CrossRef]

45. Baker, C. Tidal Power. *Energy Policy* **1991**, *19*, 792–797. [CrossRef]

46. Lewis, M.J.; Angeloudis, A.; Robins, P.E.; Evans, P.S.; Neill, S.P. Influence of Storm Surge on Tidal Range Energy. *Energy* **2017**, *122*, 25–36. [CrossRef]
47. University of Graz Wegener Center for Climate and Global Change. Available online: https://wegc203116.uni-graz.at/meted/oceans/tides_intro/print.htm (accessed on 11 January 2019).
48. Neumann, D. *Ocean Currents*; Elsevier: Amsterdam, The Netherlands, 1968.
49. Munk, W.H. On the Wind-driven Ocean Circulation. *J. Meteorol.* **1950**, *7*, 80–93. [CrossRef]
50. Fofonoff, N.P.; Webster, F.; Deacon, E. A Discussion on Ocean Currents and Their Dynamics—Current Measurements in the Western Atlantic. *Philos. Trans. R. Soc. Lond.* **1971**, *270*, 423–436. [CrossRef]
51. Rintoul, S.R.; Hughes, C.W.; Olbers, D. Chapter 4.6. The Antarctic Circumpolar Current System. *Int. Geophys.* **2001**, *77*, 271–302.
52. Bajah, A.S.; Myers, L.E. Fundamentals Applicable to the Utilization of Marine Current Turbines for Energy Production. *Renew. Energy* **2003**, *28*, 2205–2211. [CrossRef]
53. Bryden, I.G.; Cough, S. ME1-Marine Energy Extraction: Tidal Resource Analysis. *Renew. Energy* **2006**, *31*, 133–139. [CrossRef]
54. Lewis, A. Ocean Energy. In *IPCC Special Report on Renewable Energy Resource Sources and Climate Change Mitigation*; Edenhofer, O., Ed.; IPCC: Geneve, Switzerland, 2011.
55. Etemadi, A.; Emdadi, A.; Asefafshar, O.; Emami, Y. Electricity Generation by the Ocean Thermal Energy. *Energy Procedia* **2011**, *12*, 936–943. [CrossRef]
56. Von Jouanne, A.; Brekken, T.K.A. Ocean and Geothermal energy Systems. *IEEE* **2017**, *99*, 1–19. [CrossRef]
57. Gerstandt, K.; Peinemann, K.V.; Erik, S.; Thorsen, T.; Holt, T. Membrane Processes in Energy Supply for an Osmotic Power Plant. *Desalination* **2008**, *224*, 64–70. [CrossRef]
58. Helfer, F.; Lemckert, C.; Anissimov, Y.G. Osmotic Power with Pressure Retarded Osmosis: Theory, Performance and Trends—A review. *J. Membr. Sci.* **2014**, *453*, 337–358. [CrossRef]
59. McKendry, P. Energy Production from Biomass (part 1): Overview of Biomass. *Bioresour. Technol.* **2002**, *83*, 37–46. [CrossRef]
60. Wei, N.; Quarterman, J.; Jin, Y.S. Marine Macroalgae: An Untapped Resource for Producing Fuels and Chemicals. *Trends Biotechnol.* **2013**, *31*, 70–77. [CrossRef]
61. Ross, A.B.; Jones, J.M.; Kubacki, M.L.; Bridgeman, T. Classification of Macroalgae as Fuel and its Thermochemical Behaviour. *Bioresour. Technol.* **2008**, *99*, 6494–6504. [CrossRef]
62. Peteiro, C.; García-Tasende, M. Uso y Cultivo de Laminarias, las Grandes Algas Marinas. *Investig. Y Cienc.* **2015**, *466*, 8–9.
63. FAO. *Food and Agriculture Organization of the United Nations*; Technical Report; FAO: Rome, Italy, 2014.
64. Hughes, A.D.; Kelly, M.; Black, K.D.; Suzanne-Stanley, M. Biogas from Macroalgae: Is It Time to Revisit the Idea? *Biotechnol. Biofuels* **2012**, *5*, 1–7. [CrossRef]
65. Wikipedia. Available online: www.wikipedia.org (accessed on 15 June 2019).
66. Hiriart, G.; Prol-Ledesma, R.M.; Alcocer, S.; Espindola, S. Submarine Geothermics; Hydrothermal Vents and Electricity Generation. In Proceedings of the World Geotechnical Congress, Bali, Indonesia, 25–29 April 2010.
67. Sahu, A.; Yadav, N.; Sudhakar, K. Floating Photovoltaic Power Plant: A Review. *Renew. Sustain. Energy Rev.* **2016**, *66*, 815–824. [CrossRef]
68. Ez 2c Web Page. Available online: www.ez2c.de/ml/solar_land_area/ (accessed on 10 January 2019).
69. Tran, T.T.D.; Smith, A.D. Incorporating Performance-based Global Sensitivity and Uncertainty Analysis into LCOE Calculations for Emerging Renewable Energy Technologies. *Appl. Energy* **2018**, *2016*, 157–171. [CrossRef]
70. Montoya, F.G.; Aguilera, M.J.; Manzano-Agugliaro, F. Renewable Energy Production in Spain: A Review. *Renew. Sustain. Energy Rev.* **2014**, *33*, 509–531. [CrossRef]
71. IDAE Web Page. Available online: www.atlaseolicoidae.es (accessed on 11 January 2019).
72. IHCantabria Web Page. Available online: www.ihcantabria.com (accessed on 14 January 2019).
73. IDAE. *Evaluación del Potencial de la Energía de las Olas*; Technical Report; IDAE: Madrid, Spain, 2011.
74. Puertos del Estado Web Page. Available online: www.puertos.es (accessed on 15 January 2019).
75. Aemet Web Page. Available online: www.aemet.com (accessed on 16 January 2019).

Journal of
*Marine Science
and Engineering*

Article

Evaluation of Lebanon's Offshore-Wind-Energy Potential

Gabriel Ibarra-Berastegi [1,2,*,†], **Alain Ulazia** [3,†], **Jon Saénz** [2,4,†] and **Santos José González-Rojí** [4,5,6,†]

1   Department of NE and Fluid Mechanics, University of the Basque Country (UPV/EHU), Bilbao Engineering School, 48002 Bilbao, Spain
2   Plentzia Itsas Estazioa, PIE, University of the Basque Country (UPV/EHU), 48620 Plentzia, Spain; jon.saenz@ehu.eus
3   Department of NE and Fluid Mechanics, University of the Basque Country (UPV/EHU), Gipuzkoa Engineering School, 20600 Eibar, Spain; alain.ulazia@ehu.eus
4   Department of Applied Physics II, University of the Basque Country (UPV/EHU), 48940 Leioa, Spain; santosjose.gonzalez@ehu.eus
5   Oeschger Centre for Climate Change Research, University of Bern, 3003 Bern, Switzerland
6   Climate and Environmental Physics, University of Bern, 3003 Bern, Switzerland
*   Correspondence: gabriel.ibarra@ehu.eus
†   These authors contributed equally to this work.

Received: 21 September 2019 ; Accepted: 5 October 2019 ; Published: 10 October 2019

Abstract: The only regional evaluation of Lebanese wind-energy potential (National Wind Atlas) dates back to 2011 and was carried out by a United Nations agency. In this work, data from the most recent reanalysis (ERA5) developed at the European Center for Medium Range Weather Forecast (ECMWF), corresponding to the 2010–2017 period, were used to evaluate Lebanese offshore-wind-energy potential. In the present study, wind power density associated to a SIEMENS 154/6 turbine was calculated with a horizontal resolution of 31 km and 1 hour time steps. This work incorporated the impact of air density changes into the calculations due to the seasonal evolution of pressure, temperature, and humidity. Observed average offshore air density ρ_0 was 1.19 kg/m^3 for the 2010–2017 period, but if instead of ρ_0, hourly ρ values were used, seasonal oscillations of wind power density (WPD) represented differences in percentage terms ranging from −4% in summer to +3% in winter. ERA5 provides hourly wind, temperature, pressure, and dew-point temperature values that allowed us to calculate the hourly evolution of air density during this period and could also be used to accurately evaluate wind power density off the Lebanese coast. There was a significant gradient in wind power density along the shore, with the northern coastal area exhibiting the highest potential and reaching winter values of around 400 W/m^2. Finally, this study suggests that the initial results provided by the National Wind Atlas overestimated the true offshore-wind-energy potential, thus highlighting the suitability of ERA5 as an accurate tool for similar tasks globally.

Keywords: Lebanon; offshore wind energy; wind power density; aiRthermo; air density; ERA5

---

1. Introduction

Lebanon is located in the eastern Mediterranean with a shoreline of approximately 210 km (Figure 1). Electricity generated in the country originates mostly from fossil sources that need to be imported. Following the 2020 objective for 12% of electricity from renewable sources in Lebanon [1], some studies have pointed out that electricity from renewable sources—solar and wind energy—needs to be incorporated into the electric mix [2–5].

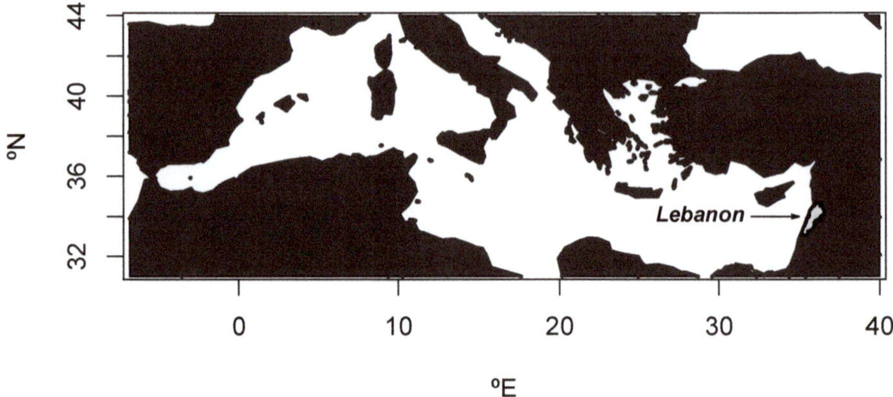

Figure 1. Lebanon, located in the eastern Mediterranean.

Younes et al. [6] suggested that a decentralized policy for energy production could contribute to the economic feasibility of incorporating renewables into the Lebanese electric system. In the case of wind energy, a study indicated that offshore wind energy could be used to support or complement existing power plants [1], but to that purpose, accurate estimation of wind resources is needed. Inland wind-energy potential has been partially assessed using a limited number of sensors [7], but the most comprehensive study carried out so far on Lebanese wind-energy potential is the National Wind Atlas, prepared by the United Nations and made public in 2011 [8]. In the case of offshore wind energy, the National Wind Atlas provides estimations of wind power density (WPD) at a height of 80 m for the Lebanese coast. The methodology applied in Atlas involves the use of wind-sensor records and a numerical model. Specific details are extensively described in the original document [9].

The objective of this paper is to characterize Lebanese offshore-wind power potential using the most recent reanalysis, ERA5 [10], freely available through the Copernicus Climate Data Store [11]. ERA5 is becoming increasingly popular for wind-energy assessment studies [12,13]. As an indicator of wind-energy potential for the Lebanese coast, wind power density WPD is used [14]. It was calculated at a height of 178 m corresponding to the hub of the SIEMENS 154/6 floating wind turbine model with a rated power of 6 MW. For comparison purposes with previous estimations gathered in the National Wind Atlas, offshore-wind power density at a height of 80 m was also calculated using ERA5 wind speed, humidity, pressure, and temperature data in order to obtain instantaneous (one-hourly) air density.

Air density is typically considered as constant during the year, with standard value ρ_0 being equal to 1.225 kg/m^3 (at sea level, 1013 mb, 15 °C) as reference for middle latitudes near the sea, or the annual mean air density for other latitudes and altitudes. The use of constant air density is usual for different wind-energy estimation methods, both at specific locations using anemometers [15–17], or over given geographical regions using mesoscale models, remote-sensing data, or reanalysis [18–29]. This is understandable since Weibull distribution is commonly fitted onto the wind-speed data of the location to be implemented on the turbine's power curve. Only wind speed is used because the power curve is provided for constant air density. However, it was shown that deformation of the power curve due to air density changes in its U^3 zone is similar to deformation due to pitch misalignment [30] or due to the presence of defective anemometers on the turbine [31].

In recent publications [13,32,33], the authors developed a technique to seasonally estimate the influence of instantaneous air density changes on the capacity factor of a given turbine. Floors et al. [34] emphasized, like us, the importance of air density, and also used ERA5 to study its effects. In this sense, the main physical magnitude that synthesizes the influence of wind speed with air density is the wind power density, the main parameter used for map representation of offshore wind energy in this work for Lebanon.

2. Data and Methodology

2.1. Data

ERA5 meteorological hourly data corresponding to the period of 2010–2017 were downloaded from the Copernicus Climate Data Store [11]. ERA5 provides a spatial resolution of 0.3° (~31 km) and geographical boundaries covering the Lebanon coastal area included 90 gridpoints with 10 positions in longitude [34.2° E, 36.9° E], and 9 positions in latitude [33° N, 35.4° N]. Since the focus of this study was the analysis of offshore wind energy, only the 30 sea gridpoints were selected for this study (Figure 2).

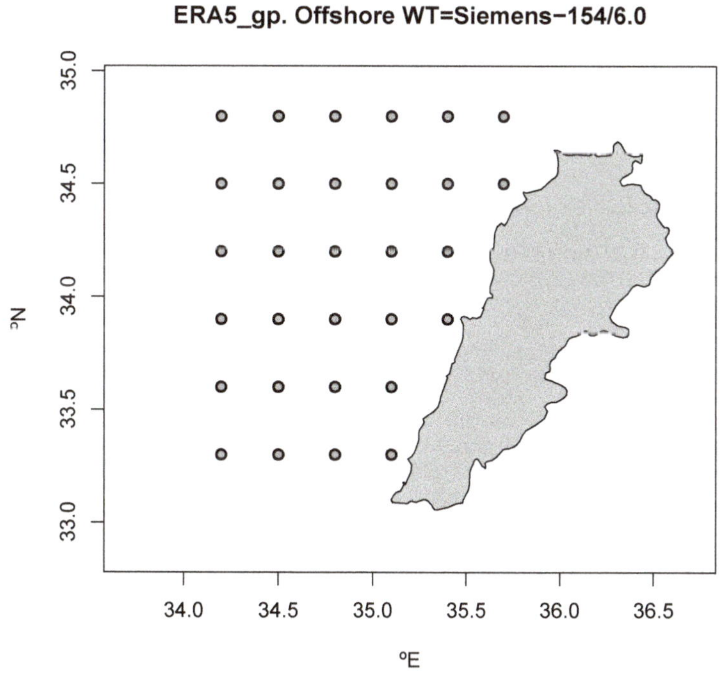

Figure 2. Location of ERA5 gridpoints on the study area.

It is to be noted that for implementation of wind-energy facilities in Lebanese waters, there were some practical constraints that deserve a brief mention even though they fall beyond the scope of this study. The Lebanese exclusive economic zone (EEZ) extends 200 nautical miles from the coast and, according to Lebanese legislation, there is a buffer area of 3 km [8,9] in which economic activities like wind energy could not be allowed. It is also worthwhile to mention that water depth increases rapidly as we move from the coast into the sea, thus making floating wind turbines the best solution for any future wind farm. For this reason, hub height h of 178 m corresponding to the SIEMENS 154/6 floating wind-turbine model was chosen to estimate wind power density. However, the current maximum technological limits are 120 km of AC cable to the coast and a mooring limit of 1000 m depth [35]. These practical constraints are represented in Figure 3 where wind turbines represent the location of the ERA5 gripdoints falling within the current administrative boundaries. It is important to mention that some of these constraints, both legal and technical, may change or disappear in the future. For this reason, they have been ignored for this study, although they should be considered for detailed analysis of wind-farm design.

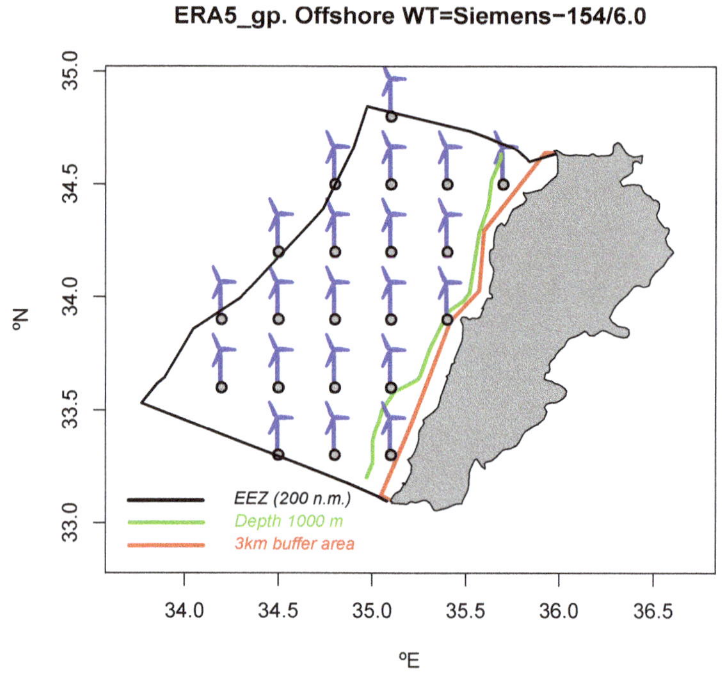

Figure 3. Lebanese exclusive economic zone, 1000 m depth boundary, and buffer area.

Although wind speed is a vectorial magnitude, for wind energy studies, wind speed values are customarily used. However, ERA5 provides the zonal (projections along the earth's parallels) and meridional (projections along the earth's meridians) of the wind speed vector (value+ incoming direction) and the wind speed value must be derived from these two components. The original ERA5 variables downloaded with the above-mentioned time and space resolution and that have been used for this study were:

(i) Zonal wind speed (u_{10}) at a height of 10 m above sea level (masl)
(ii) Meridional wind speed at 10 masl (v_{10})
(iii) Zonal wind speed at 100 masl (u_{100})
(iv) Meridional wind speed at 100 masl (v_{100})
(v) Mean sea level pressure ($mslp$)
(vi) Surface temperature (t_2)
(vii) Dew point temperature (d_2)

2.2. Methodology

First, at each gridpoint, the ERA5 zonal and meridional wind components were combined to calculate wind-speed values at 10 and 100 masl. Then, assuming a logarithmic law [13,14], surface roughness z_0 could first be estimated and then wind-speed values U_h at heights h of 80 and 178 masl could be derived (Equation (1)).

$$\frac{U_{100}}{U_{10}} = \frac{log(100/z_0)}{log(10/z_0)} \Rightarrow z_0 \Rightarrow \frac{U_h}{U_{10}} = \frac{log(h/z_0)}{log(10/z_0)} \Rightarrow U_h \tag{1}$$

wind power density (WPD_h) at hub height h of the chosen turbine provided an estimation of the energy that could be extracted from the wind (W/m^2), and it was calculated according to Equation (2) [14].

$$WPD_h = \frac{1}{2}\rho u_h^3 \tag{2}$$

Air density ρ was not directly provided by ERA5 but could be calculated by combining information on air humidity, pressure, and temperature hourly values. These calculations could be made using the regular equations for atmospheric humidity studies [36,37].

The density of moist air can be computed by means of the expression corresponding to dry air if the virtual temperature t_v is used instead of the real temperature t_2 in the equation of state of the dry air. Thus, the density of moist air is given by Equation (3), with $mslp$ pressure (Pa), t_v the virtual temperature (K) and $R_d \approx 287$ J K^{-1} kg^{-1} the constant corresponding to the mixture of gases which forms dry air.

$$\rho = \frac{mslp}{R_d t_v} \tag{3}$$

The virtual temperature t_v (K), with $t_2 \leq t_v$ is defined by Equation (4), where $\varepsilon = \frac{R_d}{R_w} \approx 0.622$ is the ratio of the gas constants corresponding to dry air R_d and water vapour $R_w \approx 487$ J K^{-1} kg^{-1}.

$$t_v = \frac{t_2}{1 - (1 - \varepsilon)\frac{e}{mslp}} \tag{4}$$

These computations can be carried out on the basis of these approximations or instead of going through these equations and solving them, a more straightforward method is to use the R [38] package called *aiRthermo* [39]. This package has been developed by the authors [40] and already incorporate the above mentioned equations. To that purpose, the following three built-in *aiRthermo* functions were applied to ERA5 hourly data:

1. ***TTdP2rh***, to calculate relative humidity rh from a given temperature t_2, dew point temperature d_2, and pressure $mslp$, all of them in SI units.
2. ***rh2w***, to calculate mixing ratio w from rh, pressure $mslp$, and temperature t_2. The mixing ratio is defined as the ratio of the mass of water vapour to the mass of dry air, and was returned as kg/kg.
3. ***densityMoistAir***, to calculate density of moist air ρ in kg/m^{-3} from $mslp$, temperature t_2, and mixing ratio w.

Although the influence of humidity on air density is known to be small and, in most cases, negligible [13,41–43], *aiRthermo* provides an exact and straightforward estimation of ρ values, including the effect of humidity. That is the reason why its effects have been now incorporated into this study. The most specific details on *aiRthermo* implementation can be found in the literature [40].

After calculating hourly ρ values, by combining Equations (1) and (2) at a height of $h = 178$ m, the hourly values of WPD_{178} corresponding to the 2010–2017 period at the selected 30 gridpoints were obtained. From WPD_{178} hourly values and with the aim of providing estimation of the variations along the year, seasonal averages $\overline{WPD_{178}}$ corresponding to the 2010–2017 period were computed. Then, according to the following monthly distribution:

- Winter: December, January, February (DJF).
- Spring: March, April, May (MAM).
- Summer: June, July, August (JJA).
- Autumn: September, October, November (SON).

hourly values of WPD_{178} corresponding to the 2000–2017 period have been used to calculate the four seasonal averages [$\overline{WPD_{178\text{-}WINTER}}$, $\overline{WPD_{178\text{-}SPRING}}$, $\overline{WPD_{178\text{-}SUMMER}}$ and $\overline{WPD_{178\text{-}AUTUMN}}$]. To illustrate some results, the seasonal averages of air density and wind speed at 178 m height ($\bar{\rho}$ and $\overline{U_{178}}$) were also computed from hourly values. Similarly, for comparison purposes with the previous National Wind Atlas [9], hourly WPD_{80} values were calculated using Equations (1) and (2) with $h = 80$ m; finally, all hourly cases corresponding to 2010–2017 were averaged to obtain $\overline{WPD_{80}}$ for the same gridpoints of the area.

3. Results

The values of WPD_{178} depend both on air density and wind speed at a height of 178 m (Equation (2)). Regarding ρ in the Lebanese offshore area, there did not exist a relevant spatial gradient, but maximum oscillations of seasonal $\bar{\rho}$ between summer and winter of $\pm3\%$ around the average value of $\rho_0 = 1.19$ kg/m^3 could be observed (Figure 4).

Coming to wind speed, the seasonal averages for the area in the 2010–2017 period, ($\overline{U_{178}}$) exhibited two major spatial gradients (Figure 5)

(i)   Higher values on the northern coast and smaller in the central coast. However, in winter, intermediate values are observed in the southern coast, while in summmer, wind speeds reach their lowest values along the southern coast.

(ii)   As we moved away from the coast into the sea, higher wind-speed values were observed.

Additionally, $\overline{U_{178}}$ showed a clear seasonal pattern with substantially lower values in summer than in winter.

However, the objective of this study was to characterize the evolution of seasonal wind power density $\overline{WPD_{178}}$ to estimate the feasibility of any future wind-farm project. It can be seen (Figure 6) that there was a major spatial gradient along the coast with a maximum in the north and a decreasing gradient southward along the shoreline. Apart from this spatial gradient, $\overline{WPD_{178}}$ showed significant seasonal oscillations driven by the above-mentioned spatial and seasonal patterns of $\bar{\rho}$ and $\overline{U_{178}}$.

As can be seen, the highest values of $\overline{WPD_{178}}$ took place in the northern part of the coast during winter, reaching values slightly below 400 W/m^2. This is because the highest seasonal values of $\bar{\rho}$ and $\overline{U_{178}}$ were also observed in that season and zone. For the whole area, the lowest seasonal $\overline{WPD_{178}}$ values were in summer, while intermediate values occurred in spring and autumn. The ratio of $\overline{WPD_{178}}$ between the maximal (winter) and minimal (summer) values at each gridpoint for the whole area was around 1.5, thus indicating strong seasonality in $\overline{WPD_{178}}$.

It is important to highlight that, if the overall average $\rho_0 = 1.19$ kg/m^3 was introduced as a constant value in Equation (2) instead of using hourly ρ values to calculate WPD_{178} (Equation (2)) and then computing seasonal averages $\overline{WPD_{178}}$, significant errors in the $\overline{WPD_{178}}$ would have taken place. More particularly, our study indicates that winter $\overline{WPD_{178}}$ averages calculated with $\rho_0 = 1.19$ kg/m^3 would underestimate the true $\overline{WPD_{178}}$ by an average of 3% throughout the whole area. Conversely, in summer, an overestimation of 4% would have taken place. This stresses the relevance of not assuming a constant ρ_0 for wind-farm feasibility studies, because it may lead to non-negligible errors at estimating wind power density.

Finally, for comparison purposes with the previous results of offshore-wind power density at 80 masl as gathered in the National Wind Atlas (p. 56) [9], $\overline{WPD_{80}}$ was also calculated with the same methodology for the 2010–2017 period (Figure 7).

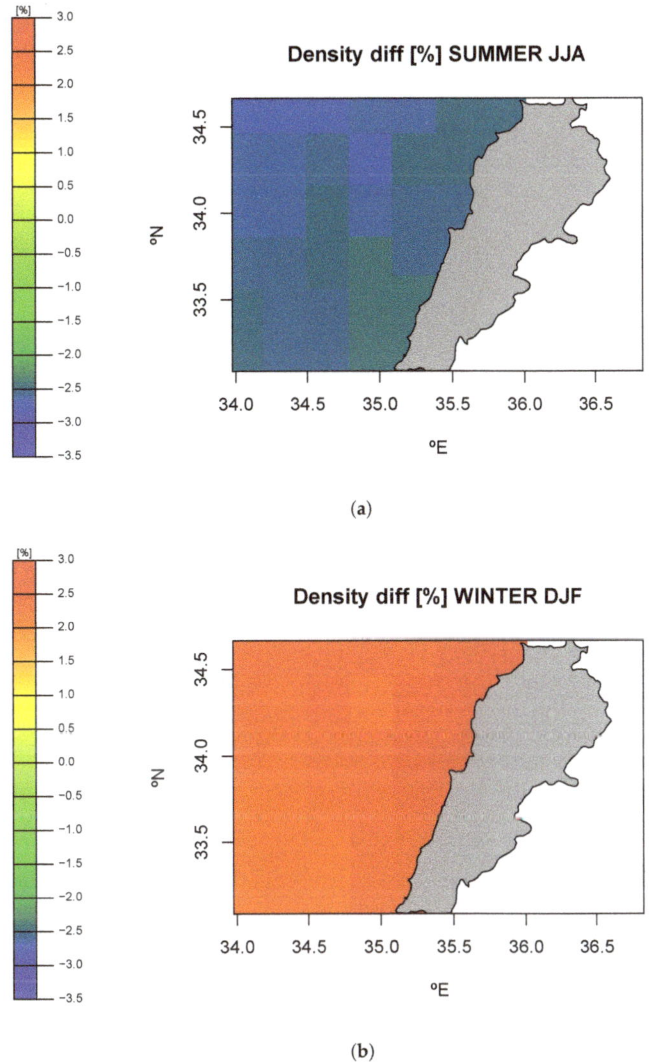

Figure 4. Seasonal air density differences of $\bar{\rho}$ [%] with respect to $\rho_0 = 1.19$ kg/m^3. (**a**) Summer. (**b**) Winter.

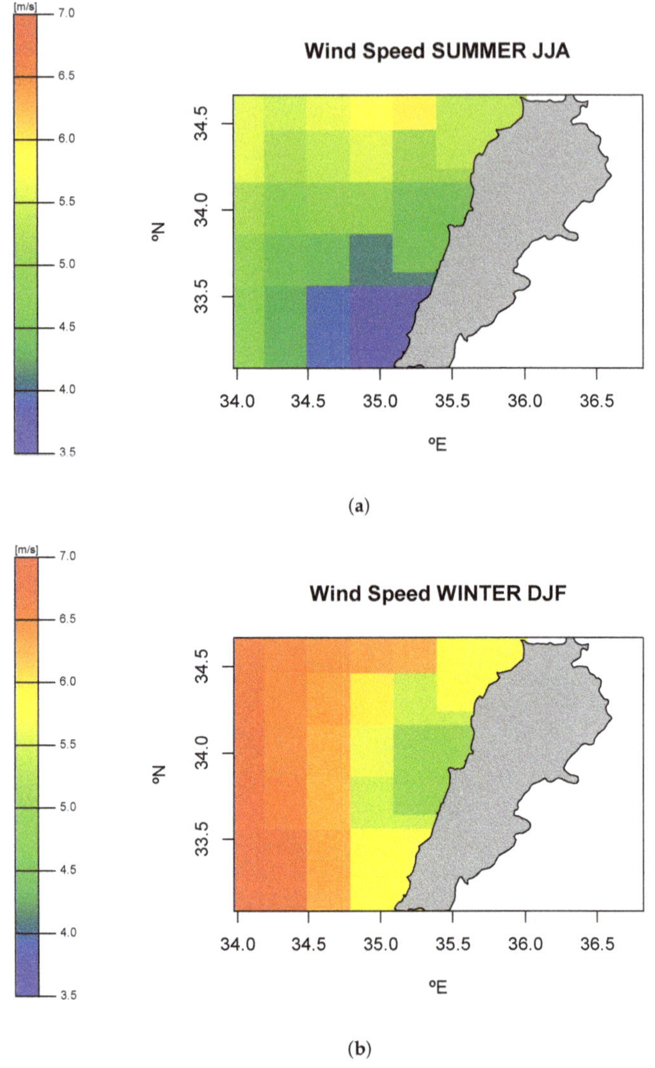

Figure 5. Lebanon offshore-wind speed $\overline{U_{178}}$ at 178 m. Seasonal averages from hourly values. (**a**) Summer. (**b**) Winter.

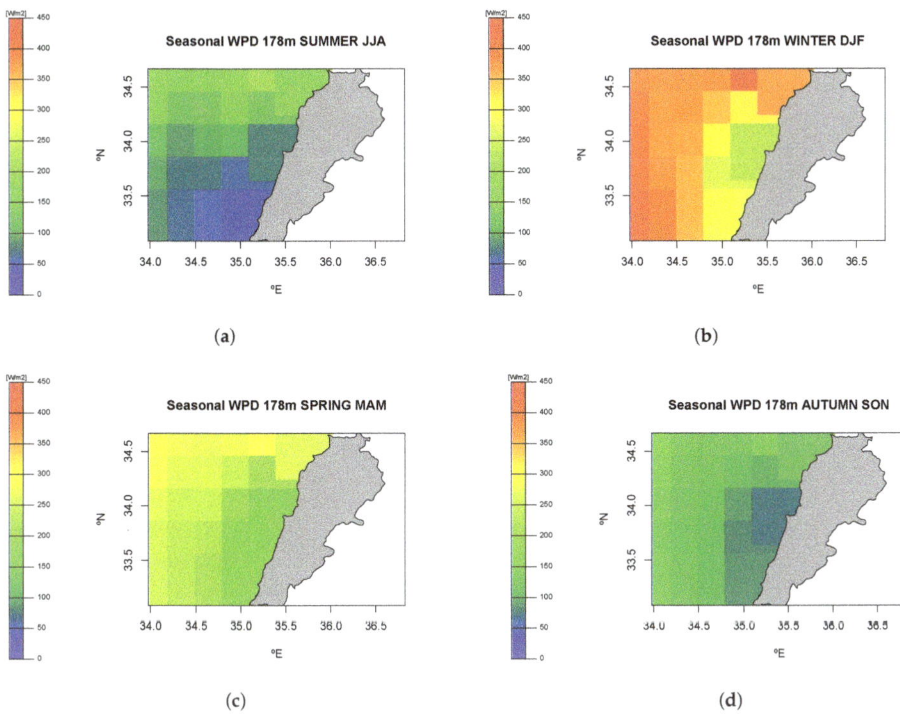

(a)

(b)

(c)

(d)

Figure 6. wind power density at 178 masl. $\overline{WPD_{178}}$. (**a**) Summer. (**b**) Winter. (**c**) Spring. (**d**) Autumn.

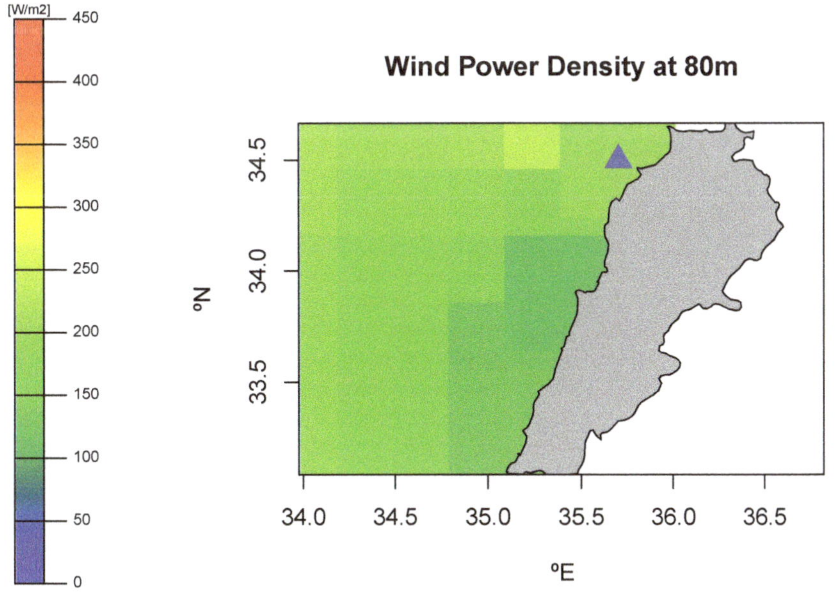

Figure 7. wind power density at 80 masl $\overline{WPD_{80}}$. Overall average. The blue triangle represents the pixel centered at coordinates [35° E, 34.5° N]

The $\overline{WPD_{80}}$ estimations in the previous United Nations report agree with our results (Figure 7) in that the northern area of the country is the zone with the highest potential. However, the attribution of $\overline{WPD_{80}}$ values in this zone reached a maximum of more than 600 W/m^2; if averaged on the same area (roughly) as the one corresponding to our pixel of 0.3° × 0.3° centered at [35° E, 34.5° N] (blue triangle in Figure 7), it would be about 450–470 W/m^2. However, the $\overline{WPD_{80}}$ value at this pixel calculated for the 2010–2017 period and derived from ERA5 was 236.76 W/m^2. This represents an overestimation in the United Nations report of roughly twice the true $\overline{WPD_{80}}$.

4. Discussion

The calculation of hourly wind power density assuming a constant value for local air density in Equation (2) is the methodological approach customarily used in many wind power feasibility studies. In those studies, either the standard value of $\rho_0 = 1.225$ kg/m^3 or local annual mean ρ_0 is adopted. In the case of the Lebanese offshore, $\rho_0 = 1.19$ kg/m^3 .

However, air density is not constant and, in the particular case of the Lebanese coast, exhibits remarkable seasonality driven by the seasonal changes in surface pressure, temperature, and humidity.

Ignoring this can lead to wrong estimations of wind-energy potential, as also pointed out in other studies [33]. In the particular case of the Lebanese offshore $\overline{WPD_{178}}$, errors moved from −3% in winter to +4% in summer. At this point, *aiRthermo* could play an important role for accurate calculation of wind power density.

In the case of the Lebanese offshore, errors in $\overline{WPD_{80}}$ with respect to previous studies were around 100%. The methodology used for the 2011 assessment is described in the UN report [9] and the differences with respect to this study can be attributed to two facts:

1. The poor spatial density of the wind observational data, obtained from a sparse meteorological network providing only surface data. In that study only surface data from 22 stations were used. An additional problem was that the records from all the stations did not cover the same observational periods
2. Generally speaking, the accuracy of the meteorological models used one decade ago was far poorer than the current assimilation algorithms used in ERA5. The model used was the MC^2 (*MesoscaleCompressibleCommunity*) computational model [44].

The use of a poor quality model (by nowadays current standards, not a decade ago) fed with the sparse and low quality data gathered before 2011, can explain the observed differences with the $\overline{WPD_{80}}$ field as obtained with ERA5 and the methodology explained in this work. This highlights the need to use state-of-the-art data and methodologies for similar studies.

Accurate values of the meteorological variables used in this study with reasonable spatial and temporal resolution are required if reliable feasibility studies are to be obtained for a given area. Along these lines, ERA5 is the most valuable tool that can be used not only to estimate Lebanese offshore-wind-energy potential, but also for any other geographical region, either offshore or onshore. These aspects explain its widespread use for wind power density estimations [12,33].

5. Conclusions

The $\overline{WPD_{178}}$ values off the Lebanese coast exhibited strong seasonality, reaching a maximal value of 400 W/m^2 in winter at the northern part of the shore. It should be noted that 400 W/m^2 is a well-known value for the limit of a good wind-energy potential [14]. Higher values were observed as we moved into the sea. Lower values were observed in summer and toward the southern part of the shoreline. Although there was an important overestimation in previous studies, the values obtained now are enough—in the frame of current legal, administrative, and technical limitations—to deploy offshore wind farms that could increase the proportion of renewable energy sources into the Lebanese electric mix. This would also contribute to a less fossil-dependent energy system and, perhaps, to the development of a local wind-energy industry. To that purpose, an accurate characterization of wind power density taking into account changes in air density is needed. The implementation of a proper methodology along with the use of ERA5 and *aiRthermo* have proven to be valuable tools to accomplish this objective, not only for the Lebanese coast, but in most other places in the world.

Author Contributions: Conceptualization, G.I.-B., A.U., J.S.; methodology, G.I.-B., A.U.; software, J.S., S.J.G.-R.; investigation, G.I.-B., J.S.; writing—original-draft preparation, G.I.-B.; writing—review and editing, all authors; supervision, all authors; project administration, G.I.-B.; funding acquisition, J.S., G.I.-B.

Funding: This work was financially supported by the Spanish Government through MINECO project CGL2016-76561-R, (MINECO/ERDF, UE) and the University of the Basque Country (UPV/EHU, GIU 17/002).

Acknowledgments: This work was financially supported by the Spanish Government through MINECO project CGL2016-76561-R, (MINECO/ERDF, UE) and the University of the Basque Country (UPV/EHU, GIU 17/002). ERA5 data were downloaded at no cost from the Copernicus Climate Data Store.

Conflicts of Interest: The authors declare no conflict of interest.

Abbreviations

The following abbreviations are used in this manuscript:

ECMWF	European Centre for Medium Range Weather Forecast
masl	meters above sea level
$mslp$	Surface pressure
u_{10}	Zonal wind speed at a height of 10 m above sea level
v_{10}	Meridional wind speed at 10 m above sea level
u_{100}	Zonal wind speed at 100 m above sea level
u_{100}	Meridional wind speed at 100 masl
rh	Relative humidity
R_d	Constant of dry air
R_w	Constant of water vapour
wp	water vapour pressure
t_2	Surface temperature
t_v	Virtual temperature
d_2	Dew point temperature
L	Latent heat of vaporization
w	Mixing ratio
e	Ratio between R_d and R_w
ρ	Hourly air density
$\bar{\rho}$	Seasonal average of air density
ρ_0	Overall average of offshore air density (1.19 kg/m^3)
U_h	Wind speed at h masl
U_{80}	Wind speed at 80 masl
U_{178}	Wind speed at 178 masl
$\overline{U_{178}}$	Seasonal average of wind speed at 178 masl
WPD	Wind Power Density
WPD_h	Wind Power Density at h masl
WPD_{80}	Hourly Wind Power Density at 80 masl
$\overline{WPD_{80}}$	2010-2017 average of Wind Power Density at 80 masl
WPD_{178}	Hourly Wind Power Density at 178 masl
$\overline{WPD_{178}}$	Seasonal average of Wind Power Density at 178 masl

References

1. Al-Kaaki, O.; Salameh, E.; Assi, A.; Arnaout, M.; Salameh, W. Offshore Wind Farms to Support Existing Power Plants Case Study: Deir Ammar Power Plant, Tripoli, Lebanon. In Proceedings of the 2018 IEEE International Multidisciplinary Conference on Engineering Technology (IMCET), Beirut, Lebanon, 14–16 November 2018; pp. 1–6.

2. Ibrahim, O.; Fardoun, F.; Younes, R.; Louahlia-Gualous, H. Energy status in Lebanon and electricity generation reform plan based on cost and pollution optimization. *Renew. Sustain. Energy Rev.* **2013**, *20*, 255–278. [CrossRef]

3. Elkhoury, M.; Nakad, Z.; Shatila, S. The assessment of wind power for electricity generation in Lebanon. *Energy Sources Part A Recovery Util. Environ. Eff.* **2010**, *32*, 1236–1247. [CrossRef]

4. El-Ali, A.; Moubayed, N.; Outbib, R. Comparison between solar and wind energy in Lebanon. In Proceedings of the 2007 9th International Conference on Electrical Power Quality and Utilisation, Barcelona, Spain, 9–11 October 2007; pp. 1–5.

5. Kinab, E.; Elkhoury, M. Renewable energy use in Lebanon: Barriers and solutions. *Renew. Sustain. Energy Rev.* **2012**, *16*, 4422–4431. [CrossRef]

6. Younes, R.; Fardoun, F.; Ibrahim, O. Electricity of Lebanon: Problems and Recommendations. *Energy Procedia* **2012**, *19*, 310–320. [CrossRef]

7. Al Zohbi, G.; Hendrick, P.; Bouillard, P. Assessment of wind energy potential in Lebanon. *Res. Mar. Sci.* **2018**, *3*, 401–414.

8. Hassan, G. *The National Wind Atlas of Lebanon: A Report*; UNDP/CEDRO: Beirut, Lebanon, 2011.

9. The National Wind Atlas of Lebanon. 2011. Available online: http://www.lb.undp.org/content/lebanon/en/home/library/environment_energy/the-national-wind-atlas-of-lebanon.html (accessed on 10 September 2019).

10. Hersbach, H. The ERA5 Atmospheric Reanalysis; AGU Fall Meeting Abstracts; 2016. Available online: http://adsabs.harvard.edu/abs/2016AGUFMNG33D..01H (accessed on 9 July 2019).

11. Copernicus Climate Data Store. 2019. Available online: https://cds.climate.copernicus.eu/ (accessed on 10 September 2019).

12. Olauson, J. ERA5: The new champion of wind power modelling? *Renew. Energy* **2018**, *126*, 322–331. [CrossRef]

13. Ulazia, A.; Nafarrate, A.; Ibarra-Berastegi, G.; Sáenz, J.; Carreno-Madinabeitia, S. The Consequences of Air Density Variations over Northeastern Scotland for Offshore Wind Energy Potential. *Energies* **2019**, *12*, 2635. [CrossRef]

14. Manwell, J.F.; McGowan, J.G.; Rogers, A.L. *Wind Energy Explained: Theory, Design and Application*; John Wiley & Sons: Hoboken, NJ, USA, 2010.

15. Monteiro, C.; Bessa, R.; Miranda, V.; Botterud, A.; Wang, J.; Conzelmann, G. *Wind Power Forecasting: State-Of-The-Art 2009*; Technical Report; Argonne National Laboratory (ANL): Lemont, IL, USA, 2009.

16. Weisser, D. A wind energy analysis of Grenada: An estimation using the Weibull density function. *Renew. Energy* **2003**, *28*, 1803–1812. [CrossRef]

17. Gökçek, M.; Bayülken, A.; Bekdemir, Ş. Investigation of wind characteristics and wind energy potential in Kirklareli, Turkey. *Renew. Energy* **2007**, *32*, 1739–1752. [CrossRef]

18. Dvorak, M.J.; Archer, C.L.; Jacobson, M.Z. California offshore wind energy potential. *Renew. Energy* **2010**, *35*, 1244–1254. [CrossRef]

19. Gross, M.S.; Magar, V. Offshore wind energy potential estimation using UPSCALE climate data. *Energy Sci. Eng.* **2015**, *3*, 342–359. [CrossRef]

20. Akdağ, S.A.; Güler, Ö. Evaluation of wind energy investment interest and electricity generation cost analysis for Turkey. *Appl. Energy* **2010**, *87*, 2574–2580. [CrossRef]

21. Fueyo, N.; Sanz, Y.; Rodrigues, M.; Montañés, C.; Dopazo, C. High resolution modelling of the on-shore technical wind energy potential in Spain. *Wind Energy* **2010**, *13*, 717–726. [CrossRef]

22. Hasager, C.B.; Barthelmie, R.J.; Christiansen, M.B.; Nielsen, M.; Pryor, S. Quantifying offshore wind resources from satellite wind maps: Study area the North Sea. *Wind Energy* **2006**, *9*, 63–74. [CrossRef]

23. Doubrawa, P.; Barthelmie, R.J.; Pryor, S.C.; Hasager, C.B.; Badger, M.; Karagali, I. Satellite winds as a tool for offshore wind resource assessment: The Great Lakes Wind Atlas. *Remote. Sens. Environ.* **2015**, *168*, 349–359. [CrossRef]

24. Carvalho, D.; Rocha, A.; Santos, C.S.; Pereira, R. Wind resource modelling in complex terrain using different mesoscale–microscale coupling techniques. *Appl. Energy* **2013**, *108*, 493–504. [CrossRef]

25. Carvalho, D.; Rocha, A.; Gómez-Gesteira, M.; Santos, C. A sensitivity study of the WRF model in wind simulation for an area of high wind energy. *Environ. Model. Softw.* **2012**, *33*, 23–34. [CrossRef]

26. Carvalho, D.; Rocha, A.; Gómez-Gesteira, M.; Santos, C.S. Sensitivity of the WRF model wind simulation and wind energy production estimates to planetary boundary layer parameterizations for onshore and offshore areas in the Iberian Peninsula. *Appl. Energy* **2014**, *135*, 234–246. [CrossRef]

27. Carvalho, D.; Rocha, A.; Gómez-Gesteira, M.; Santos, C.S. Comparison of reanalyzed, analyzed, satellite-retrieved and NWP modelled winds with buoy data along the Iberian Peninsula coast. *Remote. Sens. Environ.* **2014**, *152*, 480–492. [CrossRef]

28. Carvalho, D.; Rocha, A.; Gómez-Gesteira, M.; Santos, C.S. WRF wind simulation and wind energy production estimates forced by different reanalyses: Comparison with observed data for Portugal. *Appl. Energy* **2014**, *117*, 116–126. [CrossRef]

29. Carvalho, D.; Rocha, A.; Gómez-Gesteira, M.; Santos, C.S. Offshore wind energy resource simulation forced by different reanalyses: Comparison with observed data in the Iberian Peninsula. *Appl. Energy* **2014**, *134*, 57–64. [CrossRef]

30. Elosegui, U.; Egana, I.; Ulazia, A.; Ibarra-Berastegi, G. Pitch angle misalignment correction based on benchmarking and laser scanner measurement in wind farms. *Energies* **2018**, *11*, 3357. [CrossRef]

31. Rabanal, A.; Ulazia, A.; Ibarra-Berastegi, G.; Sáenz, J.; Elosegui, U. MIDAS: A Benchmarking Multi-Criteria Method for the Identification of Defective Anemometers in Wind Farms. *Energies* **2019**, *12*, 28. [CrossRef]

32. Ulazia, A.; Ibarra-Berastegi, G.; Sáenz, J.; Carreno-Madinabeitia, S.; González-Rojí, S.J. Seasonal correction of offshore wind energy potential due to air density: Case of the Iberian Peninsula. *Energies* **2019**, *11*, 3648. [CrossRef]

33. Ulazia, A.; Sáenz, J.; Ibarra-Berastegui, G.; González-Rojí, S.J.; Carreno-Madinabeitia, S. Global estimations of wind energy potential considering seasonal air density changes. *Energy* **2019**, *187*, 115938. [CrossRef]

34. Floors, R.; Nielsen, M. Estimating Air Density Using Observations and Re-Analysis Outputs for Wind Energy Purposes. *Energies* **2019**, *12*, 2038. [CrossRef]

35. Butterfield, C.P.; Musial, W.; Jonkman, J.; Sclavounos, P.; Wayman, L. *Engineering Challenges for Floating Offshore Wind Turbines*; National Renewable Energy Laboratory: Golden, CO, USA, 2007.

36. Bohren, C.F.; Albrecht, B.A. *Atmospheric Thermodynamics*; Oxford University Press: New York, NY, USA, 1998; p. 402.

37. Petty, G.W. *A First Course in Atmospheric Thermodynamics*; Sundog Publishing: Madison, WI, USA, 2008; p. 337.

38. R Core Team. *R: A Language and Environment for Statistical Computing*; R Foundation for Statistical Computing: Vienna, Austria, 2018. Available online: https://www.R-project.org/ (accessed on 16 July 2019).

39. Sáenz, J.; González-Rojí, S.J.; Carreno-Madinabeitia, S.; Ibarra-Berastegi, G. aiRthermo: Atmospheric Thermodynamics and Visualization. R package version 1.2.1. 2018. Available online: https://CRAN.R-project.org/package=aiRthermo (accessed on 16 July 2019).

40. Sáenz, J.; González-Rojí, S.J.; Carreno-Madinabeitia, S.; Ibarra-Berastegi, G. Analysis of atmospheric thermodynamics using the R package aiRthermo. *Comput. Geosci.* **2019**, *122*, 113–119. [CrossRef]

41. Ulazia, A.; Sáenz, J.; Ibarra-Berastegui, G. Sensitivity to the use of 3DVAR data assimilation in a mesoscale model for estimating offshore wind energy potential. A case study of the Iberian northern coastline. *Appl. Energy* **2016**, *180*, 617–627. [CrossRef]

42. Ulazia, A.; Sáenz, J.; Ibarra-Berastegui, G.; González-Rojí, S.J.; Carreno-Madinabeitia, S. Using 3DVAR data assimilation to measure offshore wind energy potential at different turbine heights in the West Mediterranean. *Appl. Energy* **2017**, *208*, 1232–1245. [CrossRef]

43. Ulazia, A.; Gonzalez-Rojí, S.J.; Ibarra-Berastegi, G.; Carreno-Madinabeitia, S.; Sáenz, J.; Nafarrate, A. Seasonal air density variations over the East of Scotland and the consequences for offshore wind energy. In Proceedings of the 2018 7th International Conference on Renewable Energy Research and Applications (ICRERA), Paris, France, 14–17 October 2018; pp. 261–265.

44. Laprise, R.; Caya, D.; Bergeron, G.; Giguère, M. The formulation of the André Robert MC2 (mesoscale compressible community) model. *Atmos.-Ocean* **1997**, *35*, 195–220. [CrossRef]

Journal of
*Marine Science
and Engineering*

Article

Categorization and Analysis of Relevant Factors for Optimal Locations in Onshore and Offshore Wind Power Plants: A Taxonomic Review

Isabel C. Gil-García [1,†], M. Socorro García-Cascales [2,†], Ana Fernández-Guillamón [3,†] and Angel Molina-García [3,*,†]

[1] Faculty of Engineering, Distance University of Madrid (UDIMA), c/ Coruña, km 38.500 28400, Collado Villalba, 28029 Madrid, Spain; isabelcristina.gil@udima.es
[2] Department of Electronics, Computer Architecture and Projects Engineering, Universidad Politécnica de Cartagena, 30202 Cartagena, Spain; socorro.garcia@upct.es
[3] Department of Electrical Engineering, Universidad Politécnica de Cartagena, 30202 Cartagena, Spain; ana.fernandez@upct.es
* Correspondence: angel.molina@upct.es; Tel.: +34-968-32-5462
† These authors contributed equally to this work.

Received: 14 October 2019; Accepted: 31 October 2019; Published: 3 November 2019

Abstract: Wind power is widely considered to be a qualified renewable, clean, ecological and inexhaustible resource that is becoming a leader in the current energy transition process. It is a mature technology solution that was quickly developed and has been massively integrated into power systems in recent years. Indeed, a remarkable number of renewable integration policies have been promoted by different governments and countries. With the aim of maximizing the power given by wind resources, the locations of both onshore and offshore wind power plants must be optimized following a sort of different criteria. Under this scenario, a number of factors and decision criteria in the evaluation and selection of locations can be identified. Moreover, the relevant wind power increasing in the power generation mix is addressed, along with a standardization of factors and decision criteria in the optimization and selection of such optimal locations. In this context, this paper describes a systematic review and meta-analysis combining most of the contributions and studies proposed during the last decade. Thus, our aim is focused on reviewing and categorizing all factors to be considered for optimal location estimation, pointing out the differences among the selected factors and the decision criteria for onshore and offshore wind power plants. In addition, our review also includes an analysis of the representative key indicators for the contributions, such as the annual frequency of publications, geographical classification, analysis by category, evaluation method and determining factors.

Keywords: wind energy; optimal selection factors; onshore-offshore wind power plant

1. Introduction

The depletion of fossil fuels, combined with the need to reduce greenhouse gas emissions, leads to a single result: energy transition [1,2]. Renewable energies, both inexhaustible and clean sources, constitute the main support for a sustainable energy transition [3]. The year 2017 culminated records for renewable energy systems installed. There was the largest increase in global capacity, reaching 2180 GW of total capacity, of which 53% belonged to hydraulic energy and 24% to wind energy (onshore + offshore) [4]. The wind industry stands out for its exponential growth during the last decade, both in accumulated MW and generated energy (See Figure 1). During 2017, 13 European and American countries reached 10% or more of their electricity consumption with wind energy, and according to the IEA Wind sources, the global generated energy amounted to 1430 T Wh. Offshore wind energy reached its best year ever, with an increase of 67% compared to 2016.

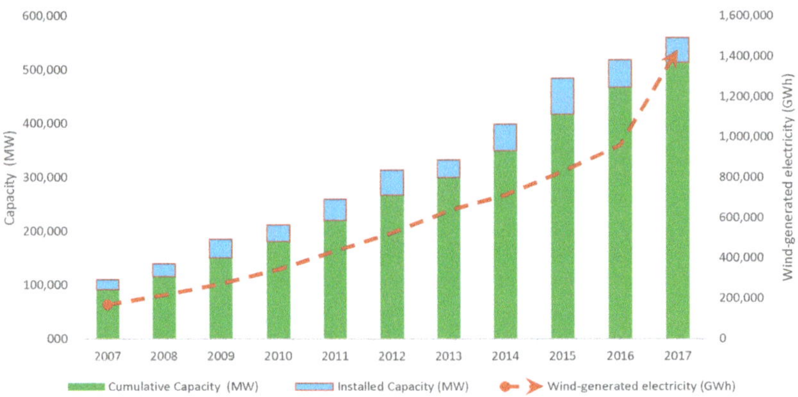

Figure 1. Comparison of accumulated, installed and generated wind energy. Data source: [5].

A significant cost reduction has been supported by an increasingly mature framework: gradual technological innovations, considerable improvements in the supply chain, reduction of the risk premium, greater qualification of developers and operators and a large market volume. Although the current wind market trend is increasingly favorable and optimistic, there are relevant barriers to overcome, mainly related to the optimal geographic location of wind power plants—both onshore and offshore. In this way, inaccurate forecasts of wind energy production can lead to the inefficiency of the wind facilities and, subsequently, to large financial losses. Presently, researchers and organizations work with the purpose of developing new optimization methodologies in the selection and evaluation of wind sites. According to the specific literature, the location of the wind power plants is associated with a group of factors that guarantee the profitability of such installations. However, and despite the extensive literature available, there is a lack of literature reviews dealing with the relational identification of the determining factors of such generation units for both onshore and offshore wind power plants. Under this framework, this paper gives an extended analysis and revision of the factors and determining decision criteria to select the optimal location of wind power plants for both onshore and offshore solutions. In addition, a categorization proposal of such factors is also provided by the authors according to the typology of the different factors. The works reviews in this paper constitute a clear evidence of the non-existence of a categorization of factors and criteria to be used for the efficient evaluations of onshore and offshore wind locations. The present analysis thus contributes to the literature by categorizing the factors and criteria—in terms of relevance, which influence the evaluation and selection of optimal locations for onshore and offshore wind plants.

The rest of the paper is structured as follows: Section 2 describes the research methodology used in the study, Section 3 presents the results and discussion on onshore wind facilities and offshore facilities with a comparison of both, and finally, Section 4 provides the conclusions and future work.

2. Proposed Methodology

The general proposed methodology of this review process is based on the Preferred Reporting Items for Systematic Reviews and Meta-Analyses (PRISMA) statement [6]. The main objective of this PRISMA statement is to address researchers in the selection of systematic review articles, guaranteeing the quality of the process. Many previous studies of different research categories have used the PRISMA statement to collect an exhaustive literature review [7–11]. In our case, the general proposed methodology based on the PRISMA statement has two main processes: systematic review and meta-analysis. Figure 2 summarizes the proposed methodology.

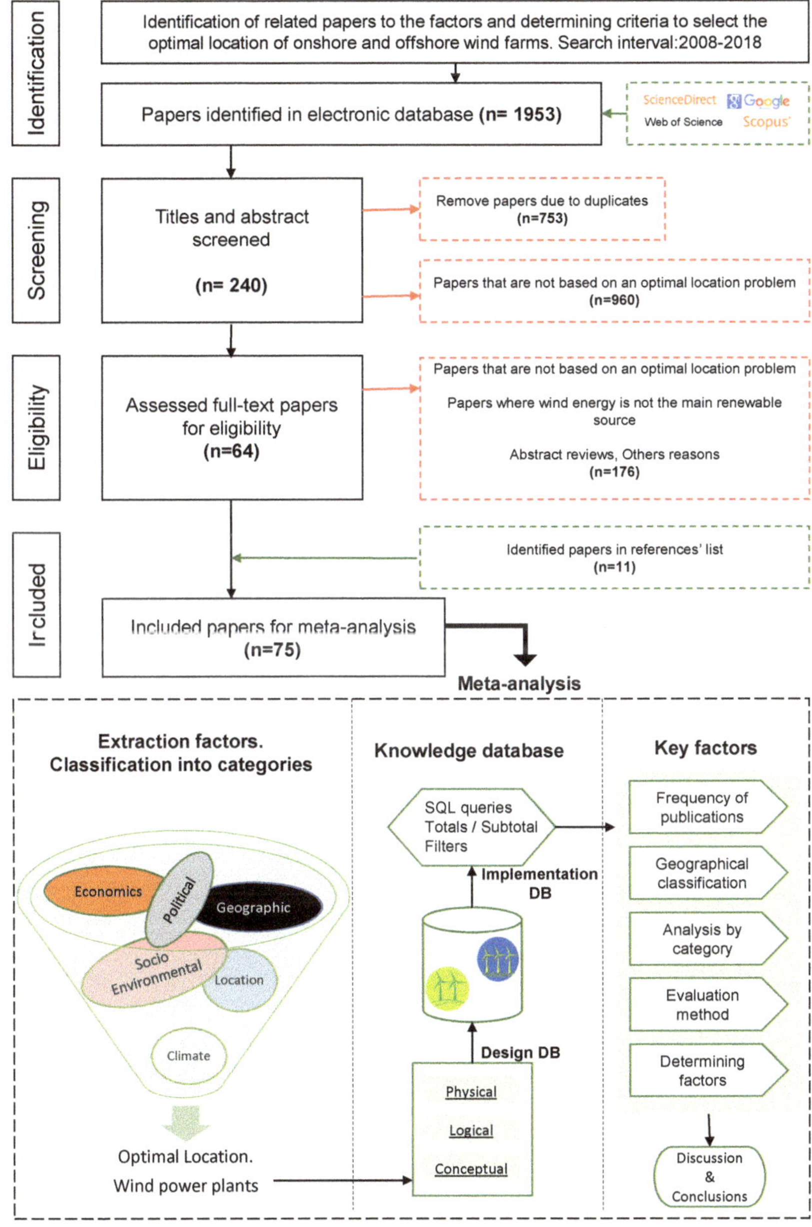

Figure 2. Proposed methodology: systematic review and meta-analysis.

2.1. Systematic Review

From the objective previously described in Section 1, this literature review aims to provide an extended analysis of the factors and determining criteria to select the optimal location of onshore and offshore wind power plants. Studies corresponding to the last decade (2008–2018) were selected for this study, in line with the evolution and exponential growth of installed wind power capacity and technological maturity during those years. The systematic review can be then divided into four stages: Identification, Screening, Eligibility and Included—see Figure 2.

According to the different contribution databases currently available, the authors included the following: ScienceDirect, Google Scholar, Web of Science and SCOPUS. Based on the keywords of the general objective and the objectives derived from the analyzed topic, a total of 1953 records were identified. Subsequently, from the contributions related to the specific problem of optimal location of power plants from the title and summary fields, we identified 753 duplicated items and 960 that were not works related to the research objectives. Finally, 240 potential contributions were identified. Once the objective of eligibility was applied, we reviewed the full text of each work, discarding revisions by the information given by the abstracts. Therefore, papers where wind resource was not the main renewable source and those that did not have the optimal location problem as their main objective were identified. In this process, 176 works were rejected, and six additional works from the reference lists were included in the total works to be considered by this analysis. Finally, 75 contributions were selected to be studied.

2.2. Meta-Analysis

Three main steps were identified in this process: extraction factors, knowledge database and key factors. First, all the factors involved in the optimal location of the specific sites were extracted. The real meaning of each factor was studied and then, we proposed a categorization based on the studies carried out by the different authors, regardless of the case study. This solution constitutes a contribution to this important barrier—'Use of the wind resources'—of the wind industry [12]. Moreover, it can be considered to be a reference for future works by providing identification and categorization factors summarized in Tables 1–6.

Table 1. Description of factors involved in the optimal location in the wind farms. Climate category.

Factor	Description
Wind speed	The wind speed that measures its kinetic energy in the site (m/s)
Power density	Power density, consider wind speed and air density (W/m^2)
Wind direction	Side where the wind blows (sexagesimal degrees)
Effective time	Occurrence of wind speed
Availability data	Accurate measurement campaign data
Turbulence	Ratio between the standard deviation of the values wind speed and its average speed, for each set of ten-minute measurements (dimensionless)
Frost periods	Duration of frost periods
Natural disasters	Probability of natural disasters
Air density	Relationship between mass and air volume (kg/m^3). Influences the kinetic energy of the air

Table 2. Description of factors involved in the optimal location in the Wind farms. Geographic category.

Factor	Description
Slope	The higher the percentage of the slope of the land, the less likely it is to install the wind farm
Altitude	At higher altitude, installation difficulties increase
Type of terrain	Soft or hard consistency
Roughness	Roughness of the terrain caused by both natural elevation and human development
Area	Area contained within the perimeter of the wind farm (m^2) or limit of the external ocean, legal marine areas of the country
Water depth	Bathymetry. Water depth in selected area of the sea (m). It is a key technical factor to decide the type of structure (fixed or floating)
Wave height	Wave height in selected area of the sea (m). It is a key technical factor to determine the effects of waves on the structure (balancing, dragging)
Water quality	It includes some properties of water such as dissolved oxygen (mg/L) to exclude areas destined for aquaculture or study co-location

Table 3. Description of factors involved in the optimal location in the wind farms. Socio-environmental category.

Factor	Description
Protected areas or distance	Completely protected areas from a legal standpoint (National and natural parks, Integral and special Natural Reserves, Special Areas of Conservation, etc.)
Agrological capacity	Suitability of the soil for certain crops
Visual impact	Visual impact according to regulations
Reduction emissions CO_2 and others	Pollution avoided compared to conventional power generation technology
Stroboscopic effect	Blinking shadow effect caused by the sun's incidence on the blades of the wind turbine
Energy-dependence contribution	Energy savings
Noise	The noise impact in quality of life
Population	The level and regularity of demand for energy in the site
Demand electricity	Sufficient electricity demand that justifies the installation
Land use	Use of land for agricultural, governmental, etc. Purposes
Flora and fauna impact	Mainly influence in birds, marine species, soil and vegetation
Shipping Routes	Ships/vessels movement routes
Fishing areas	Areas determine by the authorities for fishing

Table 4. Description of factors involved in the optimal location in the wind farms. Location category.

Factor	Description
Distance/Availability roads	Distance to roads, focused on decreasing installation and maintenance costs as well as safety in everyday transport
Distance to other wind farms	With the purpose of not exceeding the estimation of the carrying capacity of sustainable siting areas
Distance transmission lines (antennas)	Distance between any telecommunications infrastructure and the wind farm. In order to not affect the telecommunications infrastructure
Distant urban areas	Distance between urban areas, towns or cities, and location areas. In anticipation of future expansions and in compliance with the legislative framework of any country
Distances industrial/Military zones	Distance between military and industrial zones and location areas
Distance from the railway network	Distance between railway lines and possible locations. With the aim of taking advantage of the social acceptance of the zones
Distances to ports	Distance between ports and the possible sites, adaptation to the country's regulatory framework
Distances airports	Distance between the nearest airport and the different possible sites with the objective of not affecting the airspace or the future expansion of airports and facilities. Airspace restricted by the Aviation Agency
Distance to Point of Common Coupling (PCC)	Distance between nearest network or power line and the different possible sites. While this distance is smaller, the cost of the electricity infrastructure is lower and therefore, the economic and financial indicators will be better
Distances entertainment areas—historical	Distance between entertainment, historical areas and the possible sites, adaptation to the country's regulatory framework
Distance water resources (rivers, coast, lake)	Distance between water resources and the possible sites, adaptation to the country's regulatory framework, depending on whether it is a lake, river etc.
Distance of underground cables or pipes	Distance or existence of underground cables or pipes
Distance to shore	Focused on the location of offshore wind farms by regulatory measures marked by the country
Distance other point	Distance to other point as wrecks, lighthouses

Table 5. Description of factors involved in the optimal location in the wind farms. Economic category.

Factor	Description
Energy sale price	Energy sale price, very important since it is the only source of income for the installation
Energy put into the network	Energy put into the network eliminated all losses of gross energy
Infrastructure cost	Costs of the infrastructure associated with the initial investment (CAPEX)
NPV	Net present value, financial indicator
IRR	Internal rate of return, financial indicator
Payback	Recovery period in years
Interest loan	Interest of the loan requested in the initial investment
Installed capacity	Installed capacity (MW)
Exploitation	Cost focused on the exploitation phase (OPEX), example: cost of land (onshore), port activities (offshore)
Stability voltage	Voltage stability to achieve the planned energy
Economic contribution	Economic contribution focused on the creation of employment, payment of taxes in town halls etc.
Decommission cost	Include the removal of the turbines and foundations (DECEX)

Table 6. Description of factors involved in the optimal location in the wind farms. Political category.

Factor	Description
Incentives	Incentives received in compensation for producing electric power from renewable sources
Taxes	Taxes involved in the activity
Policy measures	Political measures established in favor of renewable energies

From the initial categorization process, we designed a proposed knowledge database according to the requirements and objectives of the different contributions. The conceptual model of the database was then designed, translating the entities and the relationships between them. After that, the logical design was determined, normalizing the database to avoid duplication of information. In the implementation of the database, each contribution was inserted and proceeded to program queries that respond to our objectives: filters, totals, subtotals, groupings, etc.

Finally, the key factors can be estimated by considering the following aspects:

- Frequency of publications: a first analysis of the frequency of annual publications is analyzed to identify the period in which these studies became more relevant.
- Geographical classification: to identify the geographical areas with the greatest impact of publications and their possible association with indicators from different fields (governmental measures in favor of renewable energy, social acceptance, etc.) a study is carried out by country, marine area and continent.
- Quantitative analysis: to quantitatively analyze the categories and their associated factors, it is calculated by each contribution the following aspects: the number of times such factors are used in the contributions of each technology (onshore and offshore), and their percentage of use with respect to those contributions.
- Evaluation method: in the process of searching for and selecting such optimal locations for wind power plants, it is possible to identify (i) a large amount of spatial information, and (ii) the need to cluster factors and criteria from varied nature which influence with different intensities in the multicriteria decision-making. Many researchers who tried to address the complexity of these investigations have proposed to use Geographic Information System (GIS) tools and/or Multicriteria Decision-Making (MCDM) methods. Given the importance of the methodological development of these contributions, a third indicator can be identified focused on analyzing the percentage of the researchers providing a methodology that combines geographical information systems and MCDM, or they use any of them individually.
- Determining factors: they are based on the previous analysis. The first ten most relevant determining factors are identified for each onshore and offshore technology.

3. Results and Discussion

3.1. Onshore Analysis. Categorization and Factors

In the taxonomic review of the contributions focused on the optimal selection of onshore wind power plant locations, 34 relevant works published between 2008 and 2018 were selected. From these studies, we can affirm that from 2008 to 2013, the frequency of publications was considerably low in comparison to the rest of the period. Indeed, only 10 contributions aiming to optimize onshore wind power plants were found in the specific literature until 2013 (29.4% of the total works). However, in the period 2014–2018, 24 works were published. Figure 3 shows the frequency of publications for onshore wind power plants classified by the author's origin. Therefore, and according to this contribution review, the selection of an optimal location indicator became more attractive in the early 2010s. In line with the previous results, the number of publications according to the country of the case study was obtained accordingly. Thus, the countries that have published the highest number of case studies are Spain (5), the USA (4), Greece, Iran, Turkey (3), Brazil, and China (2). Clustering these contributions to provide a continental ranking, Asia leads the list with 38%, followed by Europe with 35%, North America with 12%, South America with 9% and Africa with 6%. Figure 4 shows the geographical classification of case studies for onshore wind power plant locations (2008–2018). These results conclude that studies of evaluation and selection of optimal onshore installation locations are mostly driven by developed countries during the last decade. From the categorization and qualitative analysis described in Section 2.2, both factor identification and categorization is following described for onshore wind power plant optimal location.

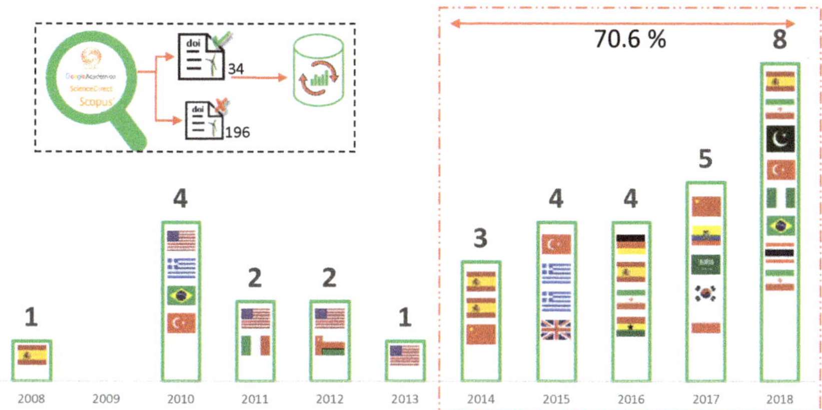

Figure 3. Onshore wind power plant optimal location. Frequency of publications (2008–2018).

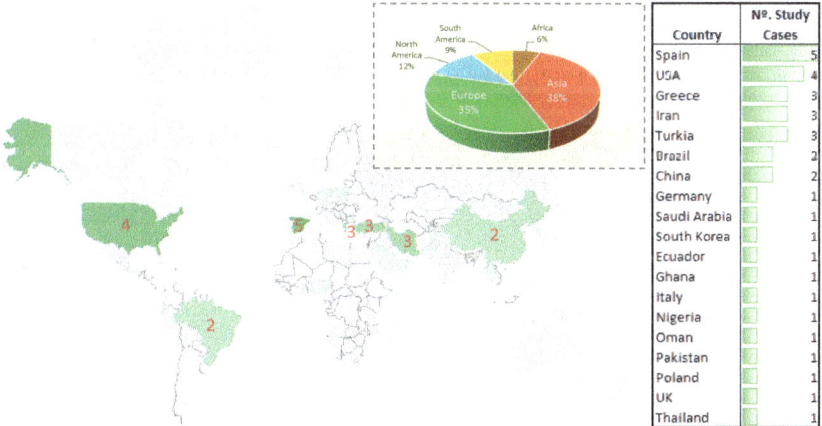

Country	Nº. Study Cases
Spain	5
USA	4
Greece	3
Iran	3
Turkia	3
Brazil	2
China	2
Germany	1
Saudi Arabia	1
South Korea	1
Ecuador	1
Ghana	1
Italy	1
Nigeria	1
Oman	1
Pakistan	1
Poland	1
UK	1
Thailand	1

Figure 4. Onshore wind power plant optimal location. Geographical case studies (2008–2018).

3.1.1. Climate Category (C_1)

In line with Table 1, a total of nine factors were included in this category, labeled from ($C_{1.1}$) to ($C_{1.9}$). Table 7 summarizes the main contributions by including (or not) the factors corresponding to the climate category for onshore optimal location methodologies. Both references and the number of works—labeled as Absolute Frequency (AF)—are included in the table, as well as the percentage of contributions where such factor is considered in the different studies. According to the results, the most relevant factor is Wind Speed ($C_{1.1}$), accounting for 32 works that directly included it in their studies for optimal location; followed by Air Density ($C_{1.9}$) with five contributions. The least relevant factors were Data availability ($C_{1.5}$), Turbulence ($C_{1.6}$) and Frost periods ($C_{1.7}$), with only one study.

Table 7. Climate Category (C_1)—Onshore optimal location. Quantitative analysis.

Nomenclature	Factor	References	AF	%
$C_{1.1}$	Wind speed	[13–44]	32	94
$C_{1.2}$	Power Density	[23,33,45,46]	4	12
$C_{1.3}$	Wind direction	[14,24]	2	6
$C_{1.4}$	Effective time	[18,23,45]	3	9
$C_{1.5}$	Availability data	[43]	1	3
$C_{1.6}$	Turbulence	[45]	1	3
$C_{1.7}$	Frost periods	[42]	1	3
$C_{1.8}$	Natural disasters	[33,37,38,46]	4	12
$C_{1.9}$	Air density	[13,35,40,42,43]	5	15

3.1.2. Geographic Category (C_2)

By considering Table 2, five factors were included into this category, labeled from ($C_{2.1}$) to ($C_{2.5}$). In a similar way to the previous categorization, the results are summarized in Table 8. Two factors stand out above the rest: Slope ($C_{2.1}$) and Altitude ($C_{2.2}$). From the specific literature, 24 and 13 contributions include these factors respectively. The rest of the geographic factors became less important.

Table 8. Geographic Category (C_2)—Onshore optimal location. Quantitative analysis.

Nomenclature	Factors	References	AF	%
$C_{2.1}$	Slope	[13,17–20,22,24,25,28–35,37–41,43–45]	24	71
$C_{2.2}$	Altitude	[13,19,24,25,29,32,35,37,39,40,42–44]	13	38
$C_{2.3}$	Type of terrain	[16,17,19,21,30,40,42,45]	8	24
$C_{2.4}$	Roughness	[13,20,37]	3	9
$C_{2.5}$	Area	[22,29,31,43,44]	5	15

3.1.3. Socio-Environmental Category (C_3)

Regarding Table 3, 11 factors were included in this category, from ($C_{3.1}$) to ($C_{3.11}$). Table 9 shows the results according to the specific literature. Four factors stand out from the rest: Protected areas ($C_{3.1}$), Land use ($C_{3.10}$), Flora and fauna impact ($C_{3.11}$) and Agrological capacity ($C_{3.2}$) with 22, 12, 12 and nine publications that included these factors in their study. The rest of the factors oscillate between absolute frequencies of 8 and 1 work, respectively.

Table 9. Socio-environmental Category (C_3)—Onshore wind energy. Quantitative analysis.

Nomenclature	Factor	References	AF	%
$C_{3.1}$	Protected areas	[13,15–18,20,22,24–26,28–31,34,39–45]	22	65
$C_{3.2}$	Agrological capacity	[14,17,20,22,30,31,35,37,43]	9	26
$C_{3.3}$	Visual impact	[13,17,19,23,37,42,43]	7	21
$C_{3.4}$	Reduction emissions	[23,33,38]	3	9
$C_{3.5}$	Stroboscopic effect	[37]	1	3
$C_{3.6}$	Energy-dependence contribution	[23,33]	2	6
$C_{3.7}$	Noise	[13,15,19,23,36,37,42,43]	8	24
$C_{3.8}$	Population	[14,16,21,46]	4	12
$C_{3.9}$	Demand electricity	[17,24,45]	3	9
$C_{3.10}$	Land use	[14,17,20,21,28,32,34,35,42–44,46]	12	35
$C_{3.11}$	Flora and fauna impact	[15,19,21,23,26,27,30,33,37,40,42,43]	12	35

3.1.4. Location Category (C_4)

Table 4 shows the location category factors, including 11 factors labeled from ($C_{4.1}$) to ($C_{4.11}$). Table 10 summarizes the presence of such factors in relevant contributions. From these results, six factors exceeded 10 references—see AF column—: Distance urban areas ($C_{4.4}$)—29, Distance/availability roads ($C_{4.1}$)—26, Point of Common Coupling ($C_{4.9}$)—22, Distance transmission lines ($C_{4.3}$) and Distance airports ($C_{4.8}$) with 17 references, and Distance water resources (rivers, coast, lake) ($C_{4.11}$)—15. The rest of the factors oscillated between one and 10 publications, being minor representative of this category.

Table 10. Location Category (C_4)—Onshore wind energy. Quantitative analysis.

Nomenclature	Factor	References	AF	%
$C_{4.1}$	D. Availability roads	[16,19–22,24–34,36–45]	26	76
$C_{4.2}$	D. to other wind farms	[26]	1	3
$C_{4.3}$	D. transmission lines	[13,14,16,21,22,25,26,30–32,34,35,37–39,42,44]	17	50
$C_{4.4}$	D. urban areas	[13–18,20–32,34–36,39–45]	29	85
$C_{4.5}$	D. industrial/Military zones	[26,39]	2	6
$C_{4.6}$	D.from the railway network	[13,20,25,29,30,34,35,39]	8	24
$C_{4.7}$	D. to ports	[26,35]	2	6
$C_{4.8}$	D. airports	[13,15,17,20–22,24–26,29,31,32,36,39,40,42,44]	17	50
$C_{4.9}$	D. Point of Common Coupling (PCC)	[14,17,19,22–24,26,27,29–31,33–36,38–43,46]	22	65
$C_{4.10}$	D. entertainment areas–historical	[16,17,23,26–30,34,39]	10	29
$C_{4.11}$	D. water resources (rivers, coast, lake)	[14,17,20,21,24–26,28–30,34,39–41,44]	15	44

3.1.5. Economic Category (C_5)

A set of 11 factors were included in this category, see Table 5, from ($C_{5.1}$) to ($C_{5.11}$). Of these factors, three stand out in this category: Exploitation ($C_{5.9}$), Energy put into the network ($C_{5.2}$) and Infrastructure cost ($C_{5.3}$) with 10, nine and eight works including these factors, see Table 11.

Table 11. Economics Category (C_5)—Onshore wind energy. Quantitative analysis.

Nomenclature	Factor	References	AF	%
$C_{5.1}$	Energy sale price	[13,20,24,33]	4	12
$C_{5.2}$	Energy put into the network	[13,14,18,20,23–25,33,41]	9	26
$C_{5.3}$	Infrastructure cost	[13,20,23,24,33,38,43,46]	8	24
$C_{5.4}$	NPV	[23]	1	3
$C_{5.5}$	IRR	[23]	1	3
$C_{5.6}$	Payback	[23,33]	2	6
$C_{5.7}$	Interest loan	[20,23]	2	6
$C_{5.8}$	Installed capacity	[33]	1	3
$C_{5.9}$	Exploitation	[13,17,19,20,33,38,41–43,46]	10	29
$C_{5.10}$	Stability voltage	[33]	1	3
$C_{5.11}$	Economic contribution	[33,43,46]	3	9

3.1.6. Political Category (C_6)

In line with Table 6, the Political category includes three factors, labeled from ($C_{6.1}$) to ($C_{6.3}$). Table 12 shows the contribution analysis according to this category. No factor exceeded the three absolute frequency publications, oscillating their values between 2 and 3.

Table 12. Political Category (C_6)—Onshore wind energy. Quantitative analysis.

Nomenclature	Factor	References	AF	%
$C_{6.1}$	Incentives	[20,33]	2	6
$C_{6.2}$	Taxes	[20,33]	2	6
$C_{6.3}$	Policy measures	[23,33,43]	3	9

3.2. Offshore Analysis. Categories and Factors

In line with the previous analysis, a similar taxonomic review corresponding to the most relevant works of offshore wind power plant optimal location was carried out by the authors. In this case, and considering the specific literature available in the database described in Section 2.1, 41 contributions published between 2008 and 2018 were selected. Between 2015 and 2018, 60.98% of the total publications were published, accounting for 25 contributions in this period. Before 2015, 16 relevant papers were published, see Figure 5. These results provide a preliminary indicator similar to the onshore analysis discussed in Section 3.1. Indeed, the evaluation and selection of optimal offshore wind power plant locations have been a remarkable interest for researchers since the 2010s, and highlighting their relevance during the recent years. In terms of the number of case studies by marine areas, the North Sea tops the list with 13 contributions, followed by the North Atlantic Ocean and the China Sea with 10 and seven studies, respectively. The continental ranking is led by Europe with 60%, followed by Asia with 33%, North America with 6% and Africa with 1%, see Figure 6. The results show that evaluation and selection of optimal locations for offshore wind power plants are mostly centralized in the northern hemisphere. The categories proposed in Section 2.2 are following discussed according to the different offshore wind power plant optimal location methodologies. With this aim, the most relevant factors taken into account in such methodologies are determined and categorized accordingly.

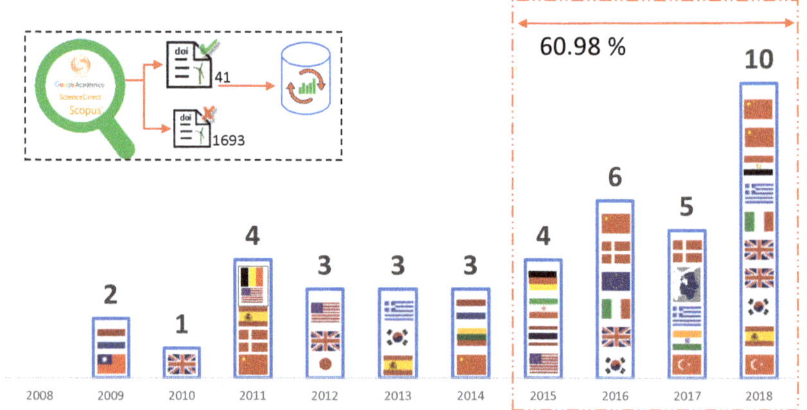

Figure 5. Offshore wind power plant optimal location. Frequency of publications (2008–2018).

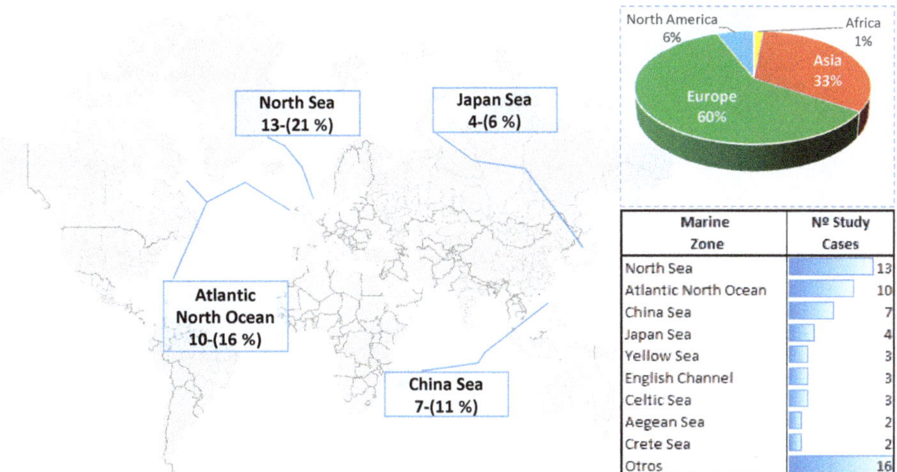

Marine Zone	Nº Study Cases	
North Sea		13
Atlantic North Ocean		10
China Sea		7
Japan Sea		4
Yellow Sea		3
English Channel		3
Celtic Sea		3
Aegean Sea		2
Crete Sea		2
Otros		16

Figure 6. Offshore wind power plant optimal location. Geographical classification of case studies (2008–2018).

3.2.1. Climate Category (C_1)

In line with the factors included in the climate category and summarized in Table 1, seven factors were identified for offshore optimal location proposals: from ($C_{1.1}$) to ($C_{1.4}$), ($C_{1.6}$), ($C_{1.8}$) and ($C_{1.9}$). The factor with the most presence in the contributions and the highest absolute frequency was Wind Speed ($C_{1.1}$). Indeed, 37 publications directly included this factor in their proposed study. Table 13 describes the climate factors as well as the number of contributions and the percentage according to the total offshore wind power plant optimal methodologies selected.

Table 13. Climate Category (C_1)—Offshore wind energy. Quantitative analysis.

Nomenclature	Factor	References	AF	%
$C_{1.1}$	Wind speed	[47–83]	37	91
$C_{1.2}$	Power Density	[51,54–57,65,67,77,81,84–86]	12	29
$C_{1.3}$	Wind direction	[47,66,68,70,76,77]	6	15
$C_{1.4}$	Effective time	[51,54,76,82]	4	10
$C_{1.6}$	Turbulence	[54,68,83]	3	7
$C_{1.8}$	Natural disasters	[54,55]	2	5
$C_{1.9}$	Air density	[70]	1	2

3.2.2. Geographic Category (C_2)

In this case, and according to Table 2, five factors have been used for offshore installation optimal location: ($C_{2.3}$) and from ($C_{2.5}$) to ($C_{2.8}$). One factor stood out above the rest: Water depth ($C_{2.6}$), with 24 contributions including this factor in the proposed methodology. From the rest of factors, three varied between 10 and 13 items, Type of terrain ($C_{2.3}$), Wave height ($C_{2.7}$) and Area ($C_{2.5}$) with a percentage of publications lower than 35%, see Table 14.

Table 14. Geographic Category (C_2)—Offshore wind energy. Quantitative analysis.

Nomenclature	Factor	References	AF	%
$C_{2.3}$	Type of terrain	[47,50,54,55,64,69,75,78,84,87]	10	24
$C_{2.5}$	Area	[56,66,71–76,78,79,81,82,87]	13	32
$C_{2.6}$	Water depth	[47,48,50–56,58–66,68,69,71,72,74–79,81–84,86]	33	80
$C_{2.7}$	Wave height	[51,53–55,57,59,69,74,78,79,82]	11	27
$C_{2.8}$	Water quality	[56,79]	2	5

3.2.3. Socio-Environmental Category (C_3)

From the 11 factors initially classified in Table 3 for the Socio-environmental Category, nine factors were selected by the different contributions for offshore installation optimal location: ($C_{3.1}$) ,($C_{3.3}$), ($C_{3.4}$), ($C_{3.7}$)-($C_{3.9}$),($C_{3.11}$)-($C_{3.13}$). According to the analyzed works, four factors stand out from the rest: Protected areas ($C_{3.1}$), Shipping Routes ($C_{3.12}$), Flora and fauna impact ($C_{3.11}$) and Fishing areas ($C_{3.13}$) with 30, 28, 22 and 16 studies that included these factors in their study. The rest of the factors ranged between seven and one absolute frequency and they did not exceed a 20% of the contributions, see Table 15.

Table 15. Socio-environmental Category (C_3)—Offshore wind energy. Quantitative analysis.

Nomenclature	Factor	References	AF	%
$C_{3.1}$	Protected areas	[47,50,52,53,55–58,61–63,65,67,69,71–79,81–87]	30	73
$C_{3.3}$	Visual impact	[55,62,63,73,81,85,87]	7	17
$C_{3.4}$	Reduction emissions	[54,60,64]	3	7
$C_{3.7}$	Noise	[53]	1	2
$C_{3.8}$	Population	[80,87]	2	5
$C_{3.9}$	Demand electricity	[63,71,85]	3	7
$C_{3.11}$	Flora and fauna impact	[47,53–57,59–63,65,67,69,73–75,79–81,85,86]	22	54
$C_{3.12}$	Shipping Routes	[47,49–51,54–58,60–63,67,69,71–76,78,80–82,84–86]	28	68
$C_{3.13}$	Fishing areas	[50,51,55,56,61,62,67,69,71,72,74,76,78,80,84,86]	16	39

3.2.4. Location Category (C_4)

Table 4 describes the factors to be considered in this category. For offshore installations, 10 factors were explicitly considered in this category: ($C_{4.2}$), ($C_{4.4}$), ($C_{4.5}$), ($C_{4.7}$)-($C_{4.10}$), ($C_{4.12}$)-($C_{4.14}$). According to the contributions, three factors had more than 15 absolute references: Distance shore ($C_{4.13}$)—26 works, Distance industrial/Military zones ($C_{4.5}$)—21 works, and Distance Point of Common Coupling ($C_{4.9}$)—18 works. The rest of the factors range between 1 and 12 absolute reference publications, see Table 16.

Table 16. Location Category (C_4)—Offshore wind energy. Quantitative analysis.

Nomenclature	Factor	References	AF	%
$C_{4.2}$	D. to other wind farms	[50,52,58,62–64,69]	7	17
$C_{4.4}$	D. urban areas	[47,87]	2	5
$C_{4.5}$	D. industrial/Military zones	[47,49,50,52,55,56,58,61–63,67,69,71,73–76,78,82,84,85]	21	51
$C_{4.7}$	D. to ports	[47,50,51,54,56–58,61,64,77,78,84]	12	29
$C_{4.8}$	D. airports	[49,71,75]	3	7
$C_{4.9}$	D. Point of Common Coupling (PCC)	[48,50,52–58,61,64,67,75,78,80,82,84,86]	18	44
$C_{4.10}$	D. entertainment areas—historical	[47,50,52,55,61,71,75,87]	8	20
$C_{4.12}$	D. underground cables or pipes	[47,49,50,61,69,73,76,78,81,84–86]	12	29
$C_{4.13}$	D. to shore	[47–50,52–62,64,69,71–73,76,77,80,83,84,87]	26	63
$C_{4.14}$	D. other point	[47]	1	2

3.2.5. Economic Category (C_5)

This category accounts for 12 factors, see Table 5. All of them were used for offshore wind power plant optimal location methodologies, which were labeled from ($C_{5.1}$) to ($C_{5.12}$). By considering the selected contributions, four factors stand out in this category: Installed capacity ($C_{5.8}$), Infrastructure cost ($C_{5.3}$), Exploitation ($C_{5.9}$) and Energy put into the network ($C_{5.2}$) with 20, 19, 18 and 11 absolute reference works. The rest of the factors have less than 10 absolute frequency contributions and they have a minor relevance in this category—less than 20% of the contributions used such factors—, see Table 17.

Table 17. Economics Category (C_5)—Offshore wind energy. Quantitative analysis.

Nomenclature	Factor	References	AF	%
$C_{5.1}$	Energy sale price	[54,59,71–73,82,83,85]	8	20
$C_{5.2}$	Energy put into the network	[50,52–55,59,71,74,82,83,85]	11	27
$C_{5.3}$	Infrastructure cost	[50,52–55,59,60,62,64,65,68,69,71–73,78,82,83,85]	19	46
$C_{5.4}$	NPV	[54,67,82,83]	4	10
$C_{5.5}$	IRR	[54,82]	2	5
$C_{5.6}$	Payback	[54,55,82,85]	4	10
$C_{5.7}$	Interest loan	[50,59,60,85]	4	10
$C_{5.8}$	Installed capacity	[50–52,59,60,62,64–69,71–74,76,83,85,86]	20	49
$C_{5.9}$	Exploitation	[50,52–55,59,60,62,64,65,69,71–73,78,82,83,85]	18	44
$C_{5.10}$	Stability voltage	[54]	1	2
$C_{5.11}$	Economic contribution	[53,58,59]	3	7
$C_{5.12}$	Decommission cost	[64,68,71,82]	4	10

3.2.6. Political Category (C_6)

Finally, and in line with Table 6, two factors were considered in this category for offshore installation optimal location: ($C_{6.1}$) and ($C_{6.3}$). These factors have less than 2 absolute reference publications, as can be seen in Table 18.

Table 18. Political Category (C_6)—Offshore wind energy. Quantitative analysis.

Nomenclature	Factor	References	AF	%
$C_{6.1}$	Incentives	[54,60]	2	5
$C_{6.3}$	Policy measures	[53]	1	2

3.3. Final Discussion

3.3.1. Categories: Comparison and Statistics

Figure 7 compares the presence of the different factors—divided by categories—in the analyzed contributions for both onshore and offshore optimal location. As can be seen, the most used categories in both onshore and offshore technologies are Location (C_2) and Socio-environmental (C_3), as there are many restrictive factors associated with the building stage of such wind power plants. Offshore technology is much more expensive than onshore technology. This aspect can be deduced from the relevance of the Economic (C_5) category, which is the third most relevant category in offshore optimal location methodologies. The annual trend keeps similar to the general analysis evaluation by categories, being Location (C_4) the most common category and the Political (C_6) the least–used category. By considering the annual tendencies, it can be affirmed that no-pattern were deduced to estimate the presence of any factor at a specific time in the analyzed period.

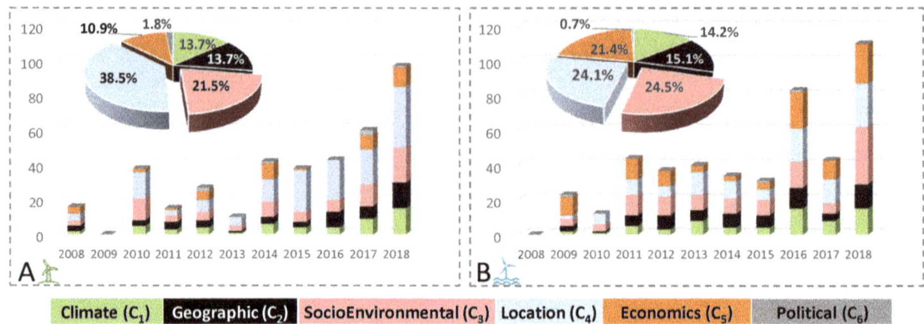

Figure 7. Categories: relevance and comparison (2008–2018). (**A**) Onshore wind power plant. (**B**) Offshore wind power plant.

3.3.2. Methodologies: Comparison and Statistics

With regard to the evaluation methods of the onshore optimal locations, and according to the specific literature, 70.6% of contributions included a combination of geospatial tools and multicriteria decision methods,—accounting for 24 works in total—. The most commonly used combination was GIS and MCDM methods, or several of them: AHP (Analytic Hierarchy Process), FAHP (Fuzzy Analytic Hierarchy Process), OWA (Ordered Weighted Average), TOPSIS, WLC (Weighted Liner Composition) [15,17,27,28,30,31,34–36,39–42,44,45]. The combination of GIS and SMCA (Spatial Multicriteria Analysis), DSS (Decision Support System), SDSS (Spatial Decision Support Systems) and MCE (Multicriteria Evaluation) can be identified in seven contributions [16,18,19,21,29,32,37]. In addition, two papers combined GIS with ELECTRE III and SMAA-TRI (Stochastic Multicriteria Acceptability Analysis) accordingly [22,25]. In terms of the optimal marine locations, it contains 63% of publications with a combination of geographic information systems and multicriteria evaluation methods (33%) or the application of only a geographic information system with an internal decision criteria (30%). The application of a GIS–MCDM combination or a combination of some of them was used in 13 studies, by considering the following processes: AHP (Analytic Hierarchy Process), FAHP (Fuzzy Analytic Hierarchy Process), OWA (Ordered Weighted Average), TOPSIS [48,52,55,57,58,63,64,67,69,71,79,80,84]. Figure 8 summarizes the methodologies proposed to estimate the optimal location for both onshore and offshore installations, as well as the relevance of each methodology according to their percentage in terms of the total contributions considered for each technology.

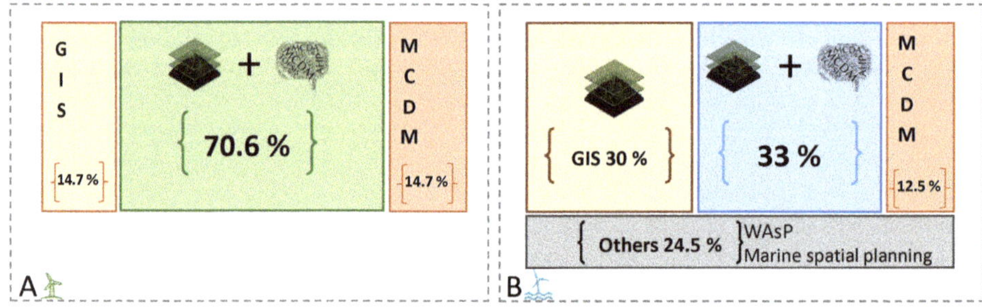

Figure 8. Statistics and comparison of location methodologies (2008–2018). (**A**) Onshore wind power plant. (**B**) Offshore wind power plant.

3.3.3. Relevant Factors: Comparison and Statistics

The most relevant factor, as expected, is Wind speed ($C_{1.1}$), directly proportional to the existence of wind power plants. Factors related to the geography of the place stand out in both technologies, such as Slope ($C_{2.1}$) and Altitude ($C_{2.2}$) in onshore plants, and Water depth ($C_{2.6}$) in offshore plants. Restrictive environmental and location factors match both technologies, such as Protected areas ($C_{3.1}$), as well as areas that are directly incompatible with this type of facilities. In addition, the Distance to Point of Common Coupling ($C_{4.9}$) is a remarkable factor due to the aim of minimizing costs and power losses. Among the determining factors of offshore optimal locations, the existence of three factors of the economic category—Infrastructure cost-CAPEX ($C_{5.3}$), Installed capacity ($C_{5.8}$) and Exploitation-OPEX ($C_{5.9}$)—and the absence of such factors in onshore optimal location processes also stands out. In addition, the Distance to shore factor ($C_{4.13}$) was proposed in 26 publications, 63% of the total offshore works. It is important to take into account the fact that most developed countries incorporate restrictive distances for industrial marine sites, and thus, some factors such as visual impact, conflicts with other industrial or tourism activities are involved. Figure 9 shows the determining factors for both technologies. The percentages provide the relative relevance of such factors for the corresponding technology, in terms of the number of contributions where each factor is considered for optimal location estimation.

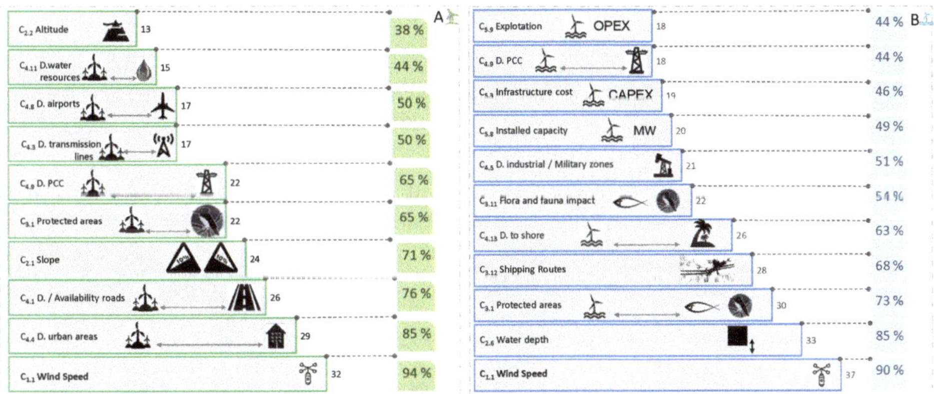

Figure 9. Relevant factors for optimal location methodologies. (**A**) Onshore wind power plant. (**B**) Offshore wind power plant.

Finally, Figure 10 graphically summarizes all the factors and categories proposed by the analyzed contributions included in this review. As can be seen, there are relevant factors only used by onshore case studies and vice-versa. From our point of view, the socio-environmental category (C_3), and more specifically Energy dependence contribution factor ($C_{3.6}$), should be considered with a higher relevance in both technologies, by considering the remarkable necessity to reduce global energy dependences. On the other hand, Factor Decommission cost ($C_{5.12}$) was only analyzed by the contributions for offshore optimal locations, although it should be considered in both technologies either in the closing of the activity or the repowering. Additionally, the existence or not of Taxes ($C_{6.2}$), belonging to the political category, is another factor to consider in both technologies, which, given the activity, should be exempted. For these reasons, and in addition to the proposed categorization, relocation of the factors, and criteria to be used in the future evaluation and selection of onshore and offshore wind optimal locations, we propose a relevant group of factors for each technology to be considered in future optimal location methodologies. These extended number of factors will allow us to estimate optimal locations from a multi-dimensional perspective, and including not only technical criteria but also environmental and energy-dependence aspects. This alternative proposal of relevant factors to be considered in onshore and offshore optimal location methodologies is depicted in Figure 11, where the additional factors to be considered for future works are highlighted in red color.

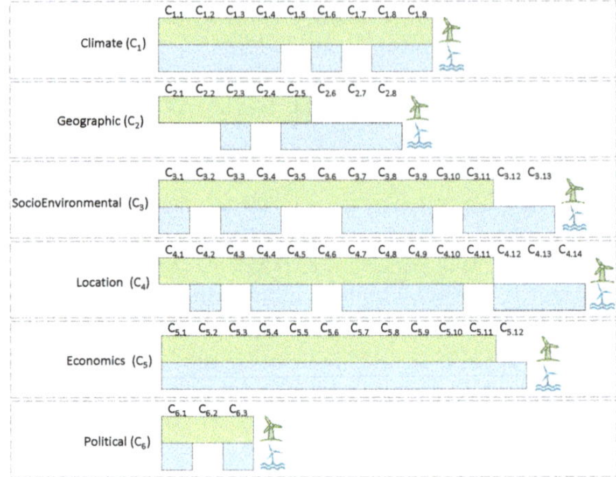

Figure 10. Categories and factors in onshore and offshore wind farm optimal location: global comparison.

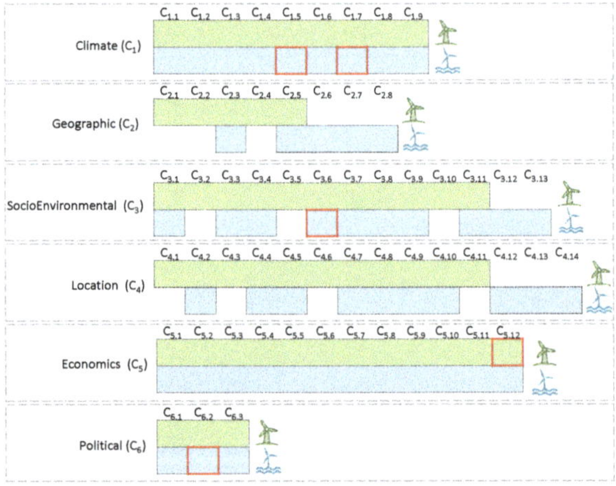

Figure 11. Proposal relevant factors in onshore and offshore wind farm optimal location.

4. Conclusions

The efficient use of renewable energy sources is extremely relevant in the global energy transition, with wind energy being the most mature technology within renewable energy sources. Each wind power plant project begins with the evaluation and selection of the optimal location. The parameters and factors to be considered for optimizing locations are different from the methodologies proposed in the specific literature. Indeed, contributions suggest different factors depending on the regulations and restrictions of each country, as well as existing data or previous studies. Under the absence of an exhaustive categorization and analysis of such factors, this paper reviews the most relevant contributions regarding the optimal location for both onshore and offshore technologies during the last decade (2008—2018). A total of 74 contributions are identified as relevant, proposing six categories to classify a total of 59 factors: climate, geographic, economic, distance, political and socio-environmental categories. Among the all factors, the wind speed factor is considered the most relevant parameter for both onshore and offshore technologies, accounting for over 90% of the contributions. The rest of the relevant factors depend on the technology—onshore or offshore—to be implemented. Economic factors also have remarkable importance, especially in offshore wind projects where 21.4% of the contributions include some factors of this category. In terms of the methodologies proposed for optimal location estimation, 50% of researchers use the combination of GIS+MCDM in their proposed methodologies. It can be considered a very successful combination, given the multiple existing spatial data as well as the wide variety of alternatives to evaluate. By considering all factors used in optimal location methodologies, we conclude that there is a lack of environmental and energy-dependence parameters which should be included in future methodologies. An extended selection of factors is proposed by the authors. This multi-dimensional perspective will be useful for future optimal location methodology, and also it is in line with current emissions and energy-dependence reduction requirements.

Author Contributions: Data curation, I.C.G.-G.; methodology and formal analysis, I.C.G.-G. and M.S.G.-C.; Resources A.F.-G.; Supervision, A.M.-G.; Writing—original draft I.C.G.-G.; and Writing—review and editing, A.F.-G. and A.M.-G.

Funding: This research received no external funding.

Acknowledgments: This work was supported by 'Ministerio de Educación, Cultura y Deporte' of Spain (grant number FPU16/04282).

Conflicts of Interest: The authors declare no conflict of interest.

Abbreviations

The following abbreviations are used in this manuscript:

AHP	Analytic Hierarchy Process
DSS	Decision Support System
FAHP	Fuzzy Analytic Hierarchy Process
GIS	Geographic Information System
OWA	Ordered Weighted Average
MCDM	Multicriteria Decision-Making
MCE	Multicriteria Evaluation
PCC	Point of Common Coupling
PRISMA	Preferred Reporting Items for Systematic Reviews and Meta-Analyses
SDSS	Spatial Decision Support Systems
SMAA	Stochastic Multicriteria Acceptability Analysis
SMCA	Spatial Multicriteria Analysis
WLC	Weighted Liner Composition

References

1. Hoehne, C.G.; Chester, M.V. Greenhouse gas and air quality effects of auto first-last mile use with transit. *Transp. Res. Part D Transp. Environ.* **2017**, *53*, 306–320. [CrossRef]
2. Zerta, M.; Schmidt, P.R.; Stiller, C.; Landinger, H. Alternative World Energy Outlook (AWEO) and the role of hydrogen in a changing energy landscape. *Int. J. Hydrog. Energy* **2008**, *33*, 3021–3025. [CrossRef]
3. Nicolini, M.; Tavoni, M. Are renewable energy subsidies effective? Evidence from Europe. *Renew. Sustain. Energy Rev.* **2017**, *74*, 412–423. [CrossRef]
4. R. Adib. *Renewable Energy Policy Network for the 21st Century (REN21)*; Renewables 2018 Global Status Report; REN21 Secretariat: Paris, France, 2018.
5. F. La Camera; *Global Energy Transformation: A Roadmap to 2050, International Renewable Energy Agency*; Technical Report; International Renewable Energy Agency (IRENA): Abu Dhabi, UAE, 2018.

6. Moher, D.; Liberati, A.; Tetzlaff, J.; Altman, D.G. Preferred reporting items for systematic reviews and meta-analyses: The PRISMA statement. *Int. J. Surg.* **2010**, *8*, 336–341. [CrossRef] [PubMed]

7. Mardani, A.; Zavadskas, E.K.; Streimikiene, D.; Jusoh, A.; Khoshnoudi, M. A comprehensive review of data envelopment analysis (DEA) approach in energy efficiency. *Renew. Sustain. Energy Rev.* **2017**, *70*, 1298–1322. [CrossRef]

8. Nicolson, M.L.; Fell, M.J.; Huebner, G.M. Consumer demand for time of use electricity tariffs: A systematized review of the empirical evidence. *Renew. Sustain. Energy Rev.* **2018**, *97*, 276–289. [CrossRef]

9. Şener, Ş.E.C.; Sharp, J.L.; Anctil, A. Factors impacting diverging paths of renewable energy: A review. *Renew. Sustain. Energy Rev.* **2018**, *81*, 2335–2342. [CrossRef]

10. Johnson, D.; Horton, E.; Mulcahy, R.; Foth, M. Gamification and serious games within the domain of domestic energy consumption: A systematic review. *Renew. Sustain. Energy Rev.* **2017**, *73*, 249–264. [CrossRef]

11. Mardani, A.; Zavadskas, E.K.; Khalifah, Z.; Zakuan, N.; Jusoh, A.; Nor, K.M.; Khoshnoudi, M. A review of multi-criteria decision-making applications to solve energy management problems: Two decades from 1995 to 2015. *Renew. Sustain. Energy Rev.* **2017**, *71*, 216–256. [CrossRef]

12. International Energy Agency Wind (IEA WIND). Task 31. Wakebench. 2019. Available online: https://community.ieawind.org/home (accessed on 2 November 2019).

13. Ramírez-Rosado, I.J.; García-Garrido, E.; Fernández-Jiménez, L.A.; Zorzano-Santamaría, P.J.; Monteiro, C.; Miranda, V. Promotion of new wind farms based on a decision support system. *Renew. Energy* **2008**, *33*, 558–566. [CrossRef]

14. Tiba, C.; Candeias, A.; Fraidenraich, N.; de S. Barbosa, E.; de Carvalho Neto, P.; de Melo Filho, J. A GIS-based decision support tool for renewable energy management and planning in semi-arid rural environments of northeast of Brazil. *Renew. Energy* **2010**, *35*, 2921–2932. [CrossRef]

15. Aydin, N.Y.; Kentel, E.; Duzgun, S. GIS-based environmental assessment of wind energy systems for spatial planning: A case study from Western Turkey. *Renew. Sustain. Energy Rev.* **2010**, *14*, 364–373. [CrossRef]

16. Janke, J.R. Multicriteria GIS modeling of wind and solar farms in Colorado. *Renew. Energy* **2010**, *35*, 2228–2234. [CrossRef]

17. Tegou, L.I.; Polatidis, H.; Haralambopoulos, D.A. Environmental management framework for wind farm siting: Methodology and case study. *J. Environ. Manag.* **2010**, *91*, 2134–2147. [CrossRef]

18. Mari, R.; Bottai, L.; Busillo, C.; Calastrini, F.; Gozzini, B.; Gualtieri, G. A GIS-based interactive web decision support system for planning wind farms in Tuscany (Italy). *Renew. Energy* **2011**, *36*, 754–763. [CrossRef]

19. Van Haaren, R.; Fthenakis, V. GIS-based wind farm site selection using spatial multi-criteria analysis (SMCA): Evaluating the case for New York State. *Renew. Sustain. Energy Rev.* **2011**, *15*, 3332–3340. [CrossRef]

20. Grassi, S.; Chokani, N.; Abhari, R.S. Large scale technical and economical assessment of wind energy potential with a GIS tool: Case study Iowa. *Energy Policy* **2012**, *45*, 73–85. [CrossRef]

21. Gorsevski, P.V.; Cathcart, S.C.; Mirzaei, G.; Jamali, M.M.; Ye, X.; Gomezdelcampo, E. A group-based spatial decision support system for wind farm site selection in Northwest Ohio. *Energy Policy* **2013**, *55*, 374–385. [CrossRef]

22. Sánchez-Lozano, J.; García-Cascales, M.; Lamata, M. Identification and selection of potential sites for onshore wind farms development in Region of Murcia, Spain. *Energy* **2014**, *73*, 311–324. [CrossRef]

23. Yunna, W.; Geng, S. Multi-criteria decision making on selection of solar-wind hybrid power station location: A case of China. *Energy Convers. Manag.* **2014**, *81*, 527–533. [CrossRef]

24. Schallenberg-Rodríguez, J.; del Pino, J.N. Evaluation of on-shore wind techno-economical potential in regions and islands. *Appl. Energy* **2014**, *124*, 117–129. [CrossRef]

25. Atici, K.B.; Simsek, A.B.; Ulucan, A.; Tosun, M.U. A GIS-based Multiple Criteria Decision Analysis approach for wind power plant site selection. *Util. Policy* **2015**, *37*, 86–96. [CrossRef]

26. Tsoutsos, T.; Tsitoura, I.; Kokologos, D.; Kalaitzakis, K. Sustainable siting process in large wind farms case study in Crete. *Renew. Energy* **2015**, *75*, 474–480. [CrossRef]

27. Watson, J.J.; Hudson, M.D. Regional Scale wind farm and solar farm suitability assessment using GIS-assisted multi-criteria evaluation. *Landsc. Urban Plan.* **2015**, *138*, 20–31. [CrossRef]

28. Latinopoulos, D.; Kechagia, K. A GIS-based multi-criteria evaluation for wind farm site selection. A regional scale application in Greece. *Renew. Energy* **2015**, *78*, 550–560. [CrossRef]

29. Noorollahi, Y.; Yousefi, H.; Mohammadi, M. Multi-criteria decision support system for wind farm site selection using GIS. *Sustain. Energy Technol. Assess.* **2016**, *13*, 38–50. [CrossRef]

30. Hofer, T.; Sunak, Y.; Siddique, H.; Madlener, R. Wind farm siting using a spatial Analytic Hierarchy Process approach: A case study of the Städteregion Aachen. *Appl. Energy* **2016**, *163*, 222–243. [CrossRef]

31. Sánchez-Lozano, J.; García-Cascales, M.; Lamata, M. GIS-based onshore wind farm site selection using Fuzzy Multi-Criteria Decision Making methods. Evaluating the case of Southeastern Spain. *Appl. Energy* **2016**, *171*, 86–102. [CrossRef]

32. Sarpong, D.; Baffoe, P.E. Selecting Suitable Sites for Wind Energy Development in Ghana. *Ghana Min. J.* **2016**, *16*, 8–20. [CrossRef]

33. Chen, W.; Zhu, Y.; Yang, M.; Yuan, J. Optimal Site Selection of Wind-Solar Complementary Power Generation Project for a Large-Scale Plug-In Charging Station. *Sustainability* **2017**, *9*. [CrossRef]

34. Ali, S.; Lee, S.M.; Jang, C.M. Determination of the Most Optimal On-Shore Wind Farm Site Location Using a GIS-MCDM Methodology: Evaluating the Case of South Korea. *Energies* **2017**, *10*, 2072. [CrossRef]

35. Villacreses, G.; Gaona, G.; Martínez-Gómez, J.; Jijón, D.J. Wind farms suitability location using geographical information system (GIS), based on multi-criteria decision making (MCDM) methods: The case of continental Ecuador. *Renew. Energy* **2017**, *109*, 275–286. [CrossRef]

36. Baseer, M.; Rehman, S.; Meyer, J.; Alam, M.M. GIS-based site suitability analysis for wind farm development in Saudi Arabia. *Energy* **2017**, *141*, 1166–1176. [CrossRef]

37. Kazak, J.; van Hoof, J.; Szewranski, S. Challenges in the wind turbines location process in Central Europe âFIXME" The use of spatial decision support systems. *Renew. Sustain. Energy Rev.* **2017**, *76*, 425–433. [CrossRef]

38. Lotfi, R.; Mostafaeipour, A.; Mardani, N.; Mardani, S. Investigation of wind farm location planning by considering budget constraints. *Int. J. Sustain. Energy* **2018**. [CrossRef]

39. Díaz-Cuevas, P. GIS-Based Methodology for Evaluating the Wind-Energy Potential of Territories: A Case Study from Andalusia (Spain). *Energies* **2018**, *11*, 2789. [CrossRef]

40. Ayodele, T.; Ogunjuyigbe, A.; Odigie, O.; Munda, J. A multi-criteria GIS based model for wind farm site selection using interval type-2 fuzzy analytic hierarchy process: The case study of Nigeria. *Appl. Energy* **2018**, *228*, 1853–1869. [CrossRef]

41. Weiss, C.V.C.; Tagliani, P.R.A.; Espinoza, J.M.A.; de Lima, L.T.; Gandra, T.B.R. Spatial planning for wind farms: Perspectives of a coastal area in southern Brazil. *Springer* **2018**, *20*, 665–666. [CrossRef]

42. Sofuoglu, S.D.F.B.S.C. MCDM analysis of wind energy in Turkey: Decision making based on environmental impact. *Environ. Sci. Pollut. Res.* **2018**, *25*, 19753–19766. [CrossRef]

43. Solangi, Y.A.; Tan, Q.; Khan, M.W.A.; Mirjat, N.H.; Ahmed, I. The Selection ofWind Power Project Location in the Southeastern Corridor of Pakistan: A Factor Analysis, AHP, and Fuzzy-TOPSIS Application. *Energies* **2018**, *11*, 1940. [CrossRef]

44. Ali, S.; Taweekun, J.; Techato, K.; Waewsak, J.; Gyawali, S. GIS based site suitability assessment for wind and solar farms in Songkhla, Thailand. *Renew. Energy* **2018**, *132*, 1360–1372. [CrossRef]

45. Al-Yahyai, S.; Charabi, Y.; Gastli, A.; Al-Badi, A. Wind farm land suitability indexing using multi-criteria analysis. *Renew. Energy* **2012**, *44*, 80–87. [CrossRef]

46. Rezaei, M.; Mostafaeipour, A.; Qolipour, M.; Tavakkoli-Moghaddam, R. Investigation of the optimal location design of a hybrid wind-solar plant: A case study. *Int. J. Hydrog. Energy* **2018**, *43*, 100–114. [CrossRef]

47. Ntoka, C. Offshore Wind Park Sitting and Micro-Sitting in Petalioi Gulf, Greece. Ph.D. Thesis, Aalborg University, Aalborg, Denmark, 2013.

48. Waewsak, J.; Landry, M.; Gagnon, Y. Offshore wind power potential of the Gulf of Thailand. *Renew. Energy* **2015**, *81*, 609–626. [CrossRef]

49. Argin, M.; Yerci, V. Offshore wind power potential of the Black Sea region in Turkey. *Int. J. Green Energy* **2017**, *14*, 811–818, doi:10.1080/15435075.2017.1331443. [CrossRef]

50. Cavazzi, S.; Dutton, A. An Offshore Wind Energy Geographic Information System (OWE-GIS) for assessment of the UK's offshore wind energy potential. *Renew. Energy* **2016**, *87*, 212–228. [CrossRef]

51. Wu, B.; Yip, T.L.; Xie, L.; Wang, Y. A fuzzy-MADM based approach for site selection of offshore wind farm in busy waterways in China. *Ocean Eng.* **2018**, *168*, 121–132. [CrossRef]

52. Chaouachi, A.; Covrig, C.F.; Ardelean, M. Multi-criteria selection of offshore wind farms: Case study for the Baltic States. *Energy Policy* **2017**, *103*, 179–192. [CrossRef]

53. Fetanat, A.; Khorasaninejad, E. A novel hybrid MCDM approach for offshore wind farm site selection: A case study of Iran. *Ocean Coast. Manag.* **2015**, *109*, 17–28. [CrossRef]

54. Wu, Y.; Zhang, J.; Yuan, J.; Geng, S.; Zhang, H. Study of decision framework of offshore wind power station site selection based on ELECTRE-III under intuitionistic fuzzy environment: A case of China. *Energy Convers. Manag.* **2016**, *113*, 66–81. [CrossRef]

55. Kim, J.Y.; Oh, K.Y.; Kang, K.S.; Lee, J.S. Site selection of offshore wind farms around the Korean Peninsula through economic evaluation. *Renew. Energy* **2013**, *54*, 189–195. [CrossRef]

56. Kim, T.; Park, J.I.; Maeng, J. Offshore wind farm site selection study around Jeju Island, South Korea. *Renew. Energy* **2016**, *94*, 619–628. [CrossRef]

57. Cradden, L.; Kalogeri, C.; Barrios, I.M.; Galanis, G.; Ingram, D.; Kallos, G. Multi-criteria site selection for offshore renewable energy platforms. *Renew. Energy* **2016**, *87*, 791–806. [CrossRef]

58. Vasileiou, M.; Loukogeorgaki, E.; Vagiona, D.G. GIS-based multi-criteria decision analysis for site selection of hybrid offshore wind and wave energy systems in Greece. *Renew. Sustain. Energy Rev.* **2017**, *73*, 745–757. [CrossRef]

59. Punt, M.J.; Groeneveld, R.A.; van Ierland, E.C.; Stel, J.H. Spatial planning of offshore wind farms: A windfall to marine environmental protection? *Ecol. Econ.* **2009**, *69*, 93–103. [CrossRef]

60. Yue, C.D.; Yang, M.H. Exploring the potential of wind energy for a coastal state. *Energy Policy* **2009**, *37*, 3925–3940. [CrossRef]

61. Government, T.S. Draft Plan for Offshore Wind Energy in Scottish Territorial Waters. 2010. Available online: https://www2.gov.scot/resource/doc/312147/0098586.pdf (accessed on 2 November 2019).

62. Moore, A.; Price, J.; Zeyringer, M. The role of floating offshore wind in a renewable focused electricity system for Great Britain in 2050. *Energy Strategy Rev.* **2018**, *22*, 270–278. [CrossRef]

63. Vagiona, D.G.; Kamilakis, M. Sustainable Site Selection for Offshore Wind Farms in the South Aegean—Greece. *Sustainability* **2018**, *10*. [CrossRef]

64. Mytilinou, V.; Lozano-Minguez, E.; Kolios, A. A Framework for the Selection of Optimum Offshore Wind Farm Locations for Deployment. *Energies* **2018**, *11*. [CrossRef]

65. Nagababu, G.; Kachhwaha, S.S.; Savsani, V. Estimation of technical and economic potential of offshore wind along the coast of India. *Energy* **2017**, *138*, 79–91. [CrossRef]

66. Mederos, A.M.; Padrón, J.M.; Lorenzo, A.F. An offshore wind atlas for the Canary Islands. *Renew. Sustain. Energy Rev.* **2011**, *15*, 612–620. [CrossRef]

67. Kim, C.K.; Jang, S.; Kim, T.Y. Site selection for offshore wind farms in the southwest coast of South Korea. *Renew. Energy* **2018**, *120*, 151–162. [CrossRef]

68. Pillai, A.C.; Chick, J.; Khorasanchi, M.; Barbouchi, S.; Johanning, L. Application of an offshore wind farm layout optimization methodology at Middelgrunden wind farm. *Ocean Eng.* **2017**, *139*, 287–297. [CrossRef]

69. Schillings, C.; Wanderer, T.; Cameron, L.; van der Wal, J.T.; Jacquemin, J.; Veum, K. A decision support system for assessing offshore wind energy potential in the North Sea. *Energy Policy* **2012**, *49*, 541–551. [CrossRef]

70. Astariz, S.; Iglesias, G. Selecting optimum locations for co-located wave and wind energy farms. Part I: The Co-Location Feasibility index. *Energy Convers. Manag.* **2016**, *122*, 589–598. [CrossRef]

71. Schallenberg-Rodríguez, J.; Montesdeoca, N.G. Spatial planning to estimate the offshore wind energy potential in coastal regions and islands. Practical case: The Canary Islands. *Energy* **2018**, *143*, 91–103. [CrossRef]

72. Abudureyimu, A.; Hayashi, Y.; Nagasaka, K. Analyzing the Economy of Off-shore Wind Energy using GIS Technique. *APCBEE Procedia* **2012**, *1*, 182–186. [CrossRef]

73. Möller, B. Continuous spatial modelling to analyse planning and economic consequences of offshore wind energy. *Energy Policy* **2011**, *39*, 511–517. [CrossRef]

74. Sheridan, B.; Baker, S.D.; Pearre, N.S.; Firestone, J.; Kempton, W. Calculating the offshore wind power resource: Robust assessment methods applied to the U.S. Atlantic Coast. *Renew. Energy* **2012**, *43*, 224–233. [CrossRef]

75. Madsen, J.; Bates, A.; Callahan, J.; Firestone, J. *Geospatial Techniques for Managing Environmental Resources. Use of Geospatial Data in Planning for Offshore Wind Development*; Springer-Verlag GmbH: Berlin/Heidelberg, Germany, 2011. [CrossRef]

76. Gao, X.; Yang, H.; Lu, L. Study on offshore wind power potential and wind farm optimization in Hong Kong. *Appl. Energy* **2014**, *130*, 519–531. [CrossRef]

77. Veigas, M.; Carballo, R.; Iglesias, G. Wave and offshore wind energy on an island. *Energy Sustain. Dev.* **2014**, *22*, 57– 65. [CrossRef]
78. Jongbloed, R.; van der Wal, J.; Lindeboom, H. Identifying space for offshore wind energy in the North Sea. Consequences of scenario calculations for interactions with other marine uses. *Energy Policy* **2014**, *68*, 320–333. [CrossRef]
79. Gimpel, A.; Stelzenmüller, V.; Grote, B.; Buck, B.H.; Floeter, J.; Núñez-Riboni, I.; Pogoda, B.; Temming, A. A GIS modelling framework to evaluate marine spatial planning scenarios: Co-location of offshore wind farms and aquaculture in the German EEZ. *Mar. Policy* **2015**, *55*, 102–115. [CrossRef]
80. Mekonnen, A.D.; Gorsevski, P.V. A web-based participatory GIS (PGIS) for offshore wind farm suitability within Lake Erie, Ohio. *Renew. Sustain. Energy Rev.* **2015**, *41*, 162–177. [CrossRef]
81. Nie, B.; Li, J. Technical potential assessment of offshore wind energy over shallow continent shelf along China coast. *Renew. Energy* **2018**, *128*, 391–399. [CrossRef]
82. Schweizer, J.; Antonini, A.; Govoni, L.; Gottardi, G.; Archetti, R.; Supino, E.; Berretta, C.; Casadei, C.; Ozzi, C. Investigating the potential and feasibility of an offshore wind farm in the Northern Adriatic Sea. *Appl. Energy* **2016**, *177*, 449–463. [CrossRef]
83. Satir, M.; Murphy, F.; McDonnell, K. Feasibility study of an offshore wind farm in the Aegean Sea, Turkey. *Renew. Sustain. Energy Rev.* **2018**, *81*, 2552–2562. [CrossRef]
84. Mahdy, M.; Bahaj, A.S. Multi criteria decision analysis for offshore wind energy potential in Egypt. *Renew. Energy* **2018**, *118*, 278–289. [CrossRef]
85. Hong, L.; Möller, B. Offshore wind energy potential in China: Under technical, spatial and economic constraints. *Energy* **2011**, *36*, 4482–4491. [CrossRef]
86. Magar, V.; Gross, M.; González-García, L. Offshore wind energy resource assessment under techno-economic and social-ecological constraints. *Ocean Coast. Manag.* **2018**, *152*, 77–87. [CrossRef]
87. Depellegrin, D.; Blažauskas, N.; Egarter-Vigl, L. An integrated visual impact assessment model for offshore windfarm development. *Ocean Coast. Manag.* **2014**, *98*, 95–110. [CrossRef]

Article

Offshore Wind Power Integration into Future Power Systems: Overview and Trends

Ana Fernández-Guillamón [1,†], Kaushik Das [2,†], Nicolaos A. Cutululis [2,†] and Ángel Molina-García [1,†,*]

1   Department of Electrical Eng, Universidad Politécnica de Cartagena, 30202 Cartagena, Spain;
    ana.fernandez@upct.es
2   Department of Wind Energy, Technical University of Denmark, 4000 Roskilde, Denmark;
    kdas@dtu.dk (K.D.); niac@dtu.dk (N.A.C.)
*   Correspondence: angel.molina@upct.es; Tel.: +34-968-32-5462
†   These authors contributed equally to this work.

Received: 10 October 2019; Accepted: 2 November 2019; Published: 7 November 2019

Abstract: Nowadays, wind is considered as a remarkable renewable energy source to be implemented in power systems. Most wind power plant experiences have been based on onshore installations, as they are considered as a mature technological solution by the electricity sector. However, future power scenarios and roadmaps promote offshore power plants as an alternative and additional power generation source, especially in some regions such as the North and Baltic seas. According to this framework, the present paper discusses and reviews trends and perspectives of offshore wind power plants for massive offshore wind power integration into future power systems. Different offshore trends, including turbine capacity, wind power plant capacity as well as water depth and distance from the shore, are discussed. In addition, electrical transmission high voltage alternating current (HVAC) and high voltage direct current (HVDC) solutions are described by considering the advantages and technical limitations of these alternatives. Several future advancements focused on increasing the offshore wind energy capacity currently under analysis are also included in the paper.

Keywords: offshore wind energy; HVAC; HVDC; P2X; hydrogen storage; CAES

1. Introduction

Energy demand has been increasing non-stop during the last decades [1]. Nowadays, fossil fuel sources (i.e., coal, oil and natural gas) provide around 85% of the world energy demand, according to the BP Energy Outlook of 2019 [2]. However, with the Paris climate agreement established in December 2015, this energy scenario is about to change [3]. This climate agreement aims to restrict maximum increase in the global average temperature below 2 °C above pre-industrial levels [4]. To fulfill this goal, greenhouse gas (GHG) emission trends should drastically change [5]. Consequently, the use of fossil fuels should be reduced, as they are considered as the main source of GHG emissions [6]. Actually, global GHG emissions are dominated by the emissions of CO_2 due to the combustion of fossil fuels, which has been increasing continuously since 1990 [7]. The power sector should be decarbonized by 2050 to meet the Paris agreement target [8]. Furthermore, Liddle and Sadorsky estimated that increasing by 1% the share of non-fossil fuel electricity generation can reduce by up to 0.82% the CO_2 emissions [9]. This environmental worry is one of the reasons to promote the integration of renewable energy sources (RES) into power systems [10]. Moreover, RES can also mitigate the energy dependence on fossil fuels imported from other countries [11]. Apart from the economic costs of these fossil fuel imports, decreasing energy dependence increases electricity supply security [12]. The International Energy Agency defines electricity supply security as the uninterrupted availability of energy sources at an affordable price [13]. However, political stability, market liberalization and foreign affairs are nowadays linked to energy supply security [14]. As a consequence, it is important to be energy-independent to guarantee the energy security of a country [15].

While RES provide an acceptable solution for these two problems, they also face many challenges as their integration increases into the grids, mostly based on their intermittency, variability and uncertainty due to their dependency on weather conditions [16]. Actually, they are usually considered as 'non-dispatchable' sources [17]. This fact makes them hard to integrate into power systems [18], as transmission system operators (TSOs) have to deal with not only the uncontrollable demand but also uncontrollable generation [19,20]. RES include bioenergy, geothermal energy, hydropower, ocean energy (tide and wave), PV, thermal solar energy and wind energy (onshore and offshore) [21]. Some of them (such as wind and solar installations) are connected to the grid through power electronic converters, reducing the rotational inertia of the system as they replace conventional generation units [22,23]. This fact compromises the frequency stability and alters the transient response [24]. As a result, several frequency control strategies have been proposed in the specific literature [25–30]. Other alternatives to increase the RES share in power systems and avoid the aforementioned problems are to complement one source with another (for instance, wind with solar and/or hydropower) [31–33] or to use storage systems (such as flywheels, pumped hydroelectric storage, batteries, hydrogen, etc.) [34,35].

Among these renewable technologies, wind is one of the most economic, prominent and matured RES technologies [36,37]. In fact, since 2001, global cumulative installed wind capacity has shown an exponential growth, as can be seen in Figure 1a. Among the total wind capacity, 23 GW came from offshore installations in 2018, compared to 1 GW in 2007, refer to Figure 1b [38]. Despite offshore wind energy dating back to the 1990s, its popularity started around ten years ago [39]. This increase is due to the current interest of the wind energy industry in offshore wind power [40]. For instance, offshore wind energy investments surpassed onshore investments in Europe in 2016, as presented in [41]. Moreover, nearly 40% of the total wind capacity is expected to come from offshore wind energy in Europe in 2030 [42,43].

In addition, offshore wind energy presents many advantages compared to onshore wind power plants, especially related to wind energy potential [44,45]: (*i*) Offshore mean wind speeds are higher and wind power variability is also lower than onshore wind power; (*ii*) their visual and acoustic impact is usually lower than onshore; subsequently (*iii*) larger wind turbines (WTs) can be installed [46]. Actually, on the European coasts, the available offshore wind energy is about 350 GW [47]; the USA's shores present an offshore wind power potential of more than 2000 GW [48]; the offshore wind resource in China is about 500 GW in water depth under 50 m [49]; and the east and west Indian coasts have an offshore wind potential of 4.4 GW and 6.7 GW, respectively [50].

Furthermore, offshore wind speed usually increases with distance from the shore, thus increasing the power generated, as it depends on the cube of the wind speed [51]. However, higher installation and maintenance costs of offshore wind power plants (OWPP) far from the shore balance the benefits of higher energy production [52]. Indeed, OWPP are around 50% more expensive than onshore wind power plants [53], but their costs are expected to decline up to 35% by 2025 [54]. The global weighted average levelized cost of energy (LCOE) in 2018 was 20% lower than in 2010. These cost reductions can be a result of [55]:

- The evolution in wind turbine technology, installation and logistics
- The economies of scale in operations and maintenance
- The improved capacity factors due to higher hub heights, better wind resources and larger rotor diameters

This paper analyzes and reviews different aspects of offshore wind power plants, including several future alternatives to increase the offshore wind power capacity. The rest of the paper is organized as follows: Section 2 presents the current status of offshore wind power plants (WTs and OWPP sizes, water depth, distance from shore and electrical transmission to shore). Future advancements possible for larger offshore wind power plant integration are analyzed in Section 3. Finally, Section 4 gives the conclusions.

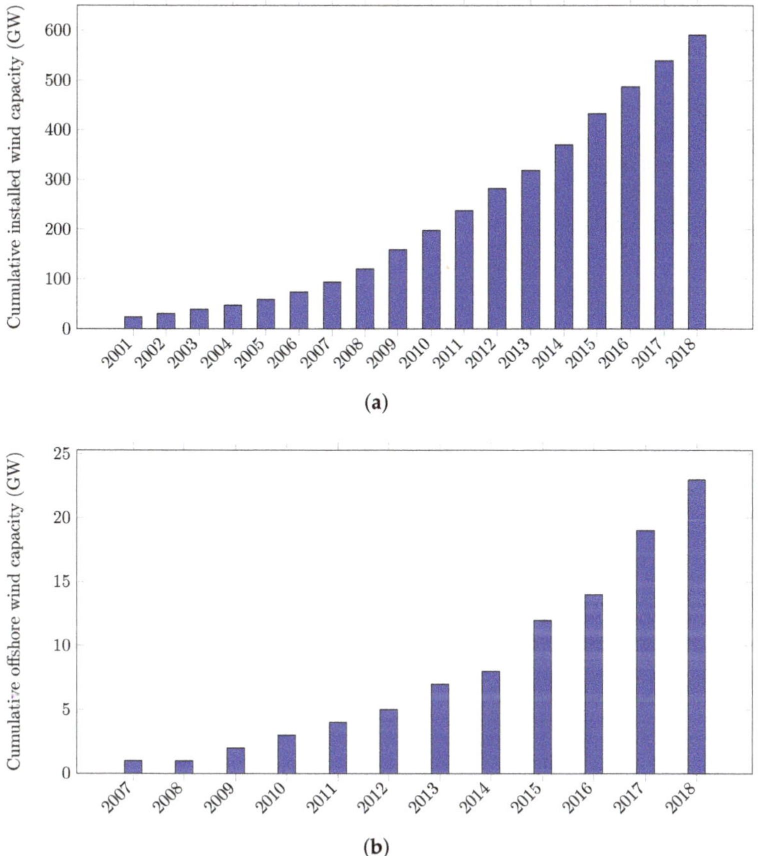

Figure 1. Global cumulative wind capacity in GW. (**a**) Global cumulative wind capacity: Onshore and offshore. (**b**) Global cumulative offshore wind capacity.

2. Current Status of Offshore Wind Power Plants

2.1. Preliminaries: Classification of Wind Turbines

WTs are usually classified as fixed speed wind turbines (FSWTs) and variable speed wind turbines (VSWTs) [56]. FSWTs work at the same rotational speed regardless of the wind speed [57]. VSWTs can operate around their optimum power point for each wind speed, using a partial or full additional power converter [58]. As a result, VSWTs are more efficient than FSWTs [59]. Moreover, WTs present different topologies depending on their generator [60]: type 1 includes a squirrel cage induction generator; type 2 includes a wound rotor induction generator; type 3 includes a doubly-fed induction generator (DFIG); and type 4 includes a full-converter synchronous generator [61]. Types 1 and 2 are FSWTs, whereas types 3 and 4 are VSWTs.

Nowadays, VSWTs are the most commonly installed WTs [62–65]. Among them, full converter generator WTs seem to be a better option than DFIG-based WTs for OWPP [66–72]. The main differences between DFIG and full-converter WTs are the following:

- The DFIG configuration needs a gearbox, generator and partial-scale power converter (around 30%), as shown in Figure 2a. The gearbox couples the blades with the generator, increasing the rotational speed from the rotor hub to the induction machine [73–75]. The stator is directly connected to the grid, whereas the rotor is connected to the power converter [76]. As a result, the converter only covers the power produced by the rotor of the DFIG [77].

- The synchronous generator of a full-converter WT is excited by an external DC source or by permanent magnets [78]. In this case, the hole generator is connected to the grid through a power converter [79]. Hence, all the generated power from a WT can be regulated accordingly [80]. They have low maintenance costs and negligible rotor losses [81]. Moreover, some type 4 WTs have no gearbox, as depicted with a dotted line in Figure 2b, using a direct driven multipole generator [82].

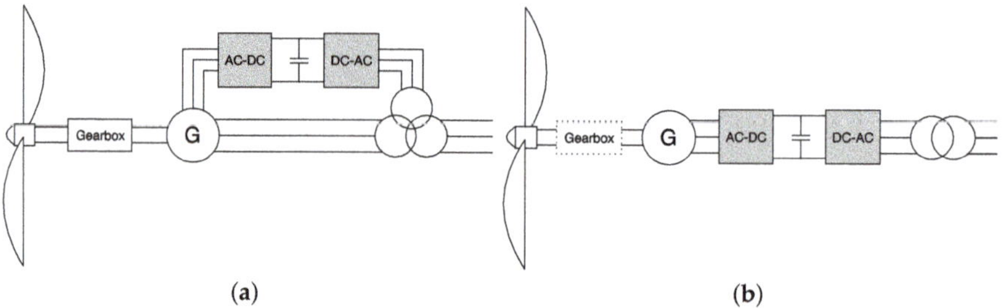

(a)

(b)

Figure 2. Variable speed wind turbines types. (**a**) Doubly-fed induction generator (DFIG) wind turbine. (**b**) Full-converter wind turbine.

2.2. Offshore Trends: Turbine Capacity, Wind Power Plant Capacity, Depth and Distance from the Shore

In Europe, the rated capacity of offshore WTs has been continuously increasing during the last decade. For instance, in 2017, the average rated capacity of WTs was 5.9 MW, compared to 3 MW in 2010 and 4.8 MW in 2016 [83]. In 2018, new offshore WTs were 6.8 MW on average, 15% larger than in 2017. Comparing 2018 to 2010, the average WT increase is more than 200%. Moreover, two 8.8 MW offshore WTs were installed in the United Kingdom in 2018, those being the largest WTs installed of the world. The commercial model of those WTs was V164-8.8 MW from MHI Vestas Offshore [84]. However, nowadays there are larger commercial offshore WTs, up to 12 MW, as presented in [85]. Table 1 shows the 10 largest WTs currently available. All of them have a rated power over 8 MW, rotor diameters between 150 and 200 m and are equipped with synchronous generators (type 4).

Table 1. Biggest wind turbines currently available.

Rated Power (MW)	Manufacturer	Reference	Diameter (m)	Generator
8.0	Siemens Gamesa	SG 8.0-167 DD	167	Synchronous permanent
8.3	MHI Vestas Offshore	V164-8.3 MW	164	Synchronous permanent
8.8	MHI Vestas Offshore	V164-8.8 MW	164	Synchronous permanent
9.0	MHI Vestas Offshore	V164-9.0 MW	164	Synchronous permanent
9.5	MHI Vestas Offshore	V164-9.5 MW	164	Permanent magnet
10.0	AMSC	wt10000dd SeaTitan	190	HTS synchronous
10.0	MHI Vestas Offshore	V164-10.0MW	164	Permanent magnet
10.0	Swiss Electric	YZ150/10.0	150	Synchronous permanent
		YZ170/10.0	170	
		YZ190/10.0	190	
10.0	Siemens Gamesa	SG 10.0-193 DD	193	Synchronous permanent
12.0	General Electric	GE HALIADE-X	220	Synchronous permanent

Regarding OWPP capacity, it has also increased dramatically in the last 10 years (around 700%), in line with the increase of average offshore WT capacity. In fact, average OWPP capacity was 79.6 MW in 2007. In contrast, 561 MW was the average capacity for OWPPs in 2018 [84]. This means that considering the average WT and OWPP capacities of the year 2018, each OWPP has between 80 and 85 WTs. On the other hand, OWPP depth and distance to the shore have not increased that much in recent years. At the end of 2013, the average water depth of OWPPs was 16 m with an average distance to the shore of 29 km [86]. In 2018, the average water depth of OWPPs under construction was 27.1 m, with an average distance to shore of 33 km. This means that water depth has increased by 170% and distance to shore by around 110%. There are some OWPPs that should be mentioned: Hornsea One (UK) and EnBW Hohe See (Germany) are the OWPPs located farthest from the shore (103 km away); Kincardine Pilot (Scotland), a floating demonstration project, has a water depth of 77 m [84]; and Hywind (Scotland), the first fully operational floating wind farm, with water depths varying between 95 and 129 m [87].

2.3. Offshore Wind Power Electrical Power Transmission

For the electrical power transmission from the OWPP plant to the shore, there are two possibilities: (*i*) High voltage alternating current (HVAC) and (*ii*) high voltage direct current (HVDC). Figure 3 depicts an overview of the current state of offshore wind power energy transmission to the shore.

$$\text{OWPP transmission} \begin{cases} \text{HVAC} \\ \text{HVDC} \begin{cases} \text{LCC} \\ \text{VSC} \\ \text{DRU} \end{cases} \end{cases}$$

Figure 3. Offshore wind power plant transmission.

2.3.1. High Voltage AC

HVAC transmissions were mostly used for OWPPs until the year 2010 [88]. Their easy protection system design and the use of transformers to change between different voltage levels were the main reasons to use them [89]. However, the high capacitance of submarine HVAC cables combined with the low resistivity of sea water caused different electromagnetic dynamic and transient problems from those of conventional overhead lines, such as distortion of the voltage's shape due to resonance problems [90,91]. This high capacitance also leads to substantial charging currents, subsequently reducing the active power transmission capacity and transmitting reactive power in long distances [92]. A possible solution could be to install reactive power compensation units along the HVAC submarine cables, but they are expensive devices and it is a difficult task to carry out [93]. an alternative found in the literature is to add compensation units only at both ends (onshore and offshore) of the underwater cables, which improve the current profile along them [94,95]. However, their effect is very limited for distances over 60–75 km [96,97]. With the aforementioned considerations, the topology for HVAC transmission from OWPPs is depicted in Figure 4 [98,99]. It consists of:

- An offshore substation to increase the offshore voltage level (usually from 30–36 kV) to the transmission voltage level at 132–400 kV.
- Three-core HVAC submarine transmission cables.
- Reactive compensation units on both ends (offshore and onshore), such as static VAR compensators (SVCs) or static synchronous compensators (STATCOMs).
- An onshore substation, if the onshore interconnecting grid voltage is different from the offshore transmission system rated voltage.

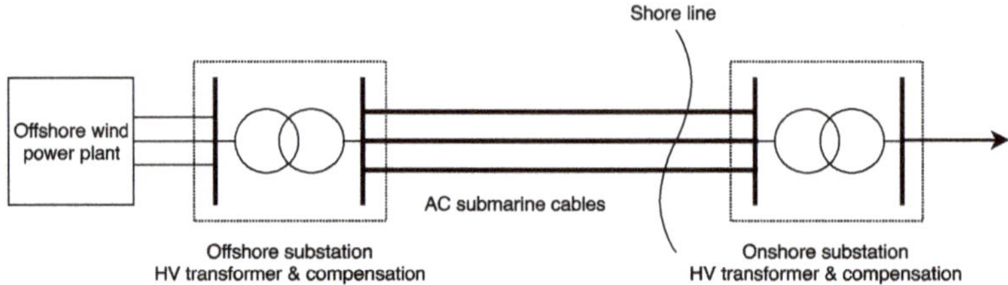

Figure 4. Offshore wind power plant high voltage alternating current (HVAC) transmission system.

As can be seen, the OWPP grid is synchronously coupled to the main onshore grid. This is another problem of HVAC links for OWPPs, as all faults in either the grid or the OWPP are propagated to the other one [100].

2.3.2. High Voltage DC

High voltage direct current (HVDC) transmission is considered as the best solution to OWPPs located far away from the land [101]. Actually, some studies conclude that HVDC links are economically viable for distances above 50–70 km [102]. A graphical comparison of costs between HVAC and HVDC transmission systems can be seen in Figure 5 [103].

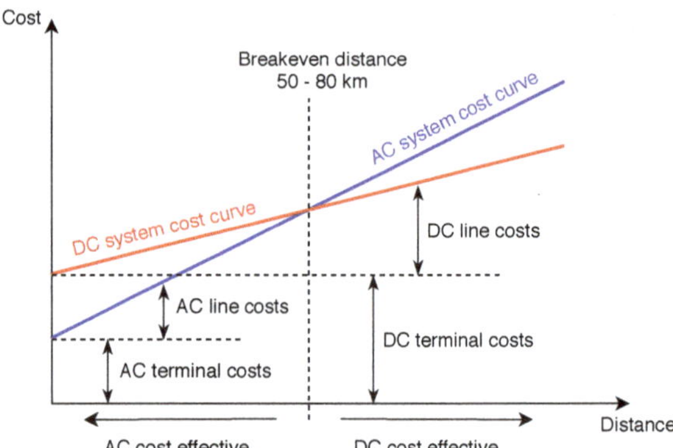

Figure 5. AC and DC system costs based on the transmission distances.

Figure 6 shows the main elements of an HVDC connection, and consists of [104,105]:

- An offshore substation to increase the voltage level to the level of the transmission line.
- AC/DC rectifier.
- AC and DC filters to cancel the low order harmonics. Furthermore, the AC filters supply some of the reactive power used by the converter, whereas the DC filters avoid the generation of circulating AC currents in the cable.
- DC current filtering reactance. This removes the possibility of a current interruption under minimum load circumstances, limiting DC fault currents and also reducing current harmonics in the DC cable.
- DC cables.
- DC/AC converter.
- An onshore substation.

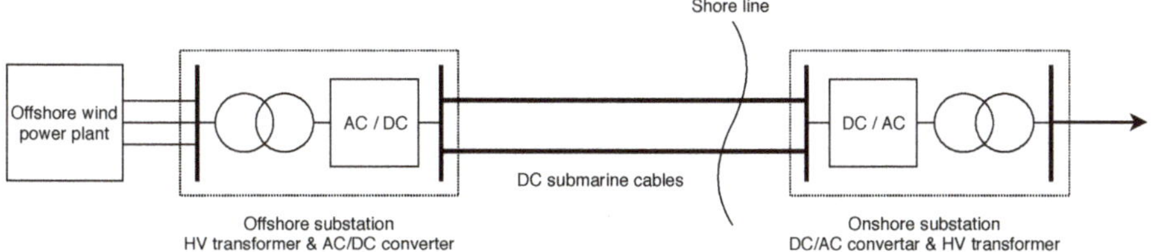

Figure 6. Offshore wind power plant HVDC transmission system.

In contrast to the HVAC topology presented in Figure 4, the HVDC link electrically decouples both the OWPP and the onshore grid, avoiding the propagation of possible disturbances between them [106,107].

Two different HVDC technologies are currently under use: Line-commutated converters (LCCs) based on thyristors and voltage-source converters (VSCs) based on insulated gate bipolar transistors (IGBTs) [108]. Among them, there is not a clear consensus about which technology is better: Some authors consider that LCCs are superior to VSCs in terms of reliability, cost and efficiency [26], whereas others affirm that the VCS–HVDC transmission system is the most promising technology [109]. a comparison between both HVDC technologies is summarized in Table 2 [110]. Recently, another technology called the diode rectifier unit (DRU) has been under discussion, though has not been implemented yet [111].

Table 2. Comparison between line-commutated converter (LCC) and voltage-source converter (VSC) HVDC technologies.

Technology	LCCs	VSCs
Semiconductor	Thyristor	IGBT
Control	Turn on	Turn on/off
Power control	Active	Active & Reactive
AC filters	Yes	No
Blackstart capability	No	Yes

Line-Commutated Converters

Traditionally, HVDC transmission systems have been based on LCCs, which use thyristors as the base technology. Actually, LCC is a trusted and mature technology [112] that links a mainland with some islands (e.g., in Northern Europe [113]). However, solutions based on thyristors usually involve the injection of some harmonics. For example, a twelve-pulse thyristor bridge, which is made up of two six-pulse bridges; the fifth and seventh harmonics can be canceled [114]. The LCC–HVDC transmission system is based on this twelve-step bridge [115,116].

The main drawbacks of the LCC–HVDC link are [116–120]:

- It can only transfer power between (at least) two active grids. As a result, an auxiliary start-up system is necessary in the OWPP.
- It demands reactive power, which needs to be supplied through reactive support devices.
- Despite most harmonics being canceled by using a twelve-pulse bridge, others still remain, thus needing additional filters.
- It requires voltage support for the OWPP AC bus. Two possible solutions can be found to overcome this requirement: (*i*) Installing a dedicated STATCOM, which increases considerably the overall cost or (*ii*) controlling the turbine inverters individually, which is technically challenging.
- The inverter is susceptible to commutation failures, especially when connected to weak AC power systems.

Blasco et al. suggest that the filter's design depends on the harmonic characteristics of the AC grid and the active power exchanged by the LCC–HVDC link, thus needing a detailed AC power system analysis [121].

Voltage-Source Converters

Since 2005, VSC–HVDC technology has been used in offshore applications. It is based on IGBTs [122]. The main characteristics of the VSC–HVDC link are summarized as [123–125]:

- It can control active and reactive power simultaneously.
- It can feed island-mode, weak AC and passive networks.
- Its station requires less space than that of an LCC (about 60% less).
- The cables are lighter.
- It does not require reactive power compensation.
- It can transmit power from zero to full-rating bidirectional, enabling OWPP start-up (black start operation) and working at low wind speeds.

Despite all these advantages, VSC–HVDC presents higher commutation losses and costs compared to the LCC–HVDC. Moreover, it can only handle limited voltage and power levels [126].

Diode Rectifier Unit

During the last years, DRU–HVDC has been under discussion. A DRU includes several diodes, a transformer and a smoothing reactor [127]. As DRU can only convert AC to DC [128], a hybrid topology combining DRU and VSC/LCC must be used, introducing the DRU as the offshore rectifier and the LCC/VSC as the onshore converter [129]. The main advantages of DRU–HVDC compared to LCC–HVDC and VSC–HVDC are [130–133]:

- Reduction of volume (80%) and weight (66%) of the platform.
- Smaller footprints.
- Reduction of power losses up to 20%.
- Reduction of total cost up to 30%.
- Capacity increased by 33%.
- Higher reliability and efficiency.
- Modular design and full encapsulation.
- Reduced operation and maintenance costs.

However, several problems have to be solved before implementing a DRU–HVDC connection [132–135]:

- As the DRU is a non-controllable passive device, the OWPP AC system must be regulated and controlled by the WT, thus requiring different WT and OWPP schemes.
- The onshore converter (LCC/VSC) controls the HVDC voltage. Subsequently, the DRU output DC voltage must be higher than the minimum voltage value to start conducting and transmit the power to the onshore station.
- Passive filters or active compensation devices are needed to remove the harmonic currents injected by the DRU.
- Voltage and frequency control stages are needed in the offshore grid for DRU commutation.
- A DRU is not able to provide auxiliary active power for the WT and OWPP substation, being then a drawback to the self-start of WTs.
- A DRU is not able to provide reactive power, needing power converters or other devices to compensate it.

3. Potential Future Advancements for Offshore Wind Energy

In 2018, ENTSO-E and ENTSO-G published their Ten Year Network Development Plan (TYNDP) scenarios. It was the first time that both European electrical and gas TSOs collaborated together. The TYNDP 2018 covers from 2020 to 2040 [136]. In 2030–2040, it is expected that between 45 and 75% of the overall European demand will be covered by RES, especially by hydro, wind and solar power. Actually, in the North Sea and Baltic Sea regions, the offshore wind power capacity is estimated to reach between 40 and 59 GW in 2030, and between 86 and 127 GW in 2040, according to the TYNDP. Other authors propose similar offshore wind power capacity scenarios in these regions; for instance, scenarios were modeled and optimized by Koivisto and Gea-Bermudez [137]. Greenpeace published in 2015 their 'Energy [R]evolution' forecast, where 148 GW of offshore wind capacity is expected to be installed in Europe in 2050 [138]. In the US, the Department of Energy considers that 22 GW of OWPPs can be installed by 2030, increasing up to 86 GW by 2050 [139]; according to the scenarios presented by the Energy Resources Institute of India for 2050, it could have 170 GW installed of offshore wind energy by then [140]; and the Chinese scenarios propose to install 200 GW of OWPPs (150 GW near offshore wind and 50 GW far offshore wind) by 2050 [141,142]. By these means, OWPPs seem to have an important energy role in the future worldwide.

However, onshore wind and other conventional fossil fuel technologies are currently cheaper than offshore wind energy [143]. As a consequence, different alternatives are being researched to reduce further costs of offshore wind power development:

- Power-to-X conversion (P2X)
- North Sea Wind Power Hub: The Hub-and-Spoke project
- Offshore storage options

These initiatives are discussed in detail in the following.

3.1. Power-to-X Conversion

P2X is based on converting power (electricity) to diverse substances (X) [144]. The different alternatives available in the P2X conversion are [145,146]:

- Power-to-heat (P2H): The electrical generation excess is linked to a heat device (electric boiler, heat pump), avoiding any intermediate energy carrier and subsequently increasing the global efficiency.
- Power-to-liquid (P2L): Different alternatives can be found in the specific literature, including the production of syngas through hydrogenation of CO_2 and reverse water gas shift; co-electrolysis of CO_2 and H_2O; or directly through the electro-reduction of CO_2 to methanol.
- Power-to-chemicals (P2C): From the syngas obtained with the power-to-liquid conversion, several compounds can be produced accordingly.
- Power-to-gas (P2G): Hydrogen is obtained from an electrolysis process and the possible subsequent conversion to methane with CO_2.
- Power-to-mobility (P2M): The electrical generation excess is used by the mobility sector through electric vehicles with an electric motor of 90% efficiency instead of an internal combustion engine (efficiency of 20%) or fuel cell (efficiency of 50%).
- Power-to-power (P2P): Electricity is converted into chemical or mechanical energy, which is stored and later reconverted into electric power.

These transformations are expected to be very relevant in future power systems, as the generated electricity excess can be stored in different ways and later used as, for instance, fuel for power plants [147]. Hence, the system's flexibility would be enhanced [148]. By these means, high capacity P2X plants could increase the RES supply by providing supply security in terms of storage facilities [149]. Moreover, as explained in [150], the P2X conversion provides a real link between different sectors, promoting the transition towards a future urban smart energy system. As an example, Figure 7 depicts the power-to-heat conversion joint [151]. an exhaustive analysis of 128 P2X demo projects in operation in Europe is discussed in [152]. These projects aim to gain experience with system integration of P2X components. Moreover, Denmark is interested in the conversion of electricity to hydrogen and liquid fuels through P2X solutions. Thus, they can become a front-runner in this technology as large Danish companies already work with it [153].

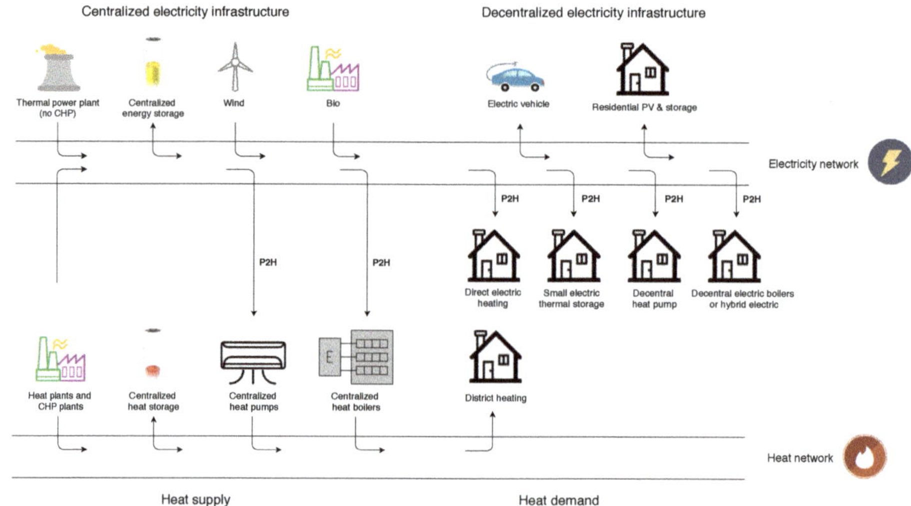

Figure 7. Power-to-heat conversion.

Several authors have already analyzed these technologies combined with the wind resource. Different flexibility options for wind power plants are analyzed in [154], concluding that the P2H solutions provide the most cost-effective scenarios with the lowest CO_2 emissions. Pursiheimo et al. focused on the feasibility of the P2G technology in Nordic countries to achieve a 100% RES system. The main applications of P2G are focused on supplying gas to transport and industrial sectors [155]. Furthermore, the use of P2G has been proved in Denmark to be a successful tool to complement wind power plants [156]. However, both investment costs of facilities and energy losses (due to the low efficiency in the conversion process) are high. Hence, the hydrogen produced from wind power also has a high cost [157]. For instance, in [158] different energy applications of hydrogen (P2P, P2G, P2M) are considered for a hybrid offshore wind–hydrogen power plant in France, obtaining negative profits due to the high investment costs in both wind and hydrogen infrastructures. Other authors conclude that the combination of a wind–hydrogen power plant should be considered to sell hydrogen directly, as re-powering hydrogen for electricity is extremely expensive [159]. Consequently, future works should be focused on reducing investment and maintenance costs for such power conversion solutions.

3.2. North Sea Wind Power Hub: The Hub-and-Spoke Project

To meet the Paris Agreement and the GHG reduction goals (refer to Section 1) in the countries around the North Sea, the North Sea Wind Power Hub (NSWPH) consortium was created. TenneT (a Dutch–German electricity TSO), Port of Rotterdam (the biggest port in Europe), Energinet (a Danish TSO) and Gasunie (a European energy infrastructure company) are the partners of the consortium [160]. NSWPH aims to facilitate the deployment of large scale OWPPs in the North Sea, evaluating and developing the Hub-and-Spoke project. The project consists on several central platforms, called hubs, which are in charge of supporting the power transport infrastructure by using the P2X conversion instead of the offshore converter platforms used currently. According to TenneT, the offshore wind power capacities will be in the range of 70 to 150 GW by the year 2040 and up to 180 GW by 2045 [161], similar values to those proposed in the TYNDP. Two main challenges are identified in NSWPH, (*i*) a strong power transmission infrastructure and (*ii*) high flexibility requirements. Mainly due to onshore surface constraints, onshore wind power plants and PV installations are not enough to decarbonize the power systems of this area [162]. The hub-and-spoke concept proposed by NSWPH is made up of several modular hubs located in different zones of the North Sea, which connect OWPPs with bordering North Sea countries. This can be seen in the figure of page 6 of [163]. By using high capacity DC cables, the power generated by OWPPs is transmitted to onshore grids in different locations connected in a smart and coordinated manner. These DC connections also provide high interconnection capacity among the different countries. Moreover, the hub-and-spoke concept can promote onshore OWPP integration through P2G transformation. Power systems thus become flexible through such P2X conversion [163].

Apart from defining the hub-and-spoke project, the NSWPH consortium also aims to demonstrate the technical feasibility of the project. So far, the consortium have concluded that [164]:

- The optimal capacity of the OWPP is estimated to be between 10 and 15 GW.
- Hub substructures can be based on four different foundation types: Ciasson island, sand island, platform and gravity-based structure. A comparison among them is presented in Table 3, as presented in [164].
- Both the spatial requirements and investment costs of the hubs are similar regardless of being all-electric, all-hydrogen or combining electricity and hydrogen:

    - All-electric hub-and-spoke: The electricity generated by the OWPP is transmitted to the shore.
    - All-hydrogen hub-and-spoke: The electricity generated by the OWPP is transformed offshore into hydrogen, and transported through pipelines to the shore.
    - Combined electricity and hydrogen hub-and-spoke: combines the two previous concepts.

Table 3. Hub substructures under consideration.

	Caisson Island	Sand Island	Platform	Gravity Based Structure
Water depth limit (m)	<25	<40	<45	>100
Construction time (years)	3–4	6–8	3–4	3–4
Size limit (GW)	6	>36	2	<6 (each WTs)
Maturity	Middle	Middle	High	Units – High/Linking – Middle
Footprint on seabed	High	High	Low	Middle

Moreover, the lifetime savings between CAPEX (capital expenditure) and OPEX (operational expenditures) for a 12 GW hub-and-spoke project (Denmark (2 GW), Germany (6 GW) and the Netherlands (4 GW)) could rise up to 2.5 billion €, without considering the P2X conversion, compared to a radial approach. This reduction is due to the lack of additional interconnection capacity between those countries. A study compared the LCOE between hub-and-spoke projects and radial approaches, concluding that the LCOE was able to be reduced for hub sizes between 6 and 12 GW, but limited for capacity hubs between 24 and 36 GW. Furthermore, electricity prices and emissions were also reduced. The total cost saving of a hub-and-spoke project compared to a no-hub project was then estimated to be between 15 and 20 billion € [165]. The main drawback of the hub-and-spoke project is that it would take more than 10 years of development and construction to become operational. Moreover, policies, regulatory framework and market design should be reconsidered to ensure a stable market. As the Paris Agreement must be fulfilled by 2050, these issues should be urgently reconsidered in order to carry out multiple hub-and-spoke projects by then [166].

3.3. Offshore Storage Options: Hydrogen and Compressed-Air Energy Storage

As electrical generation has to be immediately sold to supply the electrical consumption, and due to the stochastic nature of RES, energy storage emerges as an important solution for these sources [167]. However, as traditional energy storage technologies are difficult to use in a marine environment, new alternatives are being developed to store offshore energy. Wang et al. provide a comprehensive review on existing marine renewable energy storage solutions [168].

3.3.1. Hydrogen Energy Storage

The surplus of electricity produced by OWPP can be stored as hydrogen, and used later to generate power in fuel cells or as fuel in hydrogen vehicles [169]. Most alternatives available are based on the P2X technology, as previously described in Section 3.1.

An example of such alternatives can be found in the Deep Purple project which is based on the important CO_2 emissions from Norwegian oil and gas production. The project involves TechnipFMC, SINTEF, Subsea Valley and Maritim Forening Sogn og Fjordane, which develop the concept and new technology. It has received funding from the Research Council of Norway [170]. The Deep Purple project aims to convert electricity from OWPP to hydrogen and store such energy on the seabed. The hydrogen can then be used for several purposes [171,172]:

- Supply stable and renewable power to oil and gas installations
- Supply stable and renewable power to remote islands
- Provide a coastal hydrogen infrastructure to maritime sector
- Provide local production of power, hydrogen and oxygen to fist farming

It is expected to have a full-scale pilot by 2025 in Norway [173]. Figure 8 depicts an overview of the Deep Purple project.

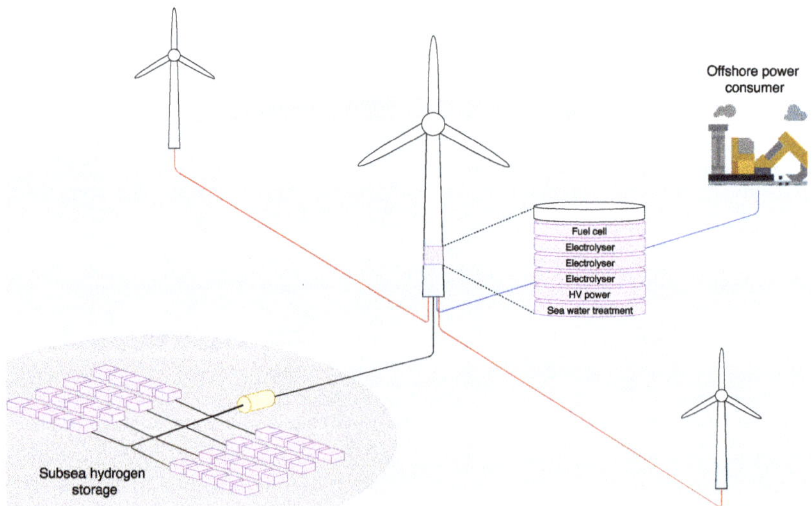

Figure 8. Deep Purple project.

3.3.2. Compressed Air Energy Storage

Compressed air energy storage (CAES) systems are a solution to energy storage based on the compression of air [174]. According to [175], the integration of CAES with wind and solar power generation can increase the RES share rate, as CAES is reported as less expensive than other storage systems, and to be large and powerful enough to store energy on a utility scale level [176]. Moreover, due to the high installation and capital cost of undersea transmission cables, offshore CAES can increase the cable's capacity factor, potentially lowering the average cost of offshore wind power while increasing the reliability and economic value of delivered power [177].

The TAKEOFF Business Incubator (University of Malta) has already patented the a storage technology called FLASC, with the aim of integrating large-scale energy storage into OWPPs. This solution is tailor-made for the offshore market, exploiting existing infrastructure and supply-chains, see Figure 9 [178]. FLASC uses compressed air for energy storage purposes, relying on the hydrostatic pressure of the deep-sea areas to maintain a stable pressure in the compressed air storage. As it uses existing infrastructure, it is considered a cost-effective solution. In [179,180], the multi-system integration and the working principle of the FLASC storage technology are described. This solution can also be used in order to: (*i*) Convert the intermittent RES supply into a stepped out one, simplifying their grid integration by allowing the TSO to schedule operations at specific time intervals, see Figure 10a; (*ii*) control the ramp rate of the generated power in case of sudden natural condition changes, see Figure 10b.

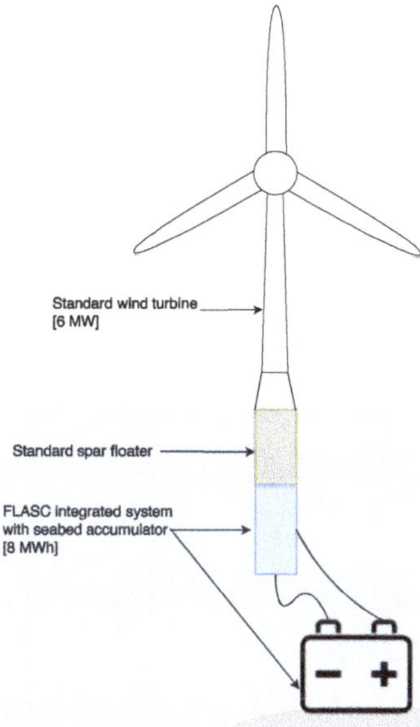

Figure 9. FLASC storage for offshore wind power plants (OWPPs).

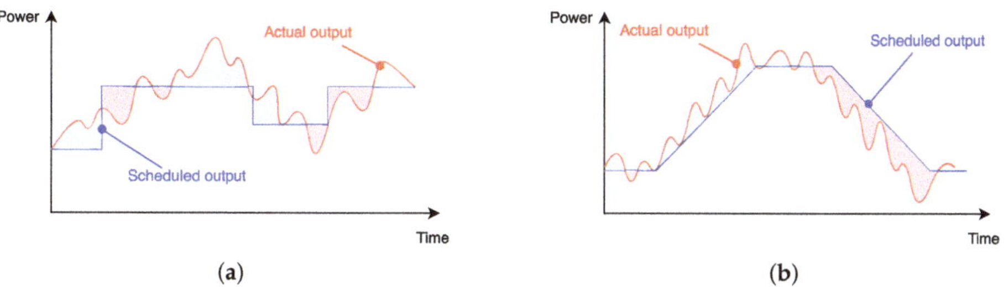

(**a**)   (**b**)

Figure 10. Further applications of FLASC storage technology. (**a**) Stepped out power control. (**b**) Ramp rate control.

A small-scale prototype was installed in Malta's Grand Harbor in May 2018. After one year, the testing campaign was completed. Results confirmed a consistently high thermal efficiency across a variety of meteorological conditions and operating regimes after hundreds of charging cycles. The prototype was removed and decommissioned, and nowadays the FLASC team is focused on developing a large-scale demonstrator in the open sea [181,182].

4. Conclusions

Future power scenarios include offshore wind energy as an important generation source. According to this framework, this paper discusses and reviews some aspects of offshore wind power plants for a massive integration into power systems. In the last decade, several characteristics such as offshore wind turbines, wind power plants, water depth and distance to shore have increased 230%, 700%, 170% and 110%, respectively. In the same way, electrical transmission has also evolved from HVAC to HVDC solutions. Moreover, HVDC technology currently offers three different possibilities: LCCs (based on thyristors), VSCs (based on IGBTs) and DRU (based on diodes). LCCs and VSCs have already been used, whereas DRU has not been implemented yet. The advantages and drawbacks of each technology have been extensively discussed in the paper. Different future advancements currently under development are also described: P2X conversion, the hub-and-spoke project as well as hydrogen and compressed air energy storage. The P2X conversion can enhance the power system's flexibility by converting the electricity surplus to other substances; however, its investment and maintenance costs should first be reduced to be economically viable. The hub-and-spoke project aims to facilitate the huge integration of offshore wind power plants in the North Sea; the total cost saving of this project, compared to a common offshore wind power plant, is estimated to be between 15 and 20 billion €. However, it is expected to take more than 10 years to become operational. Both hydrogen and compressed air energy storage systems appear as an alternative to conventional storage technologies due to the difficulty of using these traditional storage systems in the marine environment.

Author Contributions: Data curation, A.F.-G.; methodology and formal analysis, K.D. and N.A.C.; Resources A.F.-G.; Supervision, Á.M.-G.; Writing—original draft A.F.-G. and Á.M.-G.; and Writing—review and editing, K.D. and N.A.C.

Funding: This research received no external funding.

Acknowledgments: This work was supported by 'Ministerio de Educación, Cultura y Deporte' of Spain (grant numbers FPU16/04282 and EST18/00738).

Conflicts of Interest: The authors declare no conflict of interest.

Abbreviations

The following abbreviations are used in this manuscript:

AC	Alternating current
CAES	Compressed air energy storage
DC	Direct current
DFIG	Doubly fed induction generator
DRU	Diode rectifier unit
FSWTs	Fixed speed wind turbines
GHG	Greenhouse gasses
HVAC	High voltage alternating current
HVDC	High voltage direct current
IGBT	Insulated gate bipolar transistors
LCCs	Line-commutated converters
LCOE	Levelized cost of energy
OWPP	Offshore wind power plant
P2X	Power-to-X
PV	Photovoltaic

RES	Renewable energy sources
STATCOM	Static synchronous compensator
SVC	Static VAR compensator
TSO	Transmission system pperator
TYNDP	Ten Year Network Development Plan
VSCs	Voltage-source converters
VSWTs	Variable speed wind turbines
WPP	Wind power plants
WTs	Wind turbines

References

1. Wang, W.; Huang, X.; Tan, L.; Guo, J.; Liu, H. Optimization design of an inductive energy harvesting device for wireless power supply system overhead high-voltage power lines. *Energies* **2016**, *9*, 242. [CrossRef]
2. BP Energy Outlook. 2019 Edition. Available online: https://www.bp.com/content/dam/bp/business-sites/en/global/corporate/pdfs/energy-economics/energy-outlook/bp-energy-outlook-2019.pdf (accessed on 29 October 2019).
3. Clémençon, R. The two sides of the Paris climate agreement: Dismal failure or historic breakthrough? *J. Environ. Dev.* **2016**, *25*.[CrossRef]
4. Rogelj, J.; Den Elzen, M.; Höhne, N.; Fransen, T.; Fekete, H.; Winkler, H.; Schaeffer, R.; Sha, F.; Riahi, K.; Meinshausen, M. Paris Agreement climate proposals need a boost to keep warming well below 2 C. *Nature* **2016**, *534*, 631. [CrossRef] [PubMed]
5. Falkner, R. The Paris Agreement and the new logic of international climate politics. *Int. Aff.* **2016**, *92*, 1107–1125. [CrossRef]
6. Woo, J.; Choi, H.; Ahn, J. Well-to-wheel analysis of greenhouse gas emissions for electric vehicles based on electricity generation mix: A global perspective. *Transp. Res. Part D Transp. Environ.* **2017**, *51*, 340–350. [CrossRef]
7. Crippa, M.; Oreggioni, G.; Guizzardi, D.; Muntean, M.; Schaaf, E.; Lo Vullo, E.; Solazzo, E.; Monforti-Ferrario, F.; Olivier, J.G.J.; Vignati, E. *Fossil CO2 & GHG Emissions of All World Countries—2019 Report*; Publications Office of the European Union: Luxembourg, 2019.
8. Creutzig, F.; Agoston, P.; Goldschmidt, J.C.; Luderer, G.; Nemet, G.; Pietzcker, R.C. The underestimated potential of solar energy to mitigate climate change. *Nat. Energy* **2017**, *2*, 17140. [CrossRef]
9. Liddle, B.; Sadorsky, P. How much does increasing non-fossil fuels in electricity generation reduce carbon dioxide emissions? *Appl. Energy* **2017**, *197*, 212–221. [CrossRef]
10. Huber, M.; Dimkova, D.; Hamacher, T. Integration of wind and solar power in Europe: Assessment of flexibility requirements. *Energy* **2014**, *69*, 236–246, doi:10.1016/j.energy.2014.02.109. [CrossRef]
11. Pacesila, M.; Burcea, S.G.; Colesca, S.E. Analysis of renewable energies in European Union. *Renew. Sustain. Energy Rev.* **2016**, *56*, 156–170. [CrossRef]
12. Da Silva, R.C.; de Marchi Neto, I.; Seifert, S.S. Electricity supply security and the future role of renewable energy sources in Brazil. *Renew. Sustain. Energy Rev.* **2016**, *59*, 328–341. [CrossRef]
13. What is Energy Security? Available online: https://www.iea.org/topics/energysecurity/whatisenergysecurity/ (accessed on 24 September 2019).
14. Chalvatzis, K.J.; Ioannidis, A. Energy supply security in the EU: Benchmarking diversity and dependence of primary energy. *Appl. Energy* **2017**, *207*, 465–476. [CrossRef]
15. Leonavičius, V.; Genys, D.; Krikštolaitis, R. Public perception of energy security in Lithuania: Between material interest and energy independence. *J. Balt. Stud.* **2018**, *49*, 157–175. [CrossRef]
16. Wang, X.; Palazoglu, A.; El-Farra, N.H. Operational optimization and demand response of hybrid renewable energy systems. *Appl. Energy* **2015**, *143*, 324–335. [CrossRef]
17. Weitemeyer, S.; Kleinhans, D.; Vogt, T.; Agert, C. Integration of Renewable Energy Sources in future power systems: The role of storage. *Renew. Energy* **2015**, *75*, 14–20. [CrossRef]
18. Teng, F.; Mu, Y.; Jia, H.; Wu, J.; Zeng, P.; Strbac, G. Challenges on primary frequency control and potential solution from EVs in the future GB electricity system. *Appl. Energy* **2017**, *194*, 353–362. [CrossRef]

19. Zhang, W.; Fang, K. Controlling active power of wind farms to participate in load frequency control of power systems. *IET Gener. Transm. Distrib.* **2017**, *11*, 2194–2203. [CrossRef]

20. Fernandez-Guillamon, A.; Vigueras-Rodriguez, A.; Molina-Garcia, A. Analysis of power system inertia estimation in high wind power plant integration scenarios. *IET Renew. Power Gener.* **2019**.. [CrossRef]

21. Owusu, P.A.; Asumadu-Sarkodie, S. A review of renewable energy sources, sustainability issues and climate change mitigation. *Cogent Eng.* **2016**, *3*, 1167990. [CrossRef]

22. Toulabi, M.; Bahrami, S.; Ranjbar, A.M. An Input-to-State Stability Approach to Inertial Frequency Response Analysis of Doubly-Fed Induction Generator-Based Wind Turbines. *IEEE Trans. Energy Convers.* **2017**, *32*, 1418–1431. [CrossRef]

23. Fernández-Guillamón, A.; Villena-Lapaz, J.; Vigueras-Rodríguez, A.; García-Sánchez, T.; Molina-García, Á. An Adaptive Frequency Strategy for Variable Speed Wind Turbines: Application to High Wind Integration Into Power Systems. *Energies* **2018**, *11*, 1436. [CrossRef]

24. Delille, G.; Francois, B.; Malarange, G. Dynamic frequency control support by energy storage to reduce the impact of wind and solar generation on isolated power system's inertia. *IEEE Trans. Sustain. Energy* **2012**, *3*, 931–939. [CrossRef]

25. Sakamuri, J.N.; Altin, M.; Hansen, A.D.; Cutululis, N.A. Coordinated frequency control from offshore wind power plants connected to multi terminal DC system considering wind speed variation. *IET Renew. Power Gener.* **2016**, *11*, 1226–1236. [CrossRef]

26. Cardiel-Alvarez, M.A.; Rodriguez-Amenedo, J.L.; Arnaltes, S.; Montilla-DJesus, M.E. Modeling and control of LCCs rectifiers for offshore wind farms connected by HVDC links. *IEEE Trans. Energy Convers.* **2017**, *32*, 1284–1296. [CrossRef]

27. Dreidy, M.; Mokhlis, H.; Mekhilef, S. Inertia response and frequency control techniques for renewable energy sources: A review. *Renew. Sustain. Energy Rev.* **2017**, *69*, 144–155. [CrossRef]

28. Li, Y.; Xu, Z.; Østergaard, J.; Hill, D.J. Coordinated control strategies for offshore wind farm integration via VSCs-HVDC for system frequency support. *IEEE Trans. Energy Convers.* **2017**, *32*, 843–856. [CrossRef]

29. Rodriguez-Amenedo, J.L.; Arnaltes, S.; Cardiel-Alvarez, M.A.; Montilla-DJesus, M. Direct voltage and frequency control of an offshore wind farm connected through LCCs-HVDC link. In Proceedings of the 2017 19th European Conference on Power Electronics and Applications (EPE'17 ECCE Europe), Warsaw, Poland, 11–14 September 2017; pp. 1–10.

30. Fernández-Guillamón, A.; Gómez-Lázaro, E.; Muljadi, E.; Molina-García, Á. Power systems with high renewable energy sources: A review of inertia and frequency control strategies over time. *Renew. Sustain. Energy Rev.* **2019**, *115*, 109369. [CrossRef]

31. Deng, Y.Y.; Haigh, M.; Pouwels, W.; Ramaekers, L.; Brandsma, R.; Schimschar, S.; Grözinger, J.; de Jager, D. Quantifying a realistic, worldwide wind and solar electricity supply. *Glob. Environ. Chang.* **2015**, *31*, 239–252. [CrossRef]

32. Thomaidis, N.S.; Santos-Alamillos, F.J.; Pozo-Vázquez, D.; Usaola-García, J. Optimal management of wind and solar energy resources. *Comput. Oper. Res.* **2016**, *66*, 284–291. [CrossRef]

33. Barbosa, L.d.S.N.S.; Bogdanov, D.; Vainikka, P.; Breyer, C. Hydro, wind and solar power as a base for a 100% renewable energy supply for South and Central America. *PLoS ONE* **2017**, *12*, e0173820. [CrossRef]

34. Shivashankar, S.; Mekhilef, S.; Mokhlis, H.; Karimi, M. Mitigating methods of power fluctuation of photovoltaic (PV) sources—A review. *Renew. Sustain. Energy Rev.* **2016**, *59*, 1170–1184. [CrossRef]

35. Ren, G.; Liu, J.; Wan, J.; Guo, Y.; Yu, D. Overview of wind power intermittency: Impacts, measurements, and mitigation solutions. *Appl. Energy* **2017**, *204*, 47–65. [CrossRef]

36. Cardozo, C.; van Ackooij, W.; Capely, L. Cutting plane approaches for frequency constrained economic dispatch problems. *Electr. Power Syst. Res.* **2018**, *156*, 54–63. [CrossRef]

37. Snow, M.; Snow, R. The Ascent of Wind Power in Nova Scotia: Pubnico Point Wind Farm. *Int. J. Renew. Energy Sour.* **2018**, *3*, 6–9.

38. GLOBAL WIND REPORT 2018. Available online: https://gwec.net/wp-content/uploads/2019/04/GWEC-Global-Wind-Report-2018.pdf (accessed on 7 October 2019).

39. Esteban, M.D.; Diez, J.J.; López, J.S.; Negro, V. Why offshore wind energy? *Renew. Energy* **2011**, *36*, 444–450. [CrossRef]

40. Gatzert, N.; Kosub, T. Risks and risk management of renewable energy projects: The case of onshore and offshore wind parks. *Renew. Sustain. Energy Rev.* **2016**, *60*, 982–998. [CrossRef]

41. Wind Europe. *Wind in Power: 2016 European Statistics*; Wind Europe: Brussels, Belgium, 2017.

42. Chaudhary, S.K.; Teodorescu, R.; Rodriguez, P. Wind farm grid integration using vsc based hvdc transmission-an overview. In Proceedings of the 2008 IEEE Energy 2030 Conference, Atlanta, GA, USA, 17–18 November 2008; pp. 1–7.

43. Wind Europe. *Wind Energy in Europe: Scenarios for 2030*; Wind Europe: Brussels, Belgium, 2017.

44. Morgan, E.C.; Lackner, M.; Vogel, R.M.; Baise, L.G. Probability distributions for offshore wind speeds. *Energy Convers. Manag.* **2011**, *52*, 15–26. [CrossRef]

45. Pichugina, Y.L.; Banta, R.M.; Brewer, W.A.; Sandberg, S.P.; Hardesty, R.M. Doppler lidar–based wind-profile measurement system for offshore wind-energy and other marine boundary layer applications. *J. Appl. Meteorol. Climatol.* **2012**, *51*, 327–349. [CrossRef]

46. Badger, M.; Badger, J.; Nielsen, M.; Hasager, C.B.; Peña, A. Wind class sampling of satellite SAR imagery for offshore wind resource mapping. *J. Appl. Meteorol. Climatol.* **2010**, *49*, 2474–2491. [CrossRef]

47. Pérez-Collazo, C.; Greaves, D.; Iglesias, G. A review of combined wave and offshore wind energy. *Renew. Sustain. Energy Rev.* **2015**, *42*, 141–153. [CrossRef]

48. AWEA. U.S. Offshore Wind Industry. STATUS UPDATE—OCTOBER 2019. Available online: https:// www.awea.org/Awea/media/Resources/Fact%20Sheets/Offshore-Fact-Sheet-Oct-2019.pdf (accessed on 30 October 2019).

49. Dai, J.; Yang, X.; Wen, L. Development of wind power industry in China: A comprehensive assessment. *Renew. Sustain. Energy Rev.* **2018**, *97*, 156–164. [CrossRef]

50. Nagababu, G.; Kachhwaha, S.S.; Savsani, V. Estimation of technical and economic potential of offshore wind along the coast of India. *Energy* **2017**, *138*, 79–91. [CrossRef]

51. Kalmikov, A. Wind power fundamentals. In *Wind Energy Engineering*; Elsevier: Amsterdam, The Netherlands, 2017; pp. 17–24.

52. Kaldellis, J.; Kapsali, M. Shifting towards offshore wind energy—Recent activity and future development. *Energy Policy* **2013**, *53*, 136–148. [CrossRef]

53. Morthorst, P.E.; Kitzing, L. Economics of building and operating offshore wind farms. In *Offshore Wind Farms*; Elsevier: Amsterdam, The Netherlands, 2016; pp. 9–27.

54. GWEC. Global Wind Energy Outlook. 2016. Available online: http://files.gwec.net/register/success/?file= /files/GlobalWindEnergyOutlook2016 (accessed on 11 September 2019).

55. IRENA. Renewable Power Generation Costs in 2018. 2019. Available online: https://www.irena.org/-/ media/Files/IRENA/Agency/Publication/2019/May/IRENA_Renewable-Power-Generations-Costs-in- 2018.pdf (accessed on 28 October 2019).

56. Kumar, D.; Chatterjee, K. A review of conventional and advanced MPPT algorithms for wind energy systems. *Renew. Sustain. Energy Rev.* **2016**, *55*, 957–970. [CrossRef]

57. Naidu, N.S.; Singh, B. Grid-interfaced DFIG-based variable speed wind energy conversion system with power smoothening. *IEEE Trans. Sustain. Energy* **2016**, *8*, 51–58. [CrossRef]

58. Njiri, J.G.; Soeffker, D. State-of-the-art in wind turbine control: Trends and challenges. *Renew. Sustain. Energy Rev.* **2016**, *60*, 377–393. [CrossRef]

59. Bhukya, J.; Mahajan, V. The controlling of the DFIG based on variable speed wind turbine modeling and simulation. In Proceedings of the IEEE 6th International Conference on Power Systems (ICPS), New Delhi, India, 4–6 March 2016; pp. 1–6.

60. Yan, R.; Saha, T.K.; Modi, N.; Masood, N.A.; Mosadeghy, M. The combined effects of high penetration of wind and PV on power system frequency response. *Appl. Energy* **2015**, *145*, 320–330. [CrossRef]

61. Liang, X. Emerging power quality challenges due to integration of renewable energy sources. *IEEE Trans. Ind. Appl.* **2016**, *53*, 855–866. [CrossRef]

62. Edrah, M.; Lo, K.L.; Anaya-Lara, O. Impacts of high penetration of DFIG wind turbines on rotor angle stability of power systems. *IEEE Trans. Sustain. Energy* **2015**, *6*, 759–766. [CrossRef]

63. Syahputra, R.; Soesanti, I. DFIG Control Scheme of Wind Power Using ANFIS Method in Electrical Power Grid System. *Int. J. Appl. Eng. Res.* **2016**, *11*, 5256–5262.

64. Ma, Y.; Cao, W.; Yang, L.; Wang, F.F.; Tolbert, L.M. Virtual synchronous generator control of full converter wind turbines with short-term energy storage. *IEEE Trans. Ind. Electron.* **2017**, *64*, 8821–8831. [CrossRef]

65. Artigao, E.; Honrubia-Escribano, A.; Gomez-Lazaro, E. Current signature analysis to monitor DFIG wind turbine generators: A case study. *Renew. Energy* **2018**, *116*, 5–14. [CrossRef]

66. Liserre, M.; Cardenas, R.; Molinas, M.; Rodriguez, J. Overview of multi-MW wind turbines and wind parks. *IEEE Trans. Ind. Electron.* **2011**, *58*, 1081–1095. [CrossRef]

67. Blasco-Gimenez, R.; Añó-Villalba, S.; Rodríguez-D'Derlée, J.; Bernal-Pérez, S.; Morant, F. Connection of Off-Shore Wind Farms Using Diode Based HVDC Links. In *Wind Energy Conversion Systems*; Springer: Cham, Switzerland, 2012; pp. 431–464.

68. Nanou, S.I.; Papathanassiou, S.A. Grid code compatibility of VSCs-HVDC connected offshore wind turbines employing power synchronization control. *IEEE Trans. Power Syst.* **2016**, *31*, 5042–5050. [CrossRef]

69. Ruddy, J.; Meere, R.; O'Donnell, T. Low Frequency AC transmission for offshore wind power: A review. *Renew. Sustain. Energy Rev.* **2016**, *56*, 75–86. [CrossRef]

70. Jallad, J.; Mekhilef, S.; Mokhlis, H. Frequency regulation strategies in grid integrated offshore wind turbines via VSCs-HVDC technology: A review. *Energies* **2017**, *10*, 1244. [CrossRef]

71. Meere, R.; Ruddy, J.; McNamara, P.; O'Donnell, T. Variable AC transmission frequencies for offshore wind farm interconnection. *Renew. Energy* **2017**, *103*, 321–332. [CrossRef]

72. Sivalingam, K.; Sepulveda, M.; Spring, M.; Davies, P. A Review and Methodology Development for Remaining Useful Life Prediction of Offshore Fixed and Floating Wind turbine Power Converter with Digital Twin Technology Perspective. In Proceedings of the 2nd International Conference on Green Energy and Applications (ICGEA), Singapore, 24–26 March 2018; pp. 197–204.

73. Justo, J.J.; Mwasilu, F.; Jung, J.W. Doubly-fed induction generator based wind turbines: A comprehensive review of fault ride-through strategies. *Renew. Sustain. Energy Rev.* **2015**, *45*, 447–467. [CrossRef]

74. Teng, W.; Ding, X.; Zhang, X.; Liu, Y.; Ma, Z. Multi-fault detection and failure analysis of wind turbine gearbox using complex wavelet transform. *Renew. Energy* **2016**, *93*, 591–598. [CrossRef]

75. Wang, L.; Zhang, Z.; Long, H.; Xu, J.; Liu, R. Wind turbine gearbox failure identification with deep neural networks. *IEEE Trans. Ind. Inform.* **2016**, *13*, 1360–1368. [CrossRef]

76. Yang, B.; Jiang, L.; Wang, L.; Yao, W.; Wu, Q. Nonlinear maximum power point tracking control and modal analysis of DFIG based wind turbine. *Int. J. Electr. Power Energy Syst.* **2016**, *74*, 429–436. [CrossRef]

77. Serrano-González, J.; Lacal-Arántegui, R. Technological evolution of onshore wind turbines—A market-based analysis. *Wind Energy* **2016**, *19*, 2171–2187. [CrossRef]

78. Chen, Z.; Guerrero, J.M.; Blaabjerg, F. A review of the state of the art of power electronics for wind turbines. *IEEE Trans. Power Electron.* **2009**, *24*, 1859–1875. [CrossRef]

79. Blaabjerg, F.; Liserre, M.; Ma, K. Power electronics converters for wind turbine systems. *IEEE Trans. Ind. Appl.* **2011**, *48*, 708–719. [CrossRef]

80. Blaabjerg, F.; Ma, K. Future on power electronics for wind turbine systems. *IEEE J. Emerg. Sel. Top. Power Electron.* **2013**, *1*, 139–152. [CrossRef]

81. Wei, Q.; Wu, B.; Xu, D.; Zargari, N.R. A medium-frequency transformer-based wind energy conversion system used for current-source converter-based offshore wind farm. *IEEE Trans. Power Electron.* **2016**, *32*, 248–259. [CrossRef]

82. Hansen, A.D.; Iov, F.; Blaabjerg, F.; Hansen, L.H. Review of contemporary wind turbine concepts and their market penetration. *Wind Eng.* **2004**, *28*, 247–263. [CrossRef]

83. Wind Europe. *Annual Offshore Statistics in 2017*; Wind Europe: Brussels, Belgium, 2018.

84. Wind Europe. *Annual Offshore Statistics in 2018*; Wind Europe: Brussels, Belgium, 2019.

85. Windturbines database. Available online: https://en.wind-turbine-models.com/turbines (accessed on 30 September 2019).

86. Wind Europe. *European Offshore Statistics in 2013*; Wind Europe: Brussels, Belgium, 2014.

87. Equinor—the world's leading floating offshore wind developer. Available online: https://www.equinor.com/en/what-we-do/hywind-where-the-wind-takes-us.html (accessed on 3 September 2019).

88. Van Eeckhout, B.; Van Hertem, D.; Reza, M.; Srivastava, K.; Belmans, R. Economic comparison of VSCs HVDC and HVAC as transmission system for a 300 MW offshore wind farm. *Eur. Trans. Electr. Power* **2010**, *20*, 661–671. [CrossRef]

89. Bozhko, S.V.; Blasco-Gimenez, R.; Li, R.; Clare, J.C.; Asher, G.M. Control of offshore DFIG-based wind farm grid with line-commutated HVDC connection. *IEEE Trans. Energy Convers.* **2007**, *22*, 71–78. [CrossRef]

90. Perveen, R.; Kishor, N.; Mohanty, S.R. Off-shore wind farm development: Present status and challenges. *Renew. Sustain. Energy Rev.* **2014**, *29*, 780–792. [CrossRef]

91. Li, J.; Zhang, M.; Zhu, J.; Yang, Q.; Zhang, Z.; Yuan, W. Analysis of superconducting magnetic energy storage used in a submarine HVAC cable based offshore wind system. *Energy Procedia* **2015**, *75*, 691–696. [CrossRef]

92. Mitra, P.; Zhang, L.; Harnefors, L. Offshore wind integration to a weak grid by VSCs-HVDC links using power-synchronization control: A case study. *IEEE Trans. Power Deliv.* **2013**, *29*, 453–461. [CrossRef]

93. Reed, G.F.; Al Hassan, H.A.; Korytowski, M.J.; Lewis, P.T.; Grainger, B.M. Comparison of HVAC and HVDC solutions for offshore wind farms with a procedure for system economic evaluation. In Proceedings of the 2013 IEEE Energytech, Cleveland, OH, USA, 21–23 May 2013; pp. 1–7.

94. Guidi, G.; Fosso, O. Investment cost of HVAC cable reactive power compensation off-shore. In Proceedings of the 2012 IEEE International Energy Conference and Exhibition (ENERGYCON), Florence, Italy, 9–12 September 2012; pp. 299–304.

95. Hur, D. Economic considerations underlying the adoption of HVDC and HVAC for the connection of an offshore wind farm in Korea. *J. Electr. Eng. Technol.* **2012**, *7*, 157–162. [CrossRef]

96. Sharma, R.; Rasmussen, T.W.; Jensen, K.H.; Akamatov, V. Modular VSCs converter based HVDC power transmission from offshore wind power plant: Compared to the conventional HVAC system. In Proceedings of the 2010 IEEE Electrical Power & Energy Conference, Halifax, NS, Canada, 25–27 August 2010; pp. 1–6.

97. Chen, H.; Johnson, M.H.; Aliprantis, D.C. Low-frequency AC transmission for offshore wind power. *IEEE Trans. Power Deliv.* **2013**, *28*, 2236–2244. [CrossRef]

98. Negra, N.B.; Todorovic, J.; Ackermann, T. Loss evaluation of HVAC and HVDC transmission solutions for large offshore wind farms. *Electr. Power Syst. Res.* **2006**, *76*, 916–927. [CrossRef]

99. Stoutenburg, E.; Jacobson, M. Optimizing offshore transmission links for marine renewable energy farms. In Proceedings of the OCEANS 2010 MTS/IEEE SEATTLE, Seattle, WA, USA, 20–23 September 2010; pp. 1–9.

100. Bresesti, P.; Kling, W.L.; Hendriks, R.L.; Vailati, R. HVDC connection of offshore wind farms to the transmission system. *IEEE Trans. Energy Convers.* **2007**, *22*, 37–43. [CrossRef]

101. Gomis-Bellmunt, O.; Liang, J.; Ekanayake, J.; King, R.; Jenkins, N. Topologies of multiterminal HVDC-VSCs transmission for large offshore wind farms. *Electr. Power Syst. Res.* **2011**, *81*, 271–281. [CrossRef]

102. Legorburu, I.; Johnson, K.R.; Kerr, S.A. Multi-use maritime platforms-North Sea oil and offshore wind: Opportunity and risk. *Ocean Coast. Manag.* **2018**, *160*, 75–85. [CrossRef]

103. Zhang, Y.; Ravishankar, J.; Fletcher, J.; Li, R.; Han, M. Review of modular multilevel converter based multi-terminal HVDC systems for offshore wind power transmission. *Renew. Sustain. Energy Rev.* **2016**, *61*, 572–586. [CrossRef]

104. De Alegría, I.M.; Martín, J.L.; Kortabarria, I.; Andreu, J.; Ereño, P.I. Transmission alternatives for offshore electrical power. *Renew. Sustain. Energy Rev.* **2009**, *13*, 1027–1038. [CrossRef]

105. Rourke, F.O.; Boyle, F.; Reynolds, A. Marine current energy devices: Current status and possible future applications in Ireland. *Renew. Sustain. Energy Rev.* **2010**, *14*, 1026–1036. [CrossRef]

106. Chou, C.J.; Wu, Y.K.; Han, G.Y.; Lee, C.Y. Comparative evaluation of the HVDC and HVAC links integrated in a large offshore wind farm—An actual case study in Taiwan. *IEEE Trans. Ind. Appl.* **2012**, *48*, 1639–1648. [CrossRef]

107. Colmenar-Santos, A.; Perera-Perez, J.; Borge-Diez, D.; dePalacio Rodríguez, C. Offshore wind energy: A review of the current status, challenges and future development in Spain. *Renew. Sustain. Energy Rev.* **2016**, *64*, 1–18. [CrossRef]

108. Erlich, I.; Shewarega, F.; Feltes, C.; Koch, F.W.; Fortmann, J. Offshore wind power generation technologies. *Proc. IEEE* **2013**, *101*, 891–905. [CrossRef]

109. Kucuksari, S.; Erdogan, N.; Cali, U. Impact of Electrical Topology, Capacity Factor and Line Length on Economic Performance of Offshore Wind Investments. *Energies* **2019**, *12*, 3191. [CrossRef]

110. Sellick, R.; Åkerberg, M. Comparison of HVDC Light (VSCs) and HVDC Classic (LCCs) site aspects, for a 500MW 400kV HVDC transmission scheme. In Proceedings of the 10th IET International Conference on AC and DC Power Transmission (ACDC 2012), Birmingham, UK, 4–5 December 2012.

111. Saborío-Romano, O.; Bidadfar, A.; Göksu, Ö.; Zeni, L.; Cutululis, N.A. Power Oscillation Damping from Offshore Wind Farms Connected to HVDC via Diode Rectifiers. *Energies* **2019**, *12*, 3387. [CrossRef]

112. Sau-Bassols, J.; Prieto-Araujo, E.; Galceran-Arellano, S.; Gomis-Bellmunt, O. Operation and control of a Current Source Converter series tapping of an LCCs-HVDC link for integration of Offshore Wind Power Plants. *Electr. Power Syst. Res.* **2016**, *141*, 510–521. [CrossRef]

113. ENTSO-E. STATISTICAL FACTSHEET 2018. 2019. Available online: https://docstore.entsoe.eu/Documents/Publications/Statistics/Factsheet/entsoe_sfs2018_web.pdf (accessed on 26 August 2019).

114. Hart, D.W. *Power Electronics*; Tata McGraw-Hill Education: New York, NY, USA, 2011.

115. Oni, O.E.; Davidson, I.E.; Mbangula, K.N. A review of LCCs-HVDC and VSCs-HVDC technologies and applications. In Proceedings of the 2016 IEEE 16th International Conference on Environment and Electrical Engineering (EEEIC), Florence, Italy, 6–8 June 2016; pp. 1–7.

116. Barnes, M.; Van Hertem, D.; Teeuwsen, S.P.; Callavik, M. HVDC systems in smart grids. *Proc. IEEE* **2017**, *105*, 2082–2098. [CrossRef]

117. Adapa, R. High-wire act: Hvdc technology: The state of the art. *IEEE Power Energy Mag.* **2012**, *10*, 18–29. [CrossRef]

118. Liu, H.; Sun, J. Small-signal stability analysis of offshore wind farms with LCCs HVDC. In Proceedings of the 2013 IEEE Grenoble Conference, Grenoble, France, 16–20 June 2013; pp. 1–8.

119. Liu, H.; Sun, J. Voltage stability and control of offshore wind farms with AC collection and HVDC transmission. *IEEE J. Emerg. Sel. Top. Power Electron.* **2014**, *2*, 1181–1189.

120. Hamlaoui, H.; Francois, B. Interest of storage based STATCOM systems to the power quality enhancement of thyristors based LCCs HVDC links for offshore wind farm. In Proceedings of the 2018 IEEE International Conference on Industrial Technology (ICIT), Lyon, France, 20–22 February 2018; pp. 1702–1707.

121. Blasco-Gimenez, R.; Aparicio, N.; Ano-Villalba, S.; Bernal-Perez, S. LCCs-HVDC connection of offshore wind farms with reduced filter banks. *IEEE Trans. Ind. Electron.* **2012**, *60*, 2372–2380. [CrossRef]

122. Callavik, M.; Bahrman, M.; Sandeberg, P. Technology developments and plans to solve operational challenges facilitating the HVDC offshore grid. In Proceedings of the 2012 IEEE Power and Energy Society General Meeting, San Diego, CA, USA, 22–26 July 2012; pp. 1–6.

123. Henry, S.; Denis, A.; Panciatici, P. Feasibility study of off-shore HVDC grids. In Proceedings of the IEEE PES General Meeting, Minneapolis, MN, USA, 25–29 July 2010; pp. 1–5.

124. Glasdam, J.; Hjerrild, J.; Kocewiak, Ł.H.; Bak, C.L. Review on multi-level voltage source converter based HVDC technologies for grid connection of large offshore wind farms. In Proceedings of the 2012 IEEE International Conference on Power System Technology (POWERCON), Auckland, New Zealand, 23–26 October 2012; pp. 1–6.

125. Torres-Olguin, R.E.; Molinas, M.; Undeland, T. Offshore wind farm grid integration by VSCs technology with LCCs-based HVDC transmission. *IEEE Trans. Sustain. Energy* **2012**, *3*, 899–907. [CrossRef]

126. Prieto-Araujo, E.; Bianchi, F.D.; Junyent-Ferre, A.; Gomis-Bellmunt, O. Methodology for droop control dynamic analysis of multiterminal VSCs-HVDC grids for offshore wind farms. *IEEE Trans. Power Deliv.* **2011**, *26*, 2476–2485. [CrossRef]

127. Seman, S.; Trinh, N.T.; Zurowski, R.; Kreplin, S. Modeling of the diode-rectifier based HVDC transmission solution for large offshore wind power plants grid access. In Proceedings of the International Workshop on Large-Scale Integration of Wind Power into Power Systems, Vienna, Austria, 15–17 November 2016.

128. Hoffmann, M.; Rathke, C.; Menze, A.; Hemdan, N.G.; Kurrat, M. AC Fault Analysis of DRU-VSCs Hybrid HVDCTopology for Offshore Wind Farm Integration. In Proceedings of the VDE/IEEE Power and Energy Student Summit, Kaiserslautern, Germany, 2–4 July 2018.

129. Ramachandran, R.; Poullain, S.; Benchaib, A.; Bacha, S.; Francois, B. AC Grid Forming by Coordinated Control of Offshore Wind Farm connected to Diode Rectifier based HVDC Link-Review and Assessment of Solutions. In Proceedings of the 2018 20th European Conference on Power Electronics and Applications (EPE'18 ECCE Europe), Riga, Latvia, 17–21 September 2018; pp. 1–10.

130. Prignitz, C.; Eckel, H.G.; Achenbach, S.; Augsburger, F.; Schön, A. FixReF: A control strategy for offshore wind farms with different wind turbine types and diode rectifier HVDC transmission. In Proceedings of the 2016 IEEE 7th International Symposium on Power Electronics for Distributed Generation Systems (PEDG), Vancouver, BC, Canada, 27–30 June 2016; pp. 1–7.

131. Li, R.; Yu, L.; Xu, L. Offshore AC Fault Protection of Diode Rectifier Unit-Based HVdc System for Wind Energy Transmission. *IEEE Trans. Ind. Electron.* **2018**, *66*, 5289–5299. [CrossRef]

132. Li, R.; Yu, L.; Xu, L. Operation of offshore wind farms connected with DRU-HVDC transmission systems with special consideration of faults. *Glob. Energy Interconnect.* **2018**, *1*, 608–617.

133. Saborío-Romano, O.; Bidadfar, A.; Sakamuri, J.N.; Göksu, Ö.; Cutululis, N.A. Primary Frequency Response from Offshore Wind Farms Connected to HVdc via Diode Rectifiers. In Proceedings of the 2019 IEEE Milan PowerTech, Milan, Italy, 23–27 June 2019; pp. 1–6.

134. Cardiel-Alvarez, M.A.; Arnaltes, S.; Rodriguez-Amenedo, J.L.; Nami, A. Decentralized control of offshore wind farms connected to diode-based HVDC links. *IEEE Trans. Energy Convers.* **2018**, *33*, 1233–1241. [CrossRef]

135. Chang, Y.; Cai, X. Hybrid Topology of a Diode-Rectifier-based HVDC System for Offshore Wind Farms. *IEEE J. Emerg. Sel. Top. Power Electron.* **2018**, *7*, 2116–2128. [CrossRef]

136. TYNDP 2018. Available online: https://tyndp.entsoe.eu/ (accessed on 20 September 2019).

137. Koivisto, M.; Gea-Bermudez, J. *NSON-DK Energy System Scenarios—Edition 2*; DTU Wind Energy: Roskilde, Denmark, 2018.

138. Greenpeace; GWEC. Energy [r]evolution. 2015. Available online: https://storage.googleapis.com/planet4-philippines-stateless/2019/05/09ccb4ca-energy-revolution-asean.pdf (accessed on 21 September 2019).

139. U.S. Department of Energy. WindVision: A New Era for Wind Power in the United States. Available online: https://www.energy.gov/sites/prod/files/2015/03/f20/wv_executive_summary_overview_and_key_chapter_findings.pdf (accessed on 29 October 2019).

140. TERI; WWF. The Energy Report—India 100% Renewable Energy by 2050. 2013. Available online: http://awsassets.wwfindia.org/downloads/the_energy_report_india.pdf (accessed on 29 October 2019).

141. IEA; ERI. Technology Roadmap: China Wind Energy Development Roadmap 2050. 2011. Available online: https://www.china.tu-berlin.de/fileadmin/fg57/SS_2012/Umwelt/IEAchina_wind_2050.pdf (accessed on 29 October 2019).

142. China National Renewable Energy Centre. China Wind, Solar and Bioenergy Roadmap 2050. 2014. Available online: https://fch.cl/wp-content/uploads/2015/06/6.4-China-Wind-Solar-and-Bioenergy-Roadmap-2050-Short-Version.pdf (accessed on 29th October 2019).

143. Varela-Vázquez, P.; del Carmen Sánchez-Carreira, M.; Rodil-Marzábal, Ó. A novel systemic approach for analysing offshore wind energy implementation. *J. Clean. Prod.* **2019**, *212*, 1310–1318. [CrossRef]

144. De Vasconcelos, B.R.; Lavoie, J.M. Recent advances in Power-to-X technology for the production of fuels and chemicals. *Front. Chem.* **2019**, *7*, 392. [CrossRef]

145. Sternberg, A.; Bardow, A. Power-to-What?–Environmental assessment of energy storage systems. *Energy Environ. Sci.* **2015**, *8*, 389–400. [CrossRef]

146. Blanco, H.; Faaij, A. A review at the role of storage in energy systems with a focus on Power to Gas and long-term storage. *Renew. Sustain. Energy Rev.* **2018**, *81*, 1049–1086. [CrossRef]

147. Lewandowska-Bernat, A.; Desideri, U. Opportunities of power-to-gas technology in different energy systems architectures. *Appl. Energy* **2018**, *228*, 57–67. [CrossRef]

148. Koj, J.C.; Wulf, C.; Zapp, P. Environmental impacts of power-to-X systems-A review of technological and methodological choices in Life Cycle Assessments. *Renew. Sustain. Energy Rev.* **2019**, *112*, 865–879. [CrossRef]

149. Schnuelle, C.; Thoeming, J.; Wassermann, T.; Thier, P.; von Gleich, A.; Goessling-Reisemann, S. Socio-technical-economic assessment of power-to-X: Potentials and limitations for an integration into the German energy system. *Energy Res. Soc. Sci.* **2019**, *51*, 187–197. [CrossRef]

150. Nastasi, B.; Basso, G.L. Power-to-gas integration in the transition towards future urban energy systems. *Int. J. Hydrogen Energy* **2017**, *42*, 23933–23951. [CrossRef]

151. Bloess, A.; Schill, W.P.; Zerrahn, A. Power-to-heat for renewable energy integration: A review of technologies, modeling approaches, and flexibility potentials. *Appl. Energy* **2018**, *212*, 1611–1626. [CrossRef]

152. Wulf, C.; Linßen, J.; Zapp, P. Review of power-to-gas projects in Europe. *Energy Procedia* **2018**, *155*, 367–378. [CrossRef]

153. Hydrogen and Liquid Electricity Could Become Denmark's next Business Adventure. Available online: https://investindk.com/insights/hydrogen-and-liquid-electricity-could-become-denmarks-next-business-adventure (accessed on 2 October 2019).

154. Pilpola, S.; Lund, P.D. Different flexibility options for better system integration of wind power. *Energy Strategy Rev.* **2019**, *26*, 100368. [CrossRef]

155. Pursiheimo, E.; Holttinen, H.; Koljonen, T. Path toward 100% renewable energy future and feasibility of power-to-gas technology in Nordic countries. *IET Renew. Power Gener.* **2017**, *11*, 1695–1706. [CrossRef]

156. Hou, P.; Enevoldsen, P.; Eichman, J.; Hu, W.; Jacobson, M.Z.; Chen, Z. Optimizing investments in coupled offshore wind-electrolytic hydrogen storage systems in Denmark. *J. Power Sour.* **2017**, *359*, 186–197. [CrossRef]

157. Hosseini, S.E.; Wahid, M.A. Hydrogen production from renewable and sustainable energy resources: Promising green energy carrier for clean development. *Renew. Sustain. Energy Rev.* **2016**, *57*, 850–866. [CrossRef]

158. Loisel, R.; Baranger, L.; Chemouri, N.; Spinu, S.; Pardo, S. Economic evaluation of hybrid off-shore wind power and hydrogen storage system. *Int. J. Hydrogen Energy* **2015**, *40*, 6727–6739. [CrossRef]

159. Schuster, M.; Walther, T. Valuation of combined wind power plant and hydrogen storage: A decision tree approach. In Proceedings of the 2017 14th International Conference on the European Energy Market (EEM), Dresden, Germany, 6–9 June 2017; pp. 1–6.

160. North Sea Wind Power Hub. Available online: https://northseawindpowerhub.eu/ (accessed on 30 August 2019).

161. North Sea Wind Power Hub. Available online: https://www.tennet.eu/our-key-tasks/innovations/north-sea-wind-power-hub/ (accessed on 30 August 2019).

162. North Sea Wind Power Hub. 1. The Challenge. Urgent Action Is Needed to Reach a Low-Carbon Society in Time. Available online: https://northseawindpowerhub.eu/wp-content/uploads/2019/07/Concept_Paper_1-TheChallenge.pdf (accessed on 30 August 2019).

163. North Sea Wind Power Hub. 2. The Vision. The Hub-and-Spoke Concept as Modular Infrastructure Block to Scale Up Fast. Available online: https://northseawindpowerhub.eu/wp-content/uploads/2019/07/Concept_Paper_2_The-Vision.pdf (accessed on 30 August 2019).

164. North Sea Wind Power Hub. 3. Modular Hub-and-Spoke: Specific Solution Options. Case Studies Have Demonstrated Technical Feasibility. Available online: https://northseawindpowerhub.eu/wp-content/uploads/2019/07/Concept_Paper_3-Specific-solution-options.pdf (accessed on 30 August 2019).

165. North Sea Wind Power Hub. 4. The Benefits. The Modular Hub-and-Spoke Concept Has Substantial Societal Benefits and thus the Potential to Incentivise All Involved Stakeholders. Available online: https://northseawindpowerhub.eu/wp-content/uploads/2019/07/Concept_Paper_4-The-benefits.pdf (accessed on 30 August 2019).

166. North Sea Wind Power Hub. 5. Requirements to Develop. Ensure Industry Can Progress. Available online: https://northseawindpowerhub.eu/wp-content/uploads/2019/07/Concept_Paper_5-Requirements-to-develop.pdf (accessed on 30 August 2019).

167. Parra, D.; Norman, S.A.; Walker, G.S.; Gillott, M. Optimum community energy storage for renewable energy and demand load management. *Appl. Energy* **2017**, *200*, 358–369. [CrossRef]

168. Wang, Z.; Carriveau, R.; Ting, D.S.K.; Xiong, W.; Wang, Z. A review of marine renewable energy storage. *Int. J. Energy Res.* **2019**, *43*, 6108–6150. [CrossRef]

169. Shatnawi, M.; Al Qaydi, N.; Aljaberi, N.; Aljaberi, M. Hydrogen-Based Energy Storage Systems: A Review. In Proceedings of the 2018 7th International Conference on Renewable Energy Research and Applications (ICRERA), Paris, France, 14–17 October 2018; pp. 697–700.

170. Subsea Energy Storage. Available online: https://www.gceocean.no/news/posts/2018/september/subsea-energy-storage/ (accessed on 5th September 2019).

171. HYON Joins Offshore Wind-to-Hydrogen Project. Available online: https://www.offshorewind.biz/2019/06/07/hyon-joins-offshore-wind-to-hydrogen-project/ (accessed on 3rd September 2019).

172. Deep Purple. Available online: https://energyvalley.no/wp-content/uploads/2019/04/Deep-Purple-.pdf (accessed on 3rd September 2019).

173. 'Deep Purple' Seabed Hydrogen Storage for Offshore Wind Plan. Available online: https://www.rechargenews.com/wind/1802829/deep-purple-seabed-hydrogen-storage-for-offshore-wind-plan (accessed on 3rd September 2019).

174. Katsaprakakis, D. Energy storage for offshore wind farms. In *Offshore Wind Farms*; Elsevier: Amsterdam, The Netherlands, 2016; pp. 459–493.

175. Venkataramani, G.; Parankusam, P.; Ramalingam, V.; Wang, J. A review on compressed air energy storage—A pathway for smart grid and polygeneration. *Renew. Sustain. Energy Rev.* **2016**, *62*, 895–907. [CrossRef]

176. Bouman, E.A.; Øberg, M.M.; Hertwich, E.G. Environmental impacts of balancing offshore wind power with compressed air energy storage (CAES). *Energy* **2016**, *95*, 91–98. [CrossRef]

177. Li, B.; DeCarolis, J.F. A techno-economic assessment of offshore wind coupled to offshore compressed air energy storage. *Appl. Energy* **2015**, *155*, 315–322. [CrossRef]
178. FLASC. Available online: https://www.offshoreenergystorage.com/ (accessed on 4 September 2019).
179. FLASC. Available online: https://www.offshoreenergystorage.com/multi-system-integration/ (accessed on 4 September 2019).
180. FLASC. Available online: https://www.offshoreenergystorage.com/flasc-working-principle/ (accessed on 4 September 2019).
181. FLASC Prototype Official Launch. Available online: https://www.offshoreenergystorage.com/flasc-prototype-official-launch/ (accessed on 4 September 2019).
182. FLASC 1st Generation Prototype Testing Completed. Available online: https://www.offshoreenergystorage.com/flasc-1st-generation-prototype-testing-completed/ (accessed on 4th September 2019).

Journal of
*Marine Science
and Engineering*

Article

Study on Transient Overvoltage of Offshore Wind Farm Considering Different Electrical Characteristics of Vacuum Circuit Breaker

Zikai Zhou [1,†], Yaxun Guo [1,†], Xiaofeng Jiang [2], Gang Liu [1,*], Wenhu Tang [1,*], Honglei Deng [1], Xiaohua Li [1] and Ming Zheng [3]

[1]   School of Electric Power Engineering, South China University of Technology, Guangzhou 510640, China
[2]   Jieyang Power Supply Bureau of Guangdong Power Grid Co., Ltd., Jieyang 522000, China
[3]   China Energy Engineering Group Guangdong Electric Power Design Institute Co., Ltd.,
      Guangzhou 510663, China
*   Correspondence: liugang@scut.edu.cn (G.L.); wenhutang@scut.edu.cn (W.T.); Tel.: +86-20-87110613 (G.L.);
    +86-20-87111422 (W.T.)
†   These authors contributed equally to this work.

Received: 9 October 2019; Accepted: 29 October 2019; Published: 13 November 2019

Abstract: For the study of transient overvoltage (TOV) in an offshore wind farm (OWF) collector system caused by switching off vacuum circuit breakers (VCBs), a simplified experimental platform of OWF medium-voltage (MV) cable collector system was established in this paper to conduct switching operation tests of VCB and obtain the characteristic parameters for VCB, especially dielectric strength parameters; also, the effectiveness of the VCB reignition model was verified. Then, PSCAD/EMTDC was used to construct the MV collector system of the OWF, and the effects of normal switching and fault switching on TOV amplitude, steepness, and the total number of reignition of the VCB were studied, respectively, with the experimental parameters and traditional parameters of dielectric strength of the VCB. The simulation results show that when the VCB is at the tower bottom, the overvoltage amplitude generated by the normal switching is the largest, which is 1.83 p.u., and the overvoltage steepness of the fault switching is the largest, up to 142 kV/µs. The overvoltage amplitude and steepness caused by switching off VCB at the tower bottom faultily with traditional parameters are about 2 and 1.5 times of the experimental parameters under the same operating condition.

Keywords: offshore wind farm; vacuum circuit breaker; reignition characteristics; switching overvoltage

1. Introduction

Large wind farms are moving from land to sea to provide a richer and more stable source of clean energy [1,2]. Being increasingly valued by countries around the world, the operation and maintenance of offshore wind farms (OWFs) have been paid more and more attention [3,4]. Large-scale OWFs use cables as the collector system. Thus, the loads of the cables cannot exceed the limit of the cable ampacity [5,6], and the cables should be insulated reliably [7,8]. Moreover, the safety of the tower terminal transformers needs to be carefully considered. In 2004, at Horns Rev, the largest OWF in Denmark at that time, almost all the tower terminal transformers suffered insulation fault accidents [9]. In [10], the authors reported an accident of transformer insulation damage occurred by switching off a vacuum circuit breaker (VCB) after connecting with the shunt reactor in Wailuo OWF. Studies have shown that the high-frequency (HF) overvoltage generated by the frequent switching on and off for VCBs was the main factor causing insulation fault of the transformers [11–13]. Lars Lijestrand et al. simulated and calculated the HF transient process of operating overvoltage generated by switching on a no-load transformer and switching off a no-load transformer under a single-phase grounding short-circuit fault in a medium-voltage (MV) collection grid of OWF [14]. Xuezhong Liu et al. established a test circuit for the MV cable simulation system of the wind farm and a simulation calculation platform of power-frequency overvoltage [15]. Combined with the simulation and field measurement data, the influence of VCB parameters and cable length on the transient overvoltage (TOV) generated by switching off the no-load transformer in the MV collector system of OWF was studied in [16]. PSCAD/EMTDC (Manitoba HVDC Research Centre, a division of MHI Ltd., Winnipeg, Canada) and DIgSILENT/PowerFactory (DIgSILENT GmbH, Gomaringen, Germany) were used to carry out simulation calculation of switching overvoltage when the operating feeder of the OWF was a long no-load cable, and it was compared with the measured data of the actual OWF in [17]. Although there have been a lot of studies on TOV of OWFs, they mainly focus on the qualitative analysis of amplitude and steepness of the switching on TOV in the power collector system. However, there is limited literature on the study of switching-off overvoltage in a power collector system and the quantitative analysis of the overvoltage steepness. Furthermore, with the improvement of manufacturing for VCBs, the traditional electrical parameters have not adapted to today's research.

This paper mainly introduces a reignition modelling method and model verification of VCB in the OWF collector system, and also studies the characteristics of switching TOV in an OWF collector system, respectively, with the experimental parameters and traditional parameters of the dielectric strength of VCB.

The structure of this article is as follows. In Section 2, the models of double-fed induction generator (DFIG) and the VCB are established. In Section 3, the switching reignition experiment of VCB is conducted on the established experimental platform. The characteristic parameters of VCB are obtained, and the validity of the model is verified. Section 4 introduces the switching mode of the internal electrical system in OWF. In Section 5, a single-feeder MV cable collector system of OWF is constructed. The traditional parameters and experimental parameters are, respectively, used for the dielectric strength of the VCB model. The effects of normal switching and fault switching on TOV amplitude, steepness, and the total number of reignition at the high voltage side of the terminal transformer are compared. Finally, conclusions are drawn in Section 6.

2. System Component Simulation Model

2.1. DFIG Model

DFIGs are widely used in OWFs. The simplified DFIG model is composed of a wound induction motor, wind turbine components, double-pulse width modulation (PWM) converter, control system components, and filters both on the stator side and rotor side of the motor [18,19], as shown in Figure 1. The stray capacitances generally include the capacitances between stator winding and shell, stator winding and rotor, and rotor and shell. However, as the capacitance of the HF filter in the wind turbine is much larger than the capacitances mentioned above, stray capacitance has little influence on overvoltage. The study is conducted on the high-voltage side of the terminal transformer, so the stray capacitance of the induction motor can be ignored [20]. In this paper, the rated capacity of the DFIG is 4 MW.

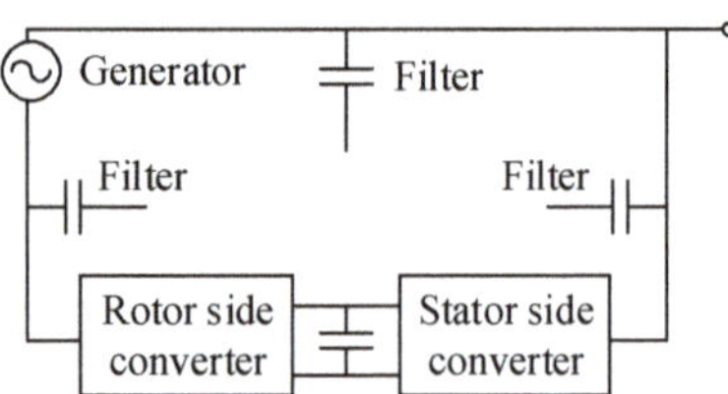

Figure 1. The double-fed induction generator (DFIG) model.

2.2. VCB Model

For inductive circuits, after the current is cut off near the zero-crossing point, an overvoltage will be caused by current chopping, which may result in the first reignition. During reignition, HF current is generated due to the influence of circuit parameters. When the HF current is cut off, the equivalent capacitance and inductance on the load side will cause electromagnetic oscillation. This oscillation will result in higher voltage that may cause the contact gap to be broken down again. Also, the HF current may couple to the other phases, producing current zeros. Thus virtual chopping occurs, causing overvoltage in other phases [10,21]. In the process of multiple reignitions, the interval between the two reignitions is extremely short. Since the second reignition is based on the previous reignition, the reignition overvoltage has the characteristics of high steepness and high amplitude,

The actual parameters of VCB are statistical and random. However, in order to study the switching overvoltage of the VCB, the parameters are set as constant values in this paper [22].

2.2.1. Chopping Current

When a VCB receives an opening instruction, the power frequency current is suddenly cut off before it reaches zero for the first time. The value of current at this time is called chopping current. When the load current varies in the range of 10 A–100 kA, the calculation for chopping current I_{ch} of VCB is as the following empirical formula:

$$I_{ch} = (2\pi I f \alpha \beta)^{\frac{1}{1-\beta}},\tag{1}$$

where f is power frequency (s^{-1}); I is current amplitude before cut-off in the first half cycle (A); and α and β are related to electrode size, material, gap distance, and circuit parameters, where $\alpha = 6.2 \times 10^{-16}$ s, $\beta = 14.3$. Normally, the chopping current of VCB is set as 3–8 A [22].

2.2.2. Dielectric Strength

When the VCB is switching to open, the dielectric strength between the contacts increases with the increase of the distance between the contacts. When the transient recovery voltage between the contacts exceeds the dielectric strength, the gap between the contacts will break down and reignite. There is an approximately linear relationship between the dielectric strength U_b and the break time during reigniting, as described in the following [23]:

$$U_b = A(t - t_{open}) + B,\tag{2}$$

where A is the rate of dielectric strength rise (kV/s); B is dielectric strength constant of VCB at the moment of contact separation (kV); t is simulation time (s); and t_{open} is breaker opening time (s).

2.2.3. HF Arc Quenching Capability

When the HF current generated by the VCB reignition is close to zero, the VCB can extinguish it. Such an arc quenching capability can be described as the rate di/dt of the time change when the HF current is crossing zero. The HF current starts with a high rate of change that the VCB cannot turn off. However, with the attenuation of HF current, when the di/dt is less than a critical value when the current is crossing zero, the VCB will cut off HF current and turn it into a disconnected state, which is generally between 100–600 A/μs [24].

2.2.4. Arcing Voltage

In practice, the arc between contacts of the VCB will generate a voltage drop, and the arcing voltage is approximately 20 V [25]. A constant arcing voltage of 20 V is achieved by changing the value of controllable resistance in the customized model, as shown below [26]:

$$R_{arc} = \frac{u_{arc}}{i_{arc}}, \tag{3}$$

where R_{arc} is arcing resistance (Ω); u_{arc} is arcing voltage (V); and i_{arc} is arcing current (A).

The VCB in the simulation is equivalent to the controlled resistance R with the parallel branch, as shown in Figure 2, where $R_s = 50\ \Omega$, $L_s = 50\ \text{mH}$, $C_s = 200\ \text{pF}$ [27]. The switching process of VCB is divided into four states in [21]. States 1–4 respectively represent the state before power frequency cut-off, transient voltage recovery, reignition, and complete switch-off. By measuring the VCB current i and the voltage u between two contacts, the C language is used to programming and solve the VCB chopping current I_{ch}, dielectric strength U_b, HF arc quenching ability, and arcing voltage u_{arc}. The program flow chart is shown in Figure 3. At the time t_{ch} when the chopping current occurring is more than 5 ms after the time t_{open} when the VCB starts operating, it can be considered that the VCB has been completely opened and there will be no reignition. The controllable resistance is realized as a real-time control by calling the program in PSCAD. Initially, the closed VCB is in State 1 and $R = 0$. Arcing occurs when the VCB starts to separate, and the program adjusts the controllable resistor R according to Equation (5) to maintain the voltage across the contacts as arcing voltage. When i is less than I_{ch}, the VCB enters State 2 after the first interruption, and $R = 1\ \text{M}\Omega$. When VCB is in State 2, if the transient voltage exceeds the dielectric strength, the VCB reignites and enters State 3, and the voltage across the contacts is the arcing voltage. When the VCB is in State 3, if the HF current quenching condition is satisfied, the current is cut off and the VCB returns to State 2, $R = 1\ \text{M}\Omega$. After multiple reignitions occur, when the transient recovery voltage cannot reach the dielectric strength, the VCB is successfully opened and remains in $R = 1\ \text{M}\Omega$.

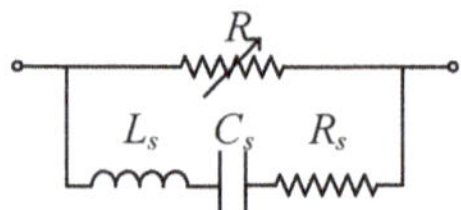

Figure 2. The vacuum circuit breaker (VCB) model.

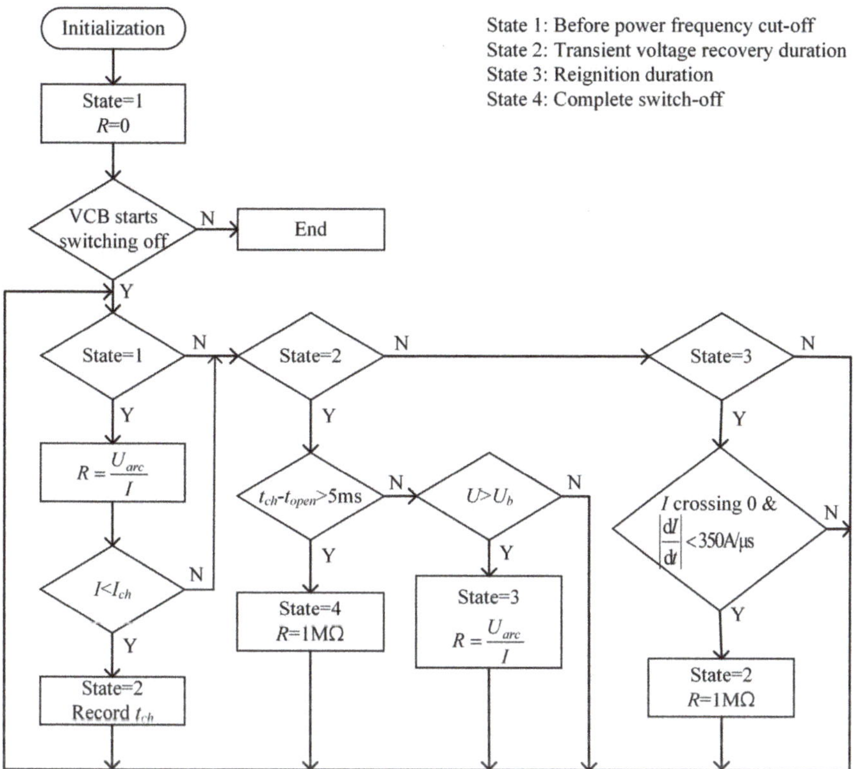

Figure 3. Block diagram of the VCB model.

2.3. Other Models

2.3.1. Transformer Model

In this paper, the unified magnetic equivalent circuit (UMEC) model is adopted for transformers, which can represent the phase coupling of the transformer in PSCAD. At the same time, capacitors are connected in parallel between the high-voltage side, low-voltage side, and high and low voltage of the transformer to simulate the HF characteristics of the transformer [28].

2.3.2. Submarine Cable Model

The frequency dependent (phase) model in PSCAD is used to model submarine cables. The three-core cable structure is used in the simulation, and the specific setting parameters are shown in [28].

3. Description of Experimental Test System and Model Verification

3.1. Description of Experimental Test System

The wiring diagram of the test system is shown in Figure 4. A transformer TX_1 with a transformation ratio of 10 kV/35 kV and a capacity of 10 MVA is used to simulate the main transformer of the offshore booster station and provides a 35 kV power supply. One kilometer-long and 80 m-long submarine cables, named $Cable_1$ and $Cable_2$ with a cross-section of 35 mm^2, are respectively connected to both sides of a 40.5 kV VCB to simulate the three-core MV cable between the 35 kV busbar to the wind turbine at the beginning of the feeder and the transformer at the top of the tower to the VCB at the tower bottom. The technical characteristics of the VCB are shown in Table 1. A transformer TX_2 with a transformation ratio of 35 kV/ 0.69 kV and a capacity of 2 MVA simulates the terminal transformer of the wind turbine. A reactor with a capacity of 1.6 Mvar is set as the load, and its capacity is approximately 80% of TX_2 to ensure that the amplitude value of power frequency current flowing through the VCB is greater than the chopping current.

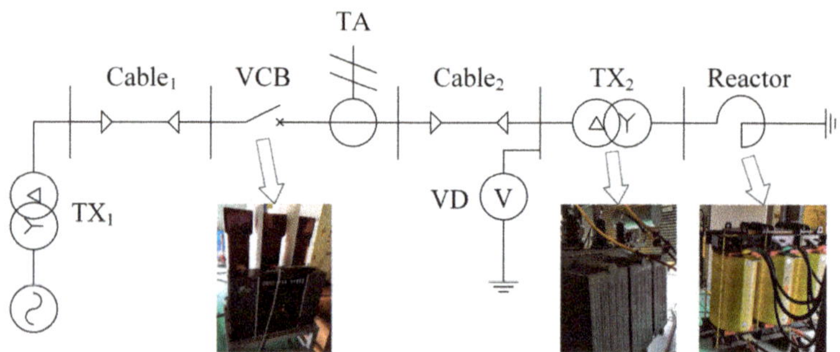

Figure 4. Wiring diagram of the test system

Table 1. Technical characteristics of the VCB.

Product Model	Rated Voltage (kV)	Rated Current (A)	Rated Short-Circuit Breaking Current (kA)	Average Opening Speed (m/s)	Clearance between Open Contacts (mm)
ZN95A-40.5	40.5	630	25	1.7	18

In order to measure and record HF signals, a 150 kV high-voltage probe VD with a ratio of 10000:1, of which the model is NRV-150, is used in the experiment. It has an accuracy of 1% when the frequency of voltage ranges from 10 Hz to 1 MHz. The model of HF current transformer TA is Pearson D101, with a frequency bandwidth of 0.25 Hz–4 MHz and 50 kA peak current. During the experiment, the sampling rate of the digital oscilloscope is set as 40 Msa/s.

3.2. Model Verification

When chopping current occurs in the VCB, the magnetic energy stored in the inductive load (such as reactor, no-load transformer, or motor, etc.) is converted into the electric field energy of the load side capacitance (usually the capacitance of the submarine cable), and thus overvoltage is generated. Therefore, the chopping current value can be calculated by the amplitude $U_{\max}$ of the first overvoltage generated by the chopping current:

$$U_{\max} = \sqrt{(U_0 + U_n)^2 + \left(I_{ch} \cdot \sqrt{\frac{L_T}{C_T}}\right)^2}, \tag{4}$$

where U_0 is cut-off transient voltage (V); U_n is power supply voltage (V); L_t is system equivalent inductance (H); and C_t is load side capacitance (F).

The reignition of the VCB is obvious when the inductive load is switched to separate. Therefore, the switching test is conducted under the condition of the inductive load to measure the three-phase voltage at the high-voltage side of TX_2 and the B-phase current at the outlet side of the VCB. The waveform obtained from the experiment is shown in Figure 5a.

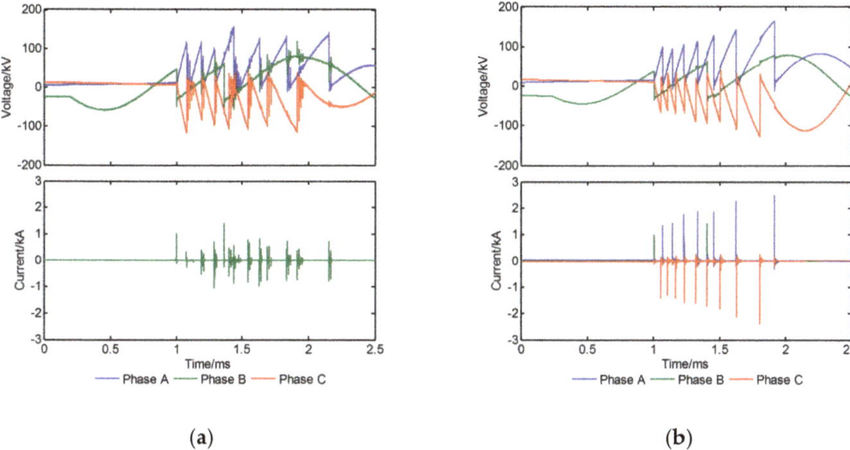

(a)   (b)

Figure 5. Voltage and current waveforms of experiment and simulation: (**a**) Waveforms of experiment voltage and current; (**b**) waveforms of simulation voltage and current.

According to Equation (4) and Figure 5a, the chopping current of VCB calculated in this experiment is 3.6 A. In Figure 5a, the value of the breakdown voltage is considered to be the value of the dielectric strength at this time. After the contacts of VCB begin to separate, each time a reignition occurs, the value of the breakdown voltage and the corresponding time are recorded and linearly fitted according to Equation (2), as shown in Figure 6. The relationship between dielectric strength and time is obtained as follows:

$$U_b = 7.355 \cdot 10^4 \cdot (t - t_{open}) + 0.69 \text{ kV}. \tag{5}$$

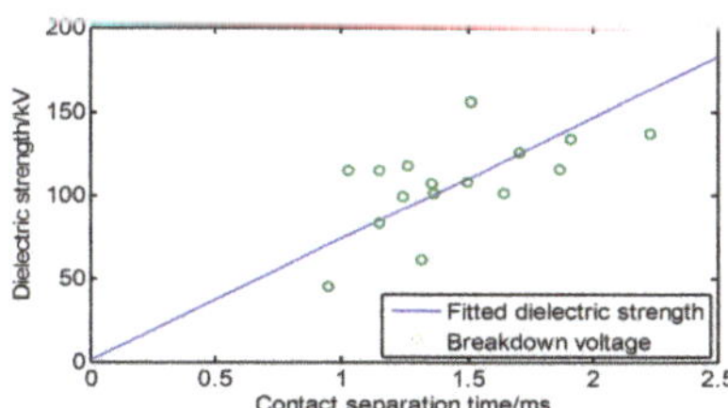

Figure 6. Dielectric strength versus contact separation time.

Since the HF arc quenching ability has little influence on the overvoltage, the critical value of di/dt is set as the average value of 350 A/μs in this paper [29], and the arcing voltage is set as 20 V. The PSCAD 4.6.2 is used to simulate and model the experimental system shown in Figure 4. The reignition model introduced in Section 2.2 of this paper is adopted in the VCB model. By cooperating with the time logic device, the switching operation is realized at any time in the model, and the waveform is obtained by simulation, shown in Figure 5b.

In the case of switching off inductive load, because of the chopping current, the rising rate of transient recovery voltage between two contacts of VCB is much faster than that of dielectric strength, which causes a lot of restrikes of VCB, as shown in Figure 5a, where B-phase is the first-opening phase. In Figure 5a,b, the amplitude and the total number of reignition of B-phase overvoltage are basically the same, and the amplitude of HF current is basically the same. However, due to the interference of external factors on the measuring equipment in the actual experiment, the measured HF current has many burrs, but its amplitude is relatively small compared with HF current generated by reignition. There is no burr in the simulation due to no external interference. As a result, the effectiveness of the VCB reignition model is well verified by the waveform of the switching TOV obtained in the experiment.

4. VCB Switching Modes of OWF Internal Electrical System

The switching modes of VCB inside the electrical system of OWF include normal switching and fault switching.

Normal switching: In this case, the no-load terminal transformer at the top of the tower is cut off; that is, after a certain wind turbine is out of operation, a corresponding VCB at tower bottom is switched off. Meanwhile, the rest wind turbines on the feeder remain in full load operation.

Fault switching: In this situation, the switching happens when wind turbines are in normal operation. Specifically, it includes two kinds of circumstances. One is that when the wind turbines are in full load operation on the whole feeder, the VCB at the beginning of the feeder is switched off. The second is that when the wind turbines are in full load operation on the whole feeder, a VCB at tower bottom is switched off.

5. Simulation of Internal Electrical System in OWF

5.1. Simulation System Setting

Wailuo OWF is located in Guangdong Province, China. Its installed capacity is 300 MW. In this section, Wailuo OWF is taken as an example to carry out simulation research. A calculation model of the internal electrical system of the OWF will be established based on the VCB and DFIG model mentioned above, as shown in Figure 7. The capacity of transformer T_0 is 180 MVA, with its ratio and leakage inductance of 220 kV/35 kV and 0.06, respectively. For transformer T_n ($n = 1, 2, \ldots, 8$), the capacity is 5 MVA, the ratio is 35 kV/0.69 kV, and the leakage inductance is 0.02 per unit. The external grid, which is connected to T_0 with a 20-km-long submarine cable, is represented by a 220 kV ideal voltage source. The length of L_1 is 80 m, and the wind turbine (WT) connects to the transformer directly. The length of L_2 between each wind turbine is 640 m, and L_0 is 5 km long. The cross-section area of submarine cables is 300 mm^2. The three dielectric strength parameters of high-, medium-, and low-voltage have been proposed in [23] and have been used in many similar simulation studies. In this paper, the dielectric strength of "high voltage VCB", which is commonly used, is compared with the parameters obtained by experiments. In the simulation, the influence of two-parameter settings on overvoltage amplitude and steepness with normal switching and fault switching is compared. The parameters used are shown in Table 2.

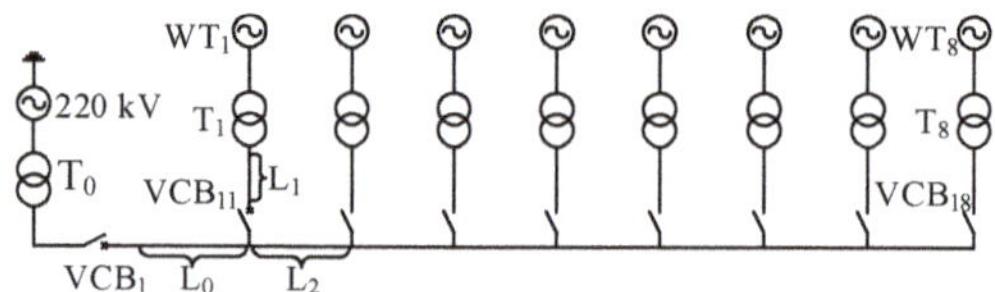

Figure 7. Wiring diagram of the simulation system.

Table 2. Traditional parameters and experimental parameters.

Parameters Type	A(kV/s)	B(kV)
traditional parameters	1.7×10^4	3.4
experimental parameters	7.355×10^4	0.69

Snapshots were taken after the stable operation of the system for 2 s, and each snapshot was started up and running for 0.08 s, with the simulation step length of 0.4 μs. In order to reduce the simulation time, a calculation model of the internal electrical system of OWF with eight DFIGs on a single feeder was built in this paper; the rated capacity of each DFIG is 4 MW. In the case of normal switching, the VCB is set to be switched off at the time of A-phase voltage zero-crossing. In the situation of fault switching, the VCB is set to be switched off at the time when A-phase current reaches the chopping current [30].

5.2. Relation Between Transformer Position and Overvoltage in Normal Switching

According to the circuit theory, when reignition occurs in the VCB, the TOV steepness on the high-voltage side of the terminal transformer is related to the current flowing through the capacitance of the high-voltage to the ground of the transformer, as shown in Equation (6):

$$\frac{\mathrm{d}u_T}{\mathrm{d}t} = -\frac{i_T}{C_H},\tag{6}$$

where u_T is the voltage at the high-voltage side of the terminal transformer (V); i_T is current in the capacitance of the high-voltage side to the ground of the terminal transformer (A); and C_H is the capacitance of the high-voltage to the ground of the transformer (F).

When any wind turbine on the feeder is out of operation, the terminal transformer and its 80 m connection cable are cut off. Then, the overvoltage on the high-voltage side of the terminal transformer on the feeder is measured, and the steepness is calculated. Since the rising rate of transient recovery voltage after switching is always lower than the rising rate of dielectric strength, they do not intersect, and there is no reignition. At this time, the amplitude and the steepness of the overvoltage at each position and the voltage waveform at the high-voltage side of transformer T_7 when switching off VCB_{17} are respectively shown in Figures 8 and 9, in which overvoltage amplitude is per unit value as follows:

$$\sqrt{2} \cdot 35 \div \sqrt{3} = 28.58 \text{ kV}.\tag{7}$$

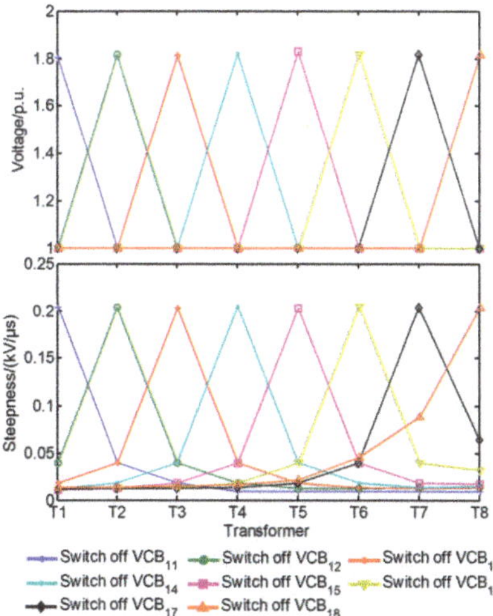

Figure 8. Overvoltage of transformers at different positions in the cases of switching off VCBs at the tower bottom normally using experimental parameters.

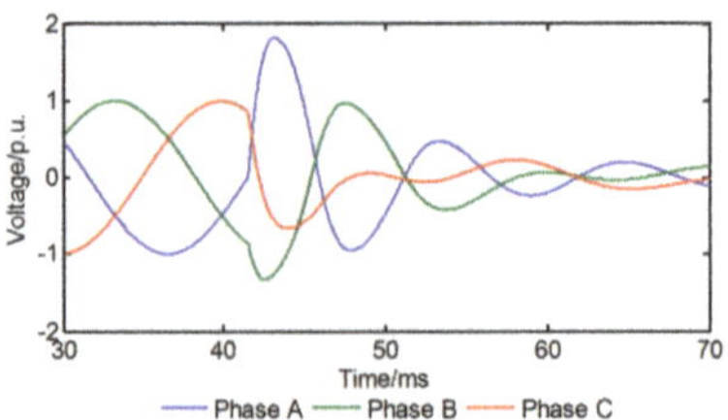

Figure 9. Overvoltage waveform of T_7 in the case of switching off VCB_{17} normally.

Figure 8 shows that the overvoltage amplitude and the maximum steepness of the transformer in the normal switching occur at the point where the VCB cuts out the transformer; they are 1.83 p.u. and 0.2 kV/μs, respectively. The overvoltage amplitude on the high-voltage side of the transformer of the other wind turbines in normal operation is basically 1 p.u., and the overvoltage steepness decreases with the increase of the propagation distance of the incident wave.

When the dielectric strength of VCB is simulated with traditional parameters, since the dielectric strength rises slowly, reignition occurs easily. The total reignition times in normal switching with different parameters are shown in Table 3. At this time, the total number of reignition normal switching is 10–14 times. The voltage amplitude and steepness of the high-voltage side of the terminal transformer at each position are shown in Figure 10. Compared with the results with experimental parameters, the maximum overvoltage amplitude drops from 1.83 p.u. to 1.58 p.u. However, the maximum overvoltage steepness reaches 76.7 kV/μs, increasing by about 380 times.

Table 3. Total reignition times in normal switching with different parameters.

Switching Case	Traditional Parameters	Experimental Parameters
normal switching	10-14	0

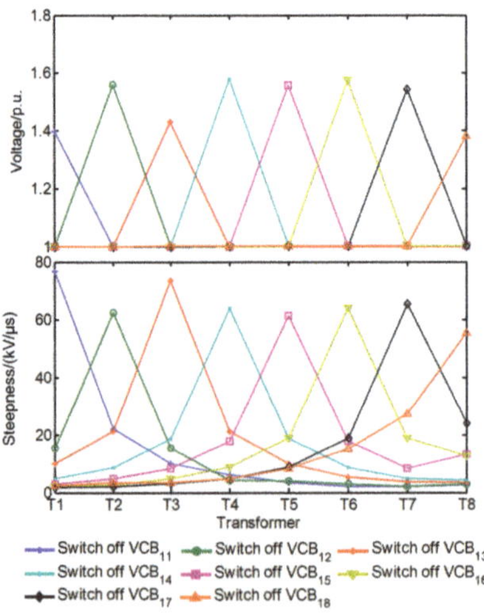

Figure 10. Overvoltage of transformers at different positions in the cases of switching off VCBs at the tower bottom normally using traditional parameters.

5.3. Relation between Transformer Position and Overvoltage in Fault Switching

5.3.1. VCB Switching at Feeder

When the wind turbines on the whole feeder line are in full-load operation, VCB_1 has no reignition in fault switching. The overvoltage waveform of the terminal transformer T_7 is shown in Figure 11. The overvoltage amplitude at the high-voltage side of the terminal transformer at all positions on the feeder is 1.27 p.u., and the maximum overvoltage steepness is about 0.1 kV/μs. When the traditional parameters are used in the simulation, VCB_1 will not have reignition. The simulation results of the two are consistent at this moment.

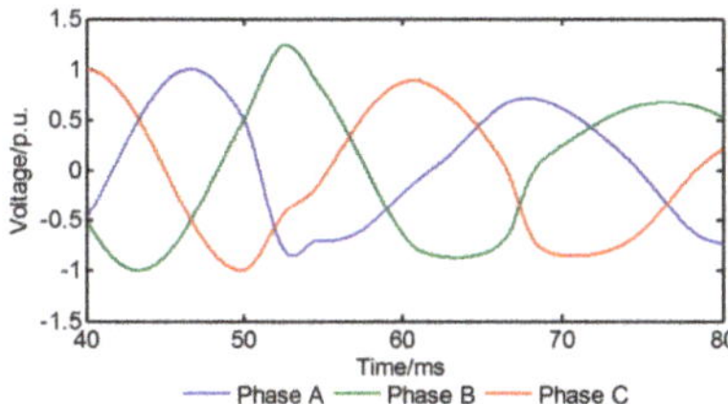

Figure 11. Overvoltage waveform of T_7 in the case of switching off VCB_1 faultily.

5.3.2. VCB Switching at Tower Bottom

When the wind turbines on the whole feeder are in full load operation, the VCB at the tower bottom has fault switching, and the simulation results are shown in Figure 12; it indicated that switching VCB only causes overvoltage with amplitude of about 1.18–1.42 p.u. on the high-voltage side of the terminal transformer where it is located, and the steepness is 83.3–142 kV/μs. The voltage amplitude of the remaining terminal transformer is 1 p.u. The voltage waveforms of terminal transform T_7 when VCB_{17} switching and its adjacent terminal transformer T_8 at the high-voltage side are respectively shown in Figure 13a,b.

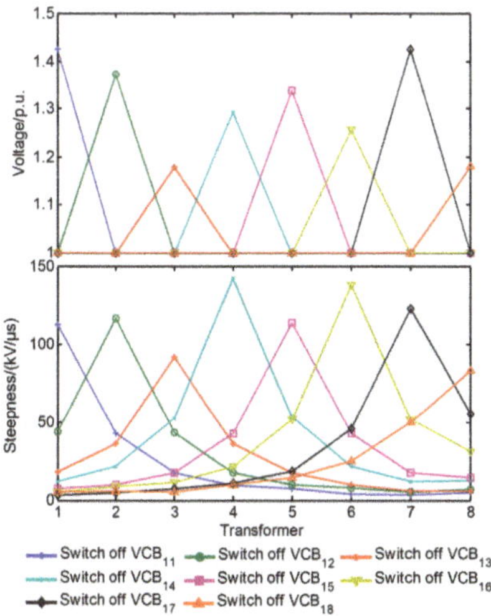

Figure 12. Overvoltage of transformers at different positions in the cases of switching off VCBs at the tower bottom faultily using experimental parameters.

During VCB_{17} switching, an overvoltage with an amplitude up to 1.42 p.u. is generated due to reignition. After reaching the high-voltage side of T_8 through the 720 m cable, the overvoltage amplitude rapidly attenuates, and the maximum steepness of voltage fluctuation also decreases from 123 kV/μs to 55.3 kV/μs. From Table 4, the total times of reignition in this condition is about 9–10.

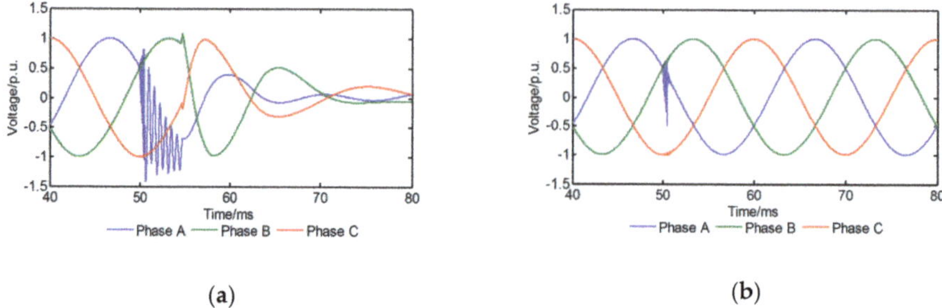

(a)   (b)

Figure 13. Voltage waveforms of T_7 and T_8 in the case of switching off VCB_{17} faultily: (a) Voltage waveform of T_7; (b) voltage waveform of T_8.

Table 4. Total reignition times in fault switching with different parameters.

Switching Case		Traditional Parameters	Experimental Parameters
fault switching	at feeder	0	0
	at tower bottom	135—171	8—10

When traditional parameters are applied for simulation, the results are obtained shown in Figure 14. Due to the slow growth rate of dielectric strength, the numbers of reignition increase significantly to 135–171 times, and the maximum overvoltage amplitude increases from 1.42 p.u to 2.96 p.u, which would cause the overvoltage amplitude of the terminal transformer at the high-voltage side of the adjacent wind turbine reaching 1.89 p.u.. Compared with the same condition when using experimental parameters, the amplitude of the switching terminal transformer increases about 2 times, and the overvoltage steepness increases about 1.5 times.

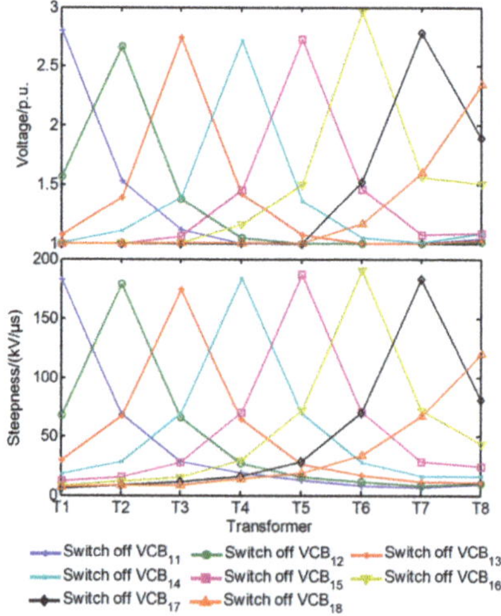

Figure 14. Overvoltage of transformers at different positions in the cases of switching off VCBs at the tower bottom faultily using traditional parameters.

6. Conclusions

In this paper, a test platform for a simplified MV cable collection system in OWF that can demonstrate the reignition phenomenon of VCB was constructed, and the parameters of VCB were calculated through experiments to verify the validity of the customized VCB model. A cable collection system of OWF was built according to the above model. The dielectric strength parameters of the VCB measured by the experiment were used in the simulation to study the TOV generated, and the results were compared with those of the traditional dielectric strength parameters. The following conclusions are drawn:

(1)   The overvoltage amplitude of the high voltage side of the transformer at different positions was basically the same in switching off feeder VCB. When the tower bottom VCB had normal or fault switching, the overvoltage steepness decreased with the increase of the propagation distance of the incident wave. The voltage amplitudes at the high-voltage side of the transformers in other positions were slightly influenced.

(2)   With experimental parameters, the critical overvoltage occurred in switching off VCBs at the tower bottom, the overvoltage amplitude in normal switching was the largest, up to 1.83 p.u., while the steepness of the overvoltage generated in fault switching was the largest, up to 142 kV/μs, and the total numbers of reignition were 9–10 times.

(3)   Due to the difference of dielectric strength of the VCB, when using the experimental parameters measured in this study, the amplitude of overvoltage of the terminal transformer was reduced to 1/2, and the steepness was reduced to 1/1.5 compared with using the traditional parameters of VCB under the same operation condition, in case of switching off VCB at the tower bottom with wind turbines full loading. Therefore, it is recommended that the actual experimental parameters of VCBs should be adopted in the following researches.

Author Contributions: Conceptualization, Z.Z. and Y.G.; methodology, X.J.; software, Z.Z. and Y.G.; validation, Y.G., X.J., and G.L.; investigation, H.D.; writing—original draft preparation, Z.Z.; writing—review and editing, Y.G. and X.J.; resources, M.Z.; visualization, X.L.; supervision, G.L.; project administration, W.T.

Funding: This research was funded in part by National Natural Science Foundation of China, grant number 51677073 and 51477054, in part by National Natural Science Fund & China-State Grid Joint Fund for Smart Grid, grant number U1766213.

Acknowledgments: The authors would like to thank Xinyu Jiang from Guangzhou Zhiguang Electric Co., Ltd. for his support on experiments.

Conflicts of Interest: The authors declare no conflict of interest.

References

1.   Esteban, M.D.; Espada, J.M.; Ortega, J.M.; López-Gutiérrez, J.-S.; Negro, V. What about Marine Renewable Energies in Spain? *J. Mar. Sci. Eng.* **2019**, *7*, 249. [CrossRef]

2.   Onea, F.; Rusu, L. A Study on the Wind Energy Potential in the Romanian Coastal Environment. *J. Mar. Sci. Eng.* **2019**, *7*, 142. [CrossRef]

3.   Gintautas, T.; Sørensen, J.D. Improved Methodology of Weather Window Prediction for Offshore Operations Based on Probabilities of Operation Failure. *J. Mar. Sci. Eng.* **2017**, *5*, 20. [CrossRef]

4.   Lei, Y.; Jian, W.; Gang, L.; Hui, M.; Ming, Z. Thermal Rating of Offshore Wind Farm Cables Installed in Ventilated J-Tubes. *Energies* **2018**, *11*, 545.

5.   Wang, P.Y.; Ma, H.; Liu, G.; Han, Z.Z.; Kang, L.Y. Dynamic Thermal Analysis of High-voltage Power Cable Insulation for Cable Dynamic Thermal Rating. *IEEE Access* **2019**, *7*, 56095–56106. [CrossRef]

6.   Wang, P.; Liu, G.; Ma, H.; Liu, Y.; Xu, T. Investigation of the Ampacity of a Prefabricated Straight-Through Joint of High Voltage Cable. *Energies* **2017**, *10*, 2050. [CrossRef]

7.   Xie, Y.; Liu, G.; Zhao, Y.; Li, L.; Ohki, Y. Rejuvenation of retired power cables by heat treatment. *IEEE Trans. Dielectr. Electr. Insul.* **2019**, *26*, 668–670. [CrossRef]

8.   Xie, Y.; Zhao, Y.; Liu, G.; Huang, J.; Li, L. Annealing Effects on XLPE Insulation of Retired High-Voltage Cable. *IEEE Access* **2019**, *7*, 104344–104353. [CrossRef]

9.   Sweet, W. Danish wind turbines take unfortunate turn. *IEEE Spectr.* **2004**, *41*, 30. [CrossRef]

10.   Zhou, J.; Xin, Y.; Tang, W.; Liu, G.; Wu, Q. Impact Factor Identification for Switching Overvoltage in an Offshore Wind Farm by Analyzing Multiple Ignition Transients. *IEEE Access* **2019**, *7*, 64651–64662. [CrossRef]

11.   Shipp, D.D.; Dionise, T.J.; Lorch, V.; Macfarlane, B.G. Transformer failure due to circuit breaker induced switching transients. *IEEE Trans. Ind. Appl.* **2011**, *47*, 707–718. [CrossRef]

12.   Xin, Y.; Tang, W.; Zhou, J.; Yang, Y.; Liu, G. Sensitivity analysis of reignition overvoltage for vacuum circuit breaker in offshore wind farm using experiment-based modeling. *Electr. Power Syst. Res.* **2019**, *172*, 86. [CrossRef]

13.   Villar, F.S.; Reza, M.; Srivastava, K.; Silva, L. High frequency transients propagation and the multiple reflections effect in collection grids for offshore wind parks. In Proceedings of the Power & Energy Society General Meeting, Detroit, MI, USA, 24–28 July 2011; pp. 1–7.

14.   Liljestrand, L.; Sannino, A.; Breder, H.; Thorburn, S. Transients in collection grids of large offshore wind parks. *Wind Energy* **2008**, *11*, 45–61. [CrossRef]

15.   Wang, X.; Liu, X.; Li, Y.; Saboor, A.; Cao, J.; Shang, Y. Study on switching transients of cable collector system in wind power plant. In Proceedings of the 2011 1st International Conference on Electric Power Equipment-Switching Technology, Xi'an, China, 23–27 October 2011; pp. 308–311.

16.   Ghafourian, S.; Arana, I.; Holboll, J.; Sorensen, T.; Popov, M.; Terzija, V. General Analysis of Vacuum Circuit Breaker Switching Overvoltages in Offshore Wind Farms. *IEEE Trans. Power Deliv.* **2016**, *31*, 2351–2359. [CrossRef]

17.   Glasdam, J.; Bak, C.L.; Hjerrild, J. Transient studies in large offshore wind farms employing detailed circuit breaker representation. *Energies* **2012**, *5*, 2214–2231. [CrossRef]

18.   Holdsworth, L.; Wu, X.G.; Jenkins, N.; Ekanayake, J.B. Dynamic modeling of doubly fed induction generator wind turbines. *IEEE Trans. Power Syst.* **2003**, *18*, 803–809.

19.   Xu, B.; Zhang, Z.; Li, C.; Zhan, P.; Wen, J. LCL filters applied in doubly fed induction generator. *Elecric Power Autom. Equip.* **2015**, *35*, 44–50.

20.   Badrzadeh, B.; Hogdahl, M.; Isabegovic, E. Transients in Wind Power Plants—Part I: Modeling Methodology and Validation. *IEEE Trans. Ind. Appl.* **2012**, *48*, 794–807. [CrossRef]

21.   Ghafourian, S.M. Switching Transients in Large Offshore Wind Farms. Ph.D. Thesis, The University of Manchester, Manchester, UK, 2015.

22.   Abdulahovic, T.; Thiringer, T.; Reza, M.; Breder, H. Vacuum Circuit Breaker Parameter Calculation and Modelling for Power System Transient Studies. *IEEE Trans. Power Deliv.* **2017**, *32*, 1165–1172. [CrossRef]

23.   Glinkowski, M.T.; Gutierrez, M.R.; Braun, D. Voltage escalation and reignition behavior of vacuum generator circuit breakers during load shedding. *IEEE Trans. Power Deliv.* **1997**, *12*, 219–226. [CrossRef]

24.   Xin, Y.; Bo, L.; Tang, W.; Wu, Q. Modeling and Mitigation for High Frequency Switching Transients Due to Energization in Offshore Wind Farms. *Energies* **2016**, *9*, 1044. [CrossRef]

25.   Leusenkamp, M.B.J. Vacuum interrupter model based on breaking tests. *IEEE Trans. Plasma Sci.* **1999**, *27*, 969–976. [CrossRef]

26.   Popov, M. Switching three-phase distribution transformers with a vacuum circuit breaker: Analysis of overvoltages and the protection of the equipment. Ph.D. Thesis, Delft University of Technology, Delft, The Netherlands, 2002.

27.   Helmer, J.; Lindmayer, M. Mathematical modeling of the high frequency behavior of vacuum interrupters and comparison with measured transients in power systems. In Proceedings of the 17th International Symposium on Discharges and Electrical Insulation in Vacuum, Berkeley, CA, USA, 21–26 July 1996; pp. 323–331.

28.   Liu, G.; Guo, Y.; Xin, Y.; Lei, Y.; Tang, W. Analysis of switching transients during energization in large offshore wind farms. *Energies* **2018**, *11*, 470. [CrossRef]

29.   Abdulahovic, T. Analysis of High-Frequency Electrical Transients in Offshore Wind Parks. Ph.D. Thesis, Chalmers University of Technology, Gothenburg, Sweden, 2011.

30.   Reza, M.; Breder, H. Cable System Transient Study: Vindforsk V-110: Experiments with Switching Transients and Their Mitigation in a Windpower Collection Grid Scale Model. Available online: https://energiforskmedia. blob.core.windows.net/media/19748/cable-system-transient-study-elforskrapport-2009-05.pdf (accessed on 30 September 2019).

Journal of
Marine Science and Engineering

Article

Analysis of Offshore Structures Based on Response Spectrum of Ice Force

Yingzhou Liu [1,2], Xin Li [1,2,*] and Youwei Zhang [3]

[1] State Key Laboratory of Coastal and Offshore Engineering, Dalian University of Technology, Dalian 116024, Liaoning, China; liuyingzhou@mail.dlut.edu.cn
[2] Institute of Earthquake Engineering, Faculty of Infrastructure Engineering, Dalian University of Technology, Dalian 116024, Liaoning, China
[3] State Key Laboratory of Structural Analysis for Industrial Equipment, Department of Engineering Mechanics, International Research Center for Computational Mechanics, Dalian University of Technology, Dalian 116024, China; ywzhang@dlut.edu.cn
* Correspondence: lixin@dlut.edu.cn; Tel.: +86-0411-84707784

Received: 19 September 2019; Accepted: 7 November 2019; Published: 14 November 2019

Abstract: With the development of large-scale offshore projects, sea ice is a potential threat to the safety of offshore structures. The main forms of damage to bottom-fixed offshore structures under sea ice are crushing failure and bending failure. Referred to as the concept of seismic response spectrums, the design response spectrum of offshore structures induced by the crushing and bending ice failure is presented. Selecting the Bohai Sea in China as an example, the sea areas were divided into different ice zones due to the different sea ice parameters. Based on the crushing and bending failure power spectral densities of ice force, a large amount of ice force time-history samples are firstly generated for each ice zone. The time-history of the maximum responses of a series of single degree of freedom systems with different natural frequencies under the ice force are calculated and subsequently, a response spectrum curve is obtained. Finally, by fitting all the response spectrum curves from different samples, the design response spectrum is generated for each ice zone. The ice force influence coefficients for crushing and bending failure are obtained, which can be used to estimate the stochastic sea ice force acting on a structure conveniently in a static way. A comparison of the proposed response spectrum method with the Monte Carlo method by a numerical example shows good agreement.

Keywords: ice force; design response spectrum; crushing failure; bending failure

---

1. Introduction

Freezing is a common natural phenomenon in winter at high latitudes of the earth. The ice force is a potential threat to bottom-fixed offshore structures due to its high magnitude and evident dynamic effects [1,2]. For instance, in 1969, the JZ20-2MSW platform was destroyed because of serious ice conditions in the Bohai Sea [3,4].

Many researchers have put their effort into the research of sea ice force and tried to present a more reasonable sea ice force model.

Su et al. [5] investigated the typical statistical characteristics of local ice loads based on the data from in-situ measurements. Neill [6] investigated the dynamic ice force on piers and piles. Torodd [7] studied model-based force and state estimation in experimental ice-induced vibrations. The feasibility and advantages of the ice-breaking cone were proven by the analytical results obtained by Ralston et al. [8] in the 1980s. Ordinary and serious ice conditions in the Bohai Sea were briefly presented by Zhang [9] and formulas for calculating the forces applied on offshore structures by ice were suggested.

From the above researches [5–9], the ice force is often regarded as a static load without considering its dynamic effects. In engineering design specifications, Nord [10], Qu [11], Barker [12], and Gravesen [13] conducted dynamic ice force experiments, respectively. Ice force acting on the Nordströmsgrund lighthouse was identified [10]. Qu [11] analyzed a random ice force for narrow conical structures through practical engineering experiments. Baker [12] and Gravesen [13] investigated the dynamic ice force on the offshore wind turbine, the dynamic characteristics of the structure under such loads were discovered.

Some dynamic ice force models have been recommended, while most of them are deterministic ones, such as the simplified dynamic ice force model presented by Kärnä and Qu [14]. Based on sea ice dynamics, Hunke et al. [15] established an elastic-viscous-plastic model. Li and Li [16] established a modified discrete element model for sea ice dynamics. Considering the redistribution process and the evolution of the ice thickness distribution, an idealized zero-dimensional model was presented by Godlovitch and Monahan [17]. Qu [18] proposed an ice dynamics model for narrow conical structures, which showed that ice-breaking cones were effective in reducing the ice force. On the other hand, the specific dynamic ice models have been developed based on local environmental conditions. Wang et al. [19] and Pedersen et al. [20] presented the proper sea ice dynamics models for the Gulf of Riga and the Greenland Sea, respectively. Besides, the dynamic and thermodynamic sea ice model for the subpolar regions was studied by Lu et al. [21]. However, sea ice breaking is a stochastic process in essence and it is usually difficult to simulate the action of ice force on offshore structures due to the low efficiency of the traditional random vibrations analysis method. Therefore, Shi [22], Zhi [23], and Ou [24] proposed the concept of sea ice force spectrums. Shi [22] recommended an ice force spectrum based on the displacement and strain responses of a single degree freedom structure, and more complicated structures were considered in the study of Zhi [23]. Ou [24] analyzed the characteristics of the random process of ice force and mechanism and established the relationship between spectral parameters and ice thickness.

The rupture forms of the sea ice can be mainly divided into several types, such as crushing failure, bending failure, and so on. Lee [25] investigated local ice load signals in ice-covered waters. Kim [26] discussed the assumptions behind rule-based ice loads of crushing failure. Through tests, the damage mechanism of sea ice was studied by Huang [27] in detail. The frequency of ice force under different ice failure modes and the occurrence probability of their magnitudes in full-scale had been studied by Suominen [28]. Zhang [29] studied the mechanism of ductile-brittle transition of sea ice damage and the influence of microcrack evolution on sea ice properties. Jones and Eylander [30] studied the ice force which acted on a vertical structure or inclined structure. Hayo et al. [31] investigated the ice-induced vibrations in the states of mixed crushing and buckling. Aksenov and Hibler [32] found that the icebreaking was highly irregular, and small cracks appeared around the broken area. Gagnon [33] established a numerical model for ice crushing failure. Sopper [34] performed a series of ice crushing tests to investigate the effects of external boundary conditions and geometric contact shapes under ice force.

In fact, ice force and the seismic effect have many similar characteristics, e.g., both of them are dynamic and stochastic with a specific frequency spectrum. Referred to as seismic design theory, the design response spectrums of sea ice force due to the crushing and bending failure are proposed. The novel method is simple and easy to analyze the response of bottom-fixed offshore structures subjected to ice.

Firstly, a single-degree of freedom (SDOF) model with different natural frequencies and damping is established to simulate different offshore structures subjected to ice force. Secondly, the crushing and bending failure power spectral densities (PSD) of the ice force, and the properties of ice are recommended. Thirdly, ice conditions in the Bohai Sea are selected as a typical investigated zone. A large amount of ice force time-history samples for crushing and bending failure are generated by applying the amplitude superposition method. Then, the maximum responses of SDOF structures with different natural frequencies subjected to each ice force time-history are obtained. The design response spectrums for both crushing and bending failure sea ice force are achieved. Finally, the numerical results validate the proposed method.

2. Analysis Model

Assume that the offshore structure can be simplified as a SDOF system, as shown in Figure 1, where m is the lumped mass; k is the shearing stiffness; and c is the damping coefficient; l_1 and l_2 indicate the heights above and under the sea level, respectively.

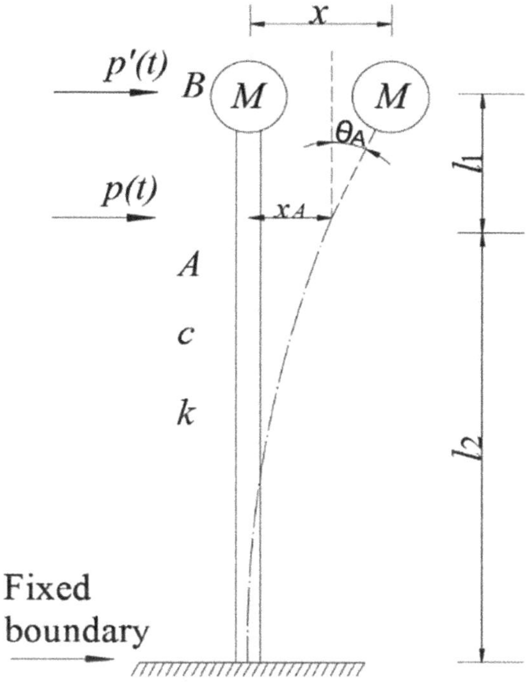

Figure 1. Single-degree of freedom (SDOF) Model.

Since the system is excited by the ice force $p(t)$ at the sea level (point A in Figure 1), rather than directly on the lumped mass (point B in Figure 1), the ice force should be replaced with an equivalent force $p'(t)$ to establish the motion equation of the system. Neglecting the inertia force, the lateral displacement x and rotation θ at point A subjected to $p(t)$ is:

$$x_A = \frac{p(t)l_2^3}{3EI}; \quad \theta_A = \frac{p(t)l_2^2}{2EI} \tag{1}$$

where EI is the flexural stiffness of the cantilever beam. The displacement at point B is then:

$$x = x_A + \theta_A l_1 = \frac{p(t)l_2^3}{3EI} + \frac{p(t)l_1 l_2^2}{2EI} \tag{2}$$

Note that:

$$k = \frac{3EI}{(l_1 + l_2)^3} \tag{3}$$

where k is the shear stiffness.

Equation (2) can be re-written as:

$$x = \frac{2l_2^3 + 3l_1 l_2^2}{2k(l_1 + l_2)^3} p(t) \tag{4}$$

According to Equation (4), the displacement at point B induced by the equivalent ice force $p'(t)$ has to be equal to that raised by $p(t)$:

$$x = \frac{p'(t)}{k} = \frac{2l_2^3 + 3l_1 l_2^2}{2k(l_1 + l_2)^3} p(t) \tag{5}$$

therefore,

$$p'(t) = \frac{2l_2^3 + 3l_1 l_2^2}{2(l_1 + l_2)^3} p(t) \tag{6}$$

The motion equation of the SDOF system subjected to ice force is then:

$$M\ddot{x} + C\dot{x} + Kx = p'(t) = \lambda p(t) \tag{7}$$

where x, $\dot{x}$, and $\ddot{x}$ are the displacement, velocity, and acceleration of the SDOF system, respectively; and

$$\lambda = \frac{2l_2^3 + 3l_1 l_2^2}{2(l_1 + l_2)^3} \tag{8}$$

where λ is the equivalent coefficient of the sea ice force.

3. PSDs of the Ice Force

3.1. Crushing Failure PSD

Based on a large amount of ice force practical measurement data of JZ9-3MDP in the Bohai Sea and the Norstromsgrund lighthouse, Kärnä and Qu [14,35] developed a crushing failure sea ice force power spectral density expression as follows:

$$S_c(f) = \frac{a\sigma^2}{1 + ka^{1.5}f^2}; \quad a = bV_{ice}^{-0.6} \tag{9}$$

where f (Hz) is the frequency of sea ice force; a and b are experimental parameters; V_{ice} is the ice velocity; and σ is the ice force standard deviation that is defined as:

$$\sigma = \frac{I_F}{1 + 3I_F} F^p; \quad F^p = \alpha\sigma_c Dh \tag{10}$$

where I_F is the interaction strength of the dynamic ice with a mean value of 0.4 MPa; F^p is the ice force amplitude when crushing failure happens; α is the comprehensive effect coefficient which lies between 0.4 and 0.7; σ_c is the ice compression strength; D is the loaded pile diameter of the structure; and h is the ice thickness.

3.2. Bending Failure PSD

In accordance with the ice load data collected on conical structures by a full-scale test in the Bohai Sea, Yue et al. [13,36] presented a bending failure sea ice force power spectral density as follows:

$$S_b(f) = \frac{10.88F_0^2 T_b^{-2.5}}{f^{3.5}} \exp\left(-5.47(fT_b)^{-0.64}\right) \tag{11}$$

where F_0 is the ice force amplitude when bending failure occurs; T_b is the ice force period. Their expressions are as follows:

$$F_0 = 3.2\sigma_f h^2 \left(\frac{D}{L_b}\right)^{0.34} ; \quad T_b = \frac{L_b}{v_{ice}} = \frac{\tau h}{v_{ice}} \tag{12}$$

where σ_f is ice bending strength; L_b is the breaking length of the ice sheet; and τ is the ratio between breaking length and ice thickness with a value around 7.3.

3.3. Compression Strength and Bending Strength

Through the experiment, Vaudrey and Li [37,38] proposed formulas of the compression and bending strength for sea ice as follows:

$$\sigma_c = 1.474 + 0.106|\theta_i| \tag{13}$$

$$\sigma_f = 0.998 - 0.063\sqrt{V_b} \tag{14}$$

where V_b is the brine volume ratio of the sea ice,

$$V_b = S_i(0.532 + 49.185/|\theta_i|) \tag{15}$$

where θ_i is the sea ice temperature; S_i is the sea ice salinity.

4. Ice Force Parameters

As shown in Equations (9)–(15), there are four important parameters, including the temperature θ_i, salinity S_i, thickness h, and velocity V_{ice}, which is related to the environment and affect the PSDs of the sea ice force. Essentially, they are all random variables and the way they are evaluated are explained in this section.

4.1. Ice Temperature and Salinity

Based on the different sea ice parameters, the Bohai Sea and the northern part of the Yellow Sea are divided into 21 zones by the China National Offshore Oil Production Research Centre and the National Marine Environmental Forecasting Center [39] as shown in Figure 2. Due to the scarcity of measured sea ice data in Zones 15–21, the ice parameters measured at a location within a certain zone is used to represent those of the zone. For instance, the data at Bayuquan Area represents that of Zone 21. The parameters of sea ice temperature and salinity corresponding to each ice zone in the Bohai Sea are listed in Table 1, which can be directly applied to Equations (12)–(15).

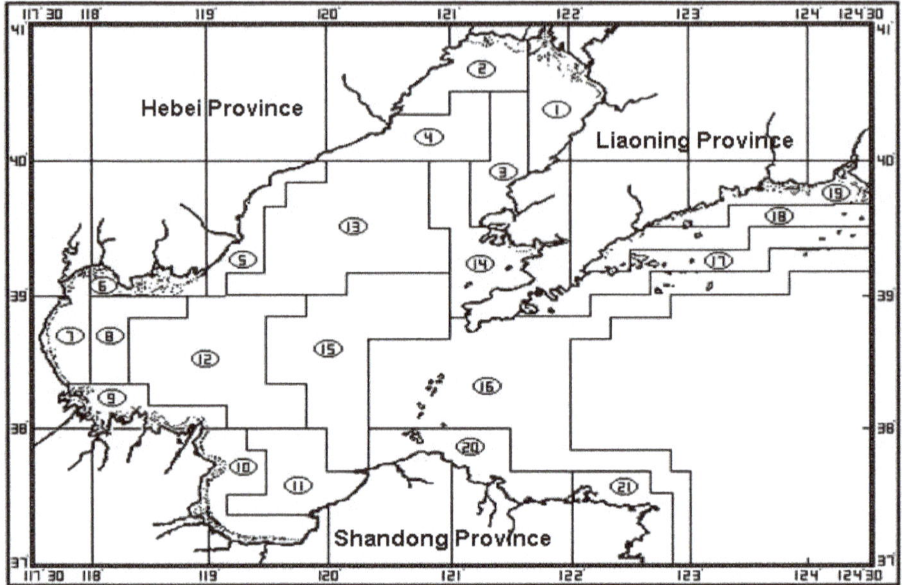

Figure 2. The divided ice zones of the Bohai Sea.

Table 1. Design values of sea ice temperature and salinity.

Zone Number	Temperature (°C)	Salinity (‰)
1	−5.4	5.50
2	−6.8	6.58
3	−5.3	7.24
4	−4.9	6.70
5	−4.5	6.70
6	−4.7	5.67
7	−4.5	4.57
8	−3.4	9.01
9	−4.5	4.29
10	−2.8	4.05
11	−3.5	4.57
12	−3.9	8.24
13	−4.2	7.50
14	−4.4	6.80
15 at Haihong Port 1	−11.1	6.3
16 at Haihong Port 2	−4.5	4.7
17 at Guanhai Trestle	−6.5	5.6
18 at Hongguang Wharf1	−2	5.6
19 at Hongguang Wharf2	−3.3	5.3
20 at Xing Cheng	−5	4.4
21 at Bayuquan Area	−9	4.8

4.2. Probability Distributions of Ice Parameters

4.2.1. Probability Distributions of Ice Thickness and Ice Velocity

Based on a large amount of data recorded from the Bohai Sea and the northern Yellow Sea during the years of 1968–1998, researchers [35,36] indicated that the annual maximum thickness and velocity of the sea ice follow the Gumbel-logistic distribution, i.e., the joint distribution function of ice thickness and velocity can be expressed as:

$$F(h, v_{ice}) = \exp\left\{-\left[(-\ln F_h(h))^m + (-\ln F_v(v_{ice}))^m\right]^{\frac{1}{m}}\right\}$$ (16)

where $F_h(h)$ and $F_v(v_{ice})$ are the marginal distribution of the random variable h and v_{ice}, respectively, which can be expressed as:

$$F_h(h) = \exp[-\exp(-\frac{h - a_h}{b_h})]; \quad F_v(v_{ice}) = \exp[-\exp(-\frac{v_{ice} - a_v}{b_v})]$$ (17)

where a_h and b_h are the estimated values of the location of Gumbel distribution for ice thickness; a_V and b_V are the estimated values of scale parameters of Gumbel distribution for ice velocity:

$$a_h = 4.51; \quad b_h = 4.56; \quad a_V = 2.90; \quad b_V = 3.41$$ (18)

In Equation (16), m ($m \geq 1$) is the correlation parameter that can be estimated as:

$$m = \frac{1}{\sqrt{(1 - \rho_{hv})}}; \quad \rho_{hv} = \frac{E[(h - \mu_h)(v_{ice} - \mu_v)]}{\sigma_h \sigma_v}$$ (19)

where μ and σ denote the mean value and the standard deviation, respectively. It is obvious that when $m = 1$, h and v_{ice} are uncorrelated; while $m \rightarrow \infty$ indicates that h and v_{ice} are perfectly correlated.

An exceedance probability would be considered in the practical structural design to define the ice load-carrying capacity of the offshore structures. In this case, the ice thickness and velocity can be determined as:

$$(h, v_{ice}) = F^{-1}\left(1 - \frac{1}{T}\right)$$ (20)

where T is the recurrence period of the sea ice; F^{-1} is the inverse function of Equation (16), which represents a spatial curved surface for specified T. The possible sea ice thickness and velocity with exceedance probabilities of 63.2% and 2% for Zone 6 in the Bohai Sea are given in Figure 3, which corresponds to frequently met sea ice force and rarely met sea ice force, respectively.

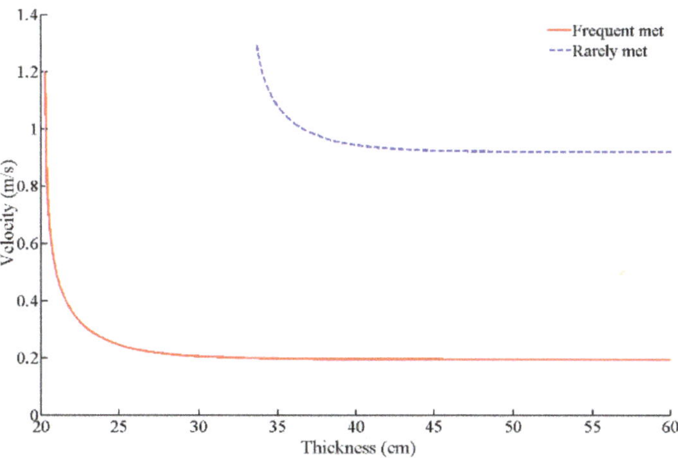

Figure 3. The possible sea ice thickness and velocity.

Since the parameters mentioned previously are given, the PSDs of sea ice force given in Equations (9) and (11) can then be calculated as well. Typical PSDs for frequent met sea ice force are given in Figure 4, whereas other parameters are listed in Table 2.

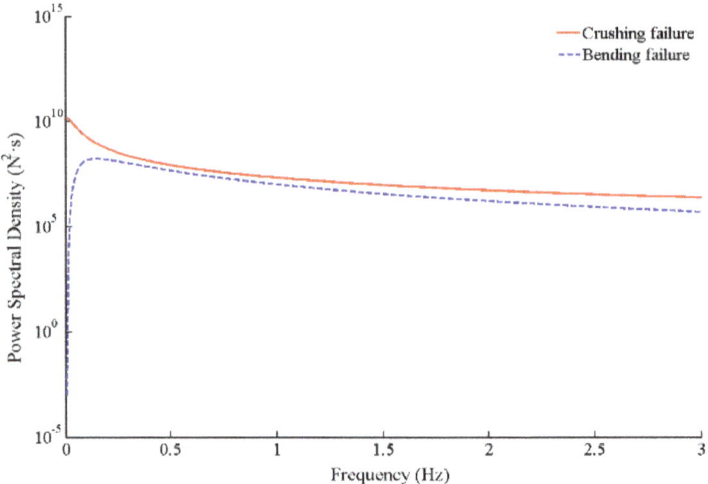

Figure 4. The power spectral densities (PSDs) of ice damage.

Table 2. Parameters for sea ice force PSDs.

k	b	I_F	β	α	h	D	V_{ice}
10.04	18.85	0.4	7.3	0.4	0.5 m	1 m	1 m/s

4.2.2. Joint Probability Density Function of Ice Thickness and Ice Velocity

The joint probability density function is usually used to describe the correlation between two variables, which is equal to the derivative of ice thickness h and ice velocity V_{ice} in Equation (16). Then, the joint probability density function of such random variables can be expressed as:

$$f_{XY}(x,y) = \frac{\partial^2 F_{XY}(x,y)}{\partial x \partial y} = \frac{F_{XY}(x,y)}{b_x b_y}\{\exp[-\frac{m(x-a_x)}{b_x}] + \exp[-\frac{m(y-a_y)}{b_y}]\}^{\frac{1-2m}{m}}$$
$$\cdot\{[\exp(-\frac{m(x-a_x)}{b_x}) + \exp(-\frac{m(y-a_y)}{b_y})]^{\frac{1}{m}} + m - 1\}.\exp[-m(\frac{x-a_x}{b_x} + \frac{y-a_y}{b_y})] \tag{21}$$

The ice zones in the Bohai Sea can be grouped into five groups through joint probability density, as listed in Table 3. The joint probability density of each zone in a group is close to each other. The independent variables x and y in the Equation (21) represent the thickness and velocity of ice, respectively. According to Equation (21), it can be discovered that the joint probability density increases with the increasing of ice thickness and ice velocity.

Table 3. Joint probability density of ice thickness and ice velocity for each ice zone.

Group Number	Ice Zone	Joint Probability Density
Group 1	Zone 1, Zone 2, Zone 6, Zone 19, Bayuquan	$0.2146e^{-4}$ to $3.4147e^{-4}$
Group 2	Zone 3, Zone 4, Zone 7, Zone 9, Haihong Port 1, Hongguang Wharf 2	$5.4056e^{-4}$ to $8.2285e^{-4}$
Group 3	Zone 5, Zone 10, Zone 18, Haihong Port 2, Guanhai Trestle, Xingcheng	1.0181^{-3} to $2.1499e^{-3}$
Group 4	Zone 8, Zone 11, Zone 14, Zone 17, Hongguang Wharf 1	$3.2114e^{-3}$ to $4.878e^{-3}$
Group 5	Zone 12, Zone 13, Zone 20	$5.1124e^{-3}$ to $5.27e^{-3}$

5. Proposed Sea Ice Response Spectrum

5.1. Generation of Ice Force Time-History Samples

For a zero-mean Gaussian Process $p(t)$ with PSD, the amplitude superposition method is used to synthesize ice force time-history samples [40]:

$$p(t) = \sum_k \sqrt{2S(f_k)\Delta f}\cos(2\pi f_k t + \phi_k) \tag{22}$$

where f_k $(k = 1, 2, 3, \cdots, M)$ denotes the k-th frequency point; Δf is the increment; and ϕ_k is the random phase that varies between $[0, 2\pi)$. Two generated samples are given in Figure 5.

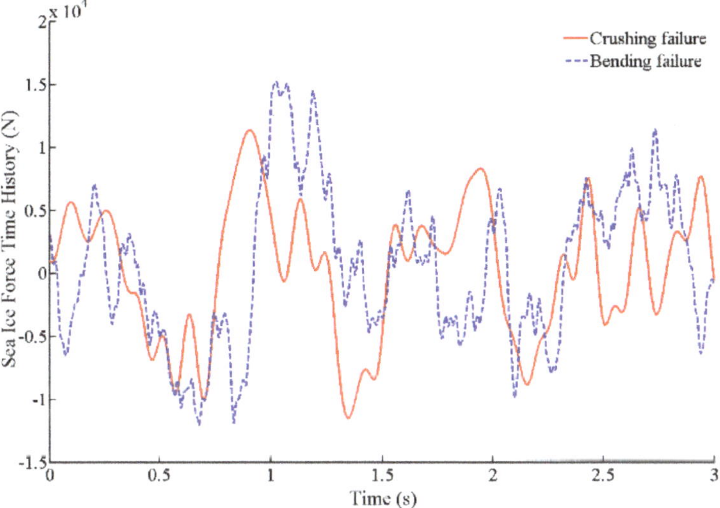

Figure 5. Ice force time-history samples.

It is noted that in Equations (10) and (12), the parameter D is a structural geometrical parameter. In order to make the proposed method generally applicable, the related item should be extracted. Substituting Equation (10) into Equation (9), the ice force crushing failure PSD can be rewritten as:

$$S_c(f) = D^2 \frac{bV_{ice}^{-0.6}I_F^2\alpha^2\sigma_c^2 h^2}{\left(1 + kb^{1.5}V_{ice}^{-0.9}f^2\right)(1 + 3I_F)^2} = D^2\overline{S}_c(f) \tag{23}$$

According to Equation (21), the time-history due to ice force crushing failure can be generated as:

$$p_c(t) = D\sum_k \sqrt{2\overline{S}_c(f_k)\Delta f}\cos(2\pi f_k t + \phi_k) = D\overline{p}_c(t) \tag{24}$$

Similarly, the ice force bending failure PSD and time-history generated can be expressed as:

$$S_b(f) = D^{0.68}\frac{111.4\sigma_f^2 h^4 T_b^{-2.5}}{(\beta h)^{0.68}f^{3.5}}\exp\left(-5.47(fT_b)^{-0.64}\right) = D^{0.68}\overline{S}_b(f) \tag{25}$$

$$p_b(t) = D^{0.34}\sum_k \sqrt{2\overline{S}_b(f_k)\Delta f}\cos(2\pi f_k t + \phi_k) = D^{0.34}\overline{p}_b(t) \tag{26}$$

5.2. Response Spectrums for Each Ice Zone

The spectral characteristics of sea ice crushing damage and bending damage are similar to those of structure under earthquakes, which have abundant frequencies. Referred to as the earthquake response spectrum theory [41–46], a similar sea ice response spectrum theory is established in this section. According to Equations (7), (24), and (26), the motion equation of a SDOF system subjected to the ice forces can be expressed as:

$$M\ddot{x} + C\dot{x} + Kx = \lambda_j \bar{p}_j(t); \quad (j = c, b) \tag{27}$$

where λ_c and λ_b are the feature coefficients of the offshore structure.

$$\lambda_c = \lambda D; \quad \lambda_b = \lambda D^{0.34} \tag{28}$$

Equation (27) can also be rewritten as:

$$\ddot{x} + 2\xi\omega\dot{x} + \omega^2 x = \frac{\lambda_j}{M}\bar{p}_j(t); \quad (j = c, b) \tag{29}$$

where ω is the natural frequency of the SDOF system; ξ is the damping ratio. By means of Duhamel integral, its solution is:

$$x(t) = \frac{\lambda_j}{M\omega'} \int_0^t e^{-\xi\omega(t-\tau)} \sin\omega'(t-\tau)\bar{p}_j(\tau)d\tau \tag{30}$$

where

$$\omega' = \omega\sqrt{1-\xi^2} \tag{31}$$

For the sake of simplifying the dynamic problem expressed by Equation (7) to a static problem, the equivalent sea ice force acting on the SDOF system would be expressed as:

$$F(t) = Kx(t) = \omega^2 Mx(t) \tag{32}$$

Ignore the tiny differences between ω' and ω, and substitute Equation (30) into Equation (32), then:

$$F(t) = \omega\lambda_j \int_0^t e^{-\xi\omega(t-\tau)} \sin\omega(t-\tau)\bar{p}_j(\tau)d\tau \tag{33}$$

The maximum absolute value of $F(t)$ is:

$$F_j^{\max} = \lambda_j \left| \omega \int_0^t e^{-\xi\omega(t-\tau)} \sin\omega(t-\tau)\bar{p}_j(\tau)d\tau \right|_{\max} = \lambda_j S_{aj} \tag{34}$$

where S_{aj} $(j = c, b)$ is the response spectrum of corresponding sea ice force time-history.

Normally speaking, a large amount of ice force time-history data based on in-situ measurements should be used as the excitation of the SDOF system to achieve the design response spectrum. Due to the scarity of the measured data, synthesized ice force time-histories have to be used. Considering the randomness of ice force due to the existence of phase φ in Equation (22), a large amount of ice force time-histories are synthesized for each zone. Based on Equation (34), a response spectrum corresponding to an ice force time-history can be calculated. Then, an envelope line covering the maximum response from the statistical data of each ice zone is fitted and normalized as the acceleration coefficient $\beta_{\max}$. Acceleration response spectrums of crushing and bending failure are shown in Figures 6 and 7 based on joint probability density of ice thickness and ice velocity in Section 4.

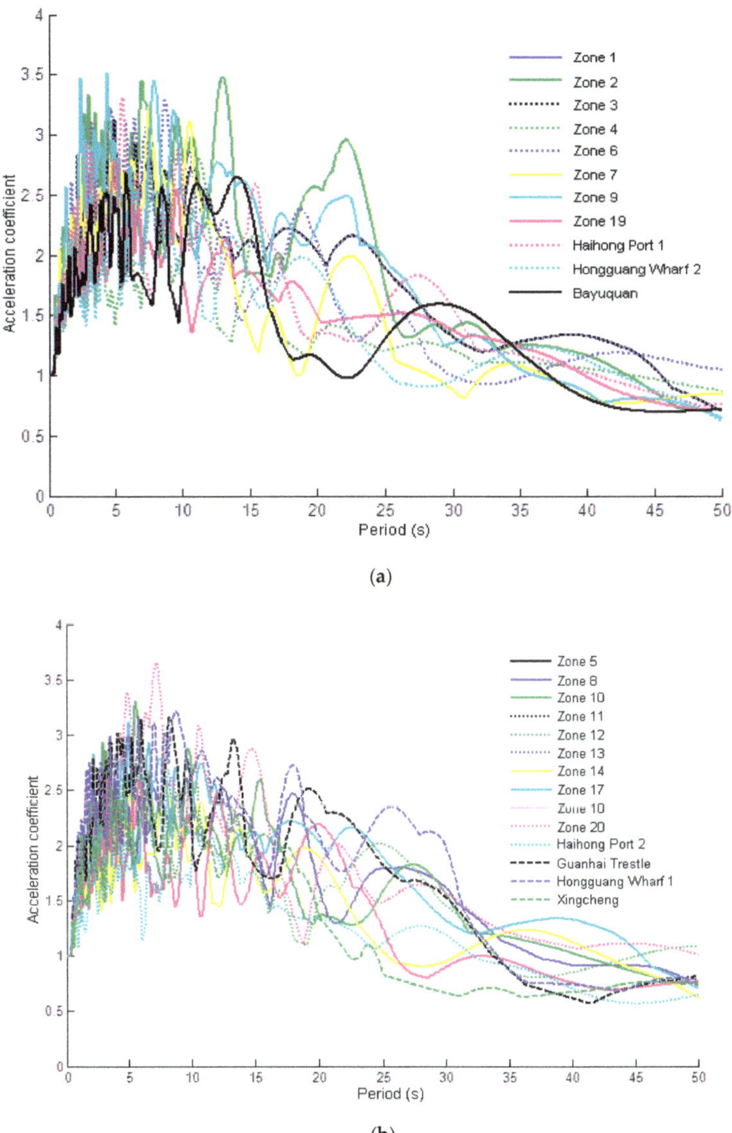

(**a**)

(**b**)

Figure 6. Response spectrums of crushing failure. (**a**) In groups 1, 2; (**b**) in groups 3, 4, and 5.

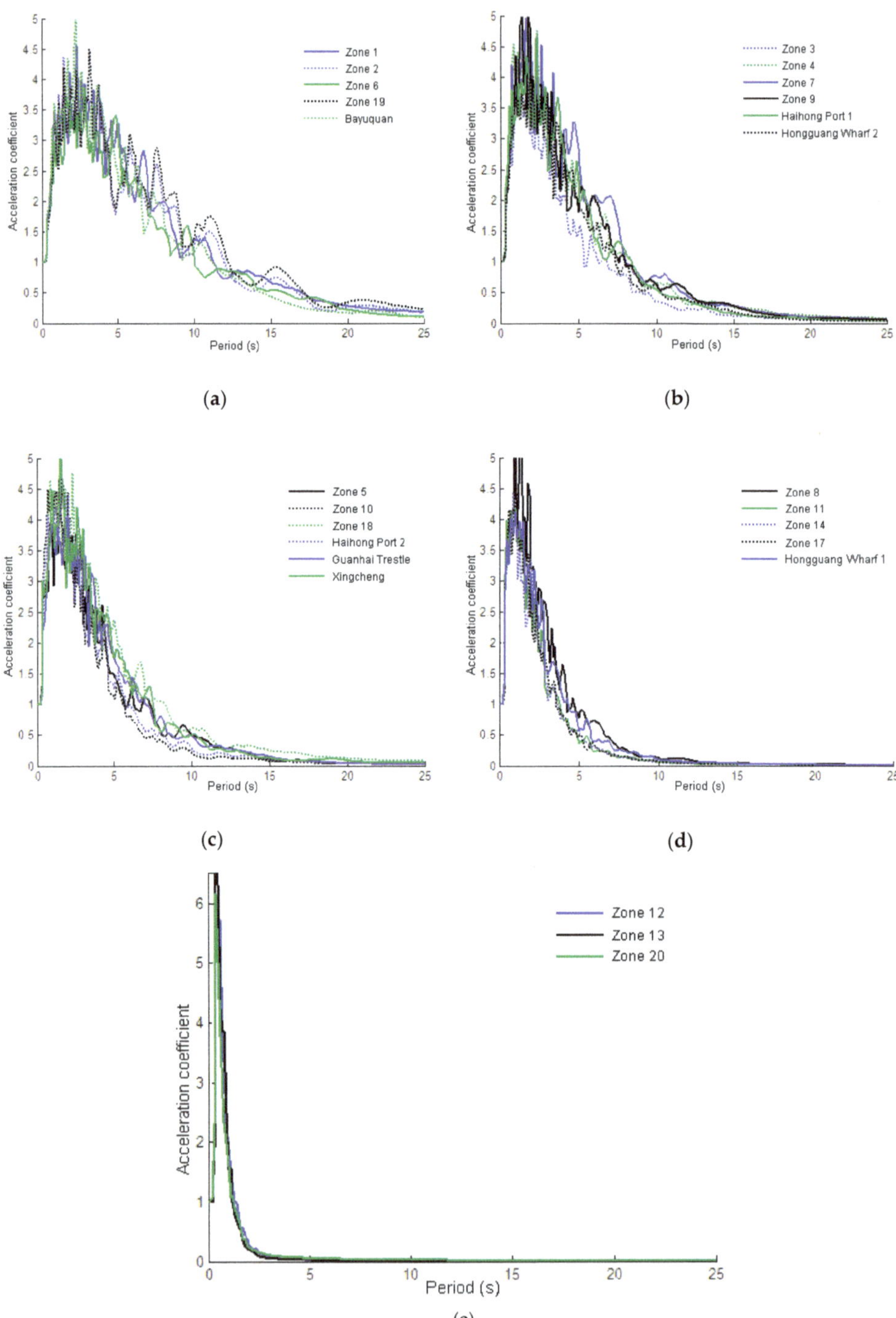

Figure 7. Response spectrums of bending failure. (**a**) In group 1; (**b**) in group 2; (**c**) in group 3; (**d**) in group 4; (**e**) in group 5.

5.3. Design Response Spectrum for the Bohai Sea in China

Based on Equation (34), it is to be noted that the value of S_{aj} changes with the ice force time-history $\bar{p}_j(t)$, the natural frequency ω, and the damping ratio ξ of the SDOF system. By taking the maximum data of all response spectrum curves obtained from different sea ice time-history samples of each ice zone, and using the piecewise fitting method, the design response spectrum of the sea ice force is achieved, which is shown in Figure 8 and denoted by α_j ($j = c, b$), where η_0, η, γ, p_1 and p_2 are the parameters related to structural damping.

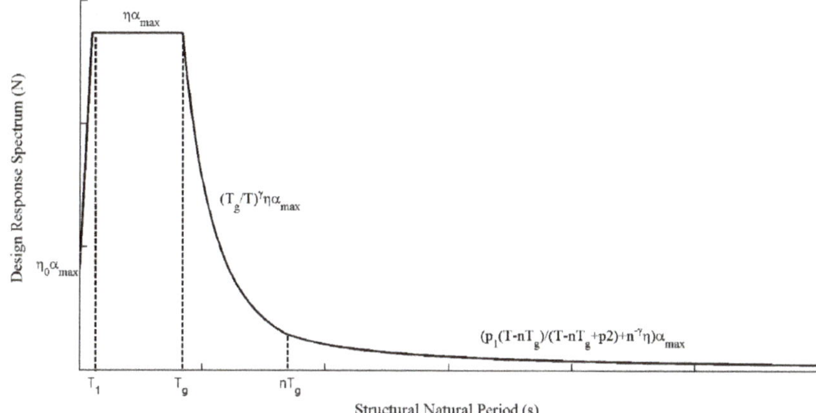

Figure 8. Design Response Spectrum.

Selecting Zone 6 as an example and taking numerical fitting, the design spectrum parameters for crushing failure sea ice force can be obtained:

$$\eta = 1 + \frac{0.05 - \xi}{0.07 + 2.1\xi}; \quad \eta_0 = 0.3\eta \tag{35}$$

$$\gamma = \frac{2.7\xi + 0.03}{\xi + 7.8 \times 10^{-3}} \tag{36}$$

$$p_1 = 0.45\xi - 0.16; \quad p_2 = 4.9\xi + 0.77 \tag{37}$$

Additionally for bending failure sea ice force:

$$\eta = 1 + \frac{0.05 - \xi}{0.05 + 3.3\xi}; \quad \eta_0 = 0.2\eta \tag{38}$$

$$\gamma = \frac{2.9\xi + 0.02}{\xi + 3.5 \times 10^{-3}} \tag{39}$$

$$p_1 = 0.46\xi - 0.13; \quad p_2 = 5.3\xi + 0.80 \tag{40}$$

Other fitting parameters combined ice velocity and thickness of exceedance probabilities for Zone 6 in the Bohai Sea are given in Tables 4 and 5.

Table 4. Fitting parameters for Zone 6.

Parameter	Crushing	Bending
$T_1(s)$	0.1	0.2
T_g (s)	0.84	0.84
n	2	2

Table 5. The value of $\alpha_{\max}$ for Zone 6.

Condition	Crushing	Bending
Frequent met	9.5×10^4	7.2×10^4
Rarely met	1.3×10^5	1.6×10^5

By replacing $F_j^{\max}$ and S_{aj} by F_{ice} and α_j in Equation (34), respectively, the stochastic sea ice force acting on a structure can then be estimated easily in a static way as:

$$F_j^{ice} = \lambda_j \alpha_j; \quad (j = c, b) \tag{41}$$

Hence, the proposed response spectrum method can be applied as follows [47]:

Step 1, Calculate η_0, η, γ, p_1 and p_2 based on Equations (35)–(40).

Step 2, Determine the value of other fitting parameters from Tables 4 and 5.

Step 3, Obtain the static sea ice force using Equation (41).

Step 4, Compute the concerned structural response.

6. Numerical Example

In this section, selecting Zone 6 as an example, the proposed method is verified by comparing the results derived from the Monte-Carlo simulation. Taking the SDOF offshore structure as the research object, whose lumped mass is 10^6 kg and damping ratio is 0.02, and which natural period varies with the changing stiffness. Assume that the loaded pile diameter of the structure is 2.7 m and the heights above and under the sea level are 20 m and 80 m, respectively. The displacements excited by crushing and bending failure sea ice force are studied, respectively.

In the Monte-Carlo simulation, 200 time-history samples are generated for both crushing and bending failure sea ice force. Based on the Monte-Carlo method and the ice force response method, Figures 9 and 10 show the displacements of the offshore structure with different structural natural periods subjected to crushing and bending failure sea ice force. It can be seen that the results obtained from the two methods display a manifested same trend.

From Figures 9 and 10, the displacement induced by ice bending failure is less than that by ice crushing failure. Meanwhile, it can be seen that the displacement under rarely met sea ice force is greater than that under frequent met sea ice force. Generally speaking, the proposed method provides an upper limit.

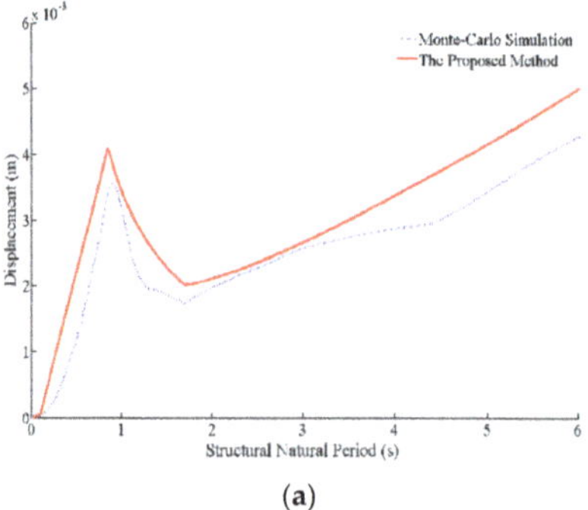

(a)

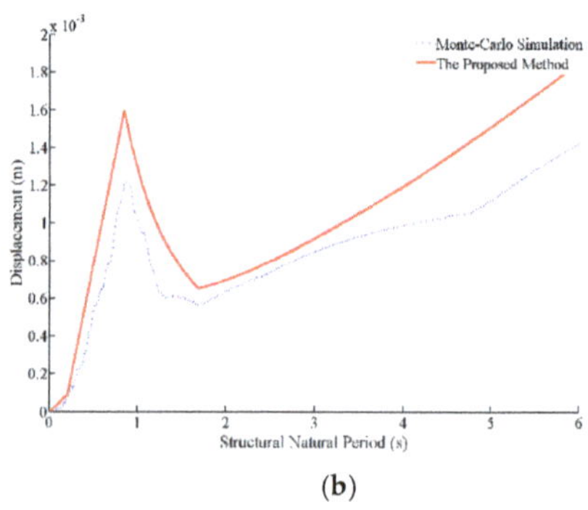

(b)

Figure 9. The displacements subjected to frequent met sea ice force. (a) Sea ice crushing failure; (b) sea ice bending failure.

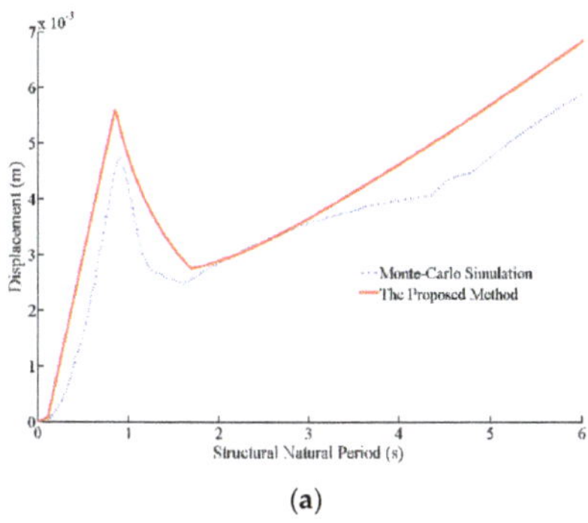
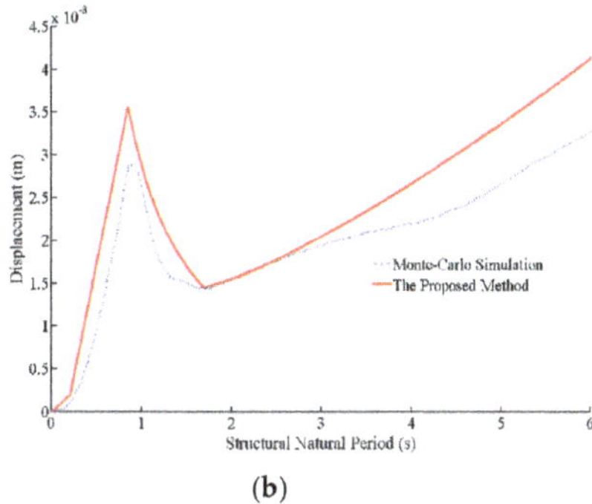

(a)

(b)

Figure 10. The displacements subjected to rarely met sea ice force. (**a**) Sea ice crushing failure; (**b**) sea ice bending failure.

As shown in Figures 9 and 10, for the offshore structures with natural periods in the interval of $[0, T_1]$, which have either large rigidity or light-weight, the response such as displacement will increase with the increasing structural natural periods. For the offshore structure with natural period in the interval of $(T_1, nT_g]$, the response will increase remarkably. It is clear that resonance will occur if the structure natural period is close to T_g. When T is larger than nT_g, the structural displacement will increase with the increase of structural natural periods due to the decreasing structural stiffness.

The computing time of the example required for the proposed method is only 0.04 s while, that for the Monte-Carlo simulation is 7537.6 s. It is obvious that the proposed method is not only easy to calculate but also matches the needs in engineering and has common applicability.

7. Conclusions

Referred to as the earthquake response spectrum theory, a new design idea to determine the maximum response of offshore structures subjected to ice forces is suggested.

(1) Considering the randomness of ice force and the complexity of structures, the theory of response spectrum suitable for offshore structures subjected to crushing and bending failure sea ice forces is established.

(2) Selecting Zone 6 in the Bohai Sea, dynamic analysis of SDOF structures subjected to synthesized ice force time-histories is performed. Then, the design response spectrums for fixed offshore structures subjected to ice forces induced by crushing and bending failure are proposed, respectively.

(3) Compared with results from the Monte-Carlo simulation and the proposed method, the proposed method is validated. Additionally, the proposed method provides an upper limit of offshore structural response subjected to ice force.

(4) For the offshore structure with a natural period in the range of $(T_1, nT_g]$, the response under ice force will increase remarkably. The maximum response will occur when the structure natural period is close to T_g.

Author Contributions: Y.L. and Y.Z.: Writing and data collection; X.L.: Data analysis, interpretation and revising; Y.L. and Y.Z.: methodology and software; Y.L.: Numerical analysis and revising. All the authors contributed to the first draft and final version of the paper.

Funding: This study was supported by the National Natural Science Foundation of China (Grant No. 51879040).

Conflicts of Interest: The authors declare no conflict of interest.

Nomenclature:

Parameter	Nomenclature	Parameter	Nomenclature
m	lumped mass of a single-degree of freedom system (SDOFS)	$p'(t)$	an equivalent force of SDOFS
k	shearing stiffness of SDOFS	F^p	ice force amplitude when crushing failure happens
c	damping coefficient of SDOFS	F_0	ice force amplitude when bending failure happens
h	sea ice thickness	$F_h(h)$ $F_v(v_{ice})$	the marginal distribution
$l_1 l_2$	heights above and under the sea level	F_{ice}	ice force
x	lateral displacement of point A	T_b	the ice force period
θ	rotation of point A	T	recurrence period of the sea ice
λ	the equivalent coefficient of the sea ice force	θ_i	sea ice temperature
f	sea ice force frequency	S_i	sea ice salinity
f_k	the k-th frequency point	ϕ_k	random phase that varies between [0, 2π)
Δf	the increment frequency	ω	natural frequency of the SDOFS
σ	ice force standard deviation	ξ	damping ratio of offshore structure
σ_c	ice compression strength	L_b	the breaking length of ice sheet
σ_f	ice flexural strength	η_0	a parameter that related to structural damping
V_{ice}	sea ice velocity	η	a parameter that related to structural damping
D	loaded pile diameter of the structure	γ	a parameter that related to structural damping
τ	the ratio between breaking length and ice thickness	p_1	a parameter that related to structural damping
$S(f)$	power spectral densities	p_2	a parameter that related to structural damping
S_{aj}	the response spectrum of corresponding sea ice force time history	I_F	the interaction strength of the dynamic ice with a mean value of 0.4
α	ice force of design response spectrum for Bohai Sea	$a\ b$	experimental parameters

References

1. Kotilainen, M. Predicting ice-induced load amplitudes on ship bow conditional on ice thickness and ship speed in the Baltic Sea. *Cold Reg. Sci. Technol.* **2017**, *135*, 116–126. [CrossRef]
2. Ranta, J.; Polojärvi, A.; Tuhkuri, J. The statistical analysis of peak ice loads in a simulated ice-structure interaction process. *Cold Reg. Sci. Technol.* **2017**, *133*, 46–55. [CrossRef]
3. Li, W.; Huang, Y.; Tian, Y. Experimental study of the ice loads on multi-piled oil piers in Bohai Sea. *Mar. Struct.* **2017**, *56*, 1–23. [CrossRef]
4. Shun, Y.; Yue, O.J.; Bi, X.J. Probability distribution of sea ice fatigue parameters in JZ20-2 sea area of the Liaodong Bay. *Ocean Eng.* **2002**, *20*, 39–44.
5. Su, B.; Riska, K.; Moan, T. Numerical simulation of local ice loads in uniform and randomly varying ice conditions. *Cold Reg. Sci. Technol.* **2011**, *65*, 145–159. [CrossRef]
6. Neill, R.C. Dynamic ice forces on piers and piles. An assessment of design guidelines in the light of recent research. *Can. J. Civ. Eng.* **1976**, *3*, 305–341. [CrossRef]
7. Nord, T.S. Model-based force and state estimation in experimental ice-induced vibrations by means of Kalman filtering. *Cold Reg. Sci. Technol.* **2015**, *111*, 13–26. [CrossRef]
8. Ralston, T.D.; Riska, K. Ice force design considerations for conical offshore structures. *Force* **1977**, *2*.
9. Zhang, F.J. The significance of sea ice forces in the design of ocean engineering structures in the Bohai Sea. *Ocean Eng.* **1987**, *5*, 59–65.

10. Nord, T.S.; Øiseth, O.; Lourens, E. Ice force identification on the Nordströmsgrund lighthouse. *Comput. Struct.* **2016**, *169*, 24–39. [CrossRef]

11. Qu, Y. Ice force spectrum on narrow conical structures. *Cold Reg. Sci. Technol.* **2007**, *49*, 161–169.

12. Barker, A. Ice loading on Danish wind turbines. *Cold Reg. Sci. Technol.* **2005**, *41*, 1–23. [CrossRef]

13. Gravesen, H. Ice loading on Danish wind turbines: Part 2. Analyses of dynamic model test results. *Cold Reg. Sci. Technol.* **2005**, *41*, 25–47. [CrossRef]

14. Kärnä, T.; Qu, Y. A New Spectral Method for Modeling Dynamic Ice Actions. In Proceedings of the 23rd International Conference on Offshore Mechanics and Arctic Engineering, Vancouver, BC, Canada, 20–25 June 2004.

15. Hunke, E.C.; Dukowicz, J. An elastic-viscous-plastic model for sea ice dynamics. *J. Phys. Oceanogr.* **1997**, *27*, 1849–1867. [CrossRef]

16. Li, B.H.; Li, H. A modified discrete element model for sea ice dynamics. *Acta Oceanol. Sin.* **2014**, *33*, 56–63. [CrossRef]

17. Godlovitch, D.; Monahan, A. An idealised stochastic model of sea ice thickness dynamics. *Cold Reg. Sci. Technol.* **2012**, *78*, 14–30. [CrossRef]

18. Qu, Y. A random ice force model for narrow conical structures. *Cold Reg. Sci. Technol.* **2006**, *45*, 148–157.

19. Wang, K.; Leppäranta, M. A sea ice dynamics model for the Gulf of Riga. *Proc. Est. Acad. Sci. Eng.* **2003**, *9*, 107–125.

20. Pedersen, L.T.; Coon, M.D. A sea ice model for the marginal ice zone with an application to the Greenland Sea. *J. Geophys. Res. Ocean.* **2004**, *109*, 1–8. [CrossRef]

21. Lu, Q.M.; Larsen, J.; Tryde, P. A dynamic and thermodynamic sea ice model for subpolar regions. *J. Geophys. Res. Ocean.* **1990**, *95*, 13433–13457. [CrossRef]

22. Shi, Q.Z. Dynamics of sea ice and spectrums of ice force. *China Ocean Eng.* **1994**, *16*, 107–111.

23. Zhi, H. Identification of random ice force spectra of offshore platforms. *Ocean Eng.* **2000**, *18*, 14–17.

24. Ou, J.P. A stochastic process model of ice pressure on a Bohai jacket platform and its parameters. *Acta Oceanol. Sin.* **1998**, *20*, 111–118.

25. Lee, J.H. Characteristics analysis of local ice load signals in ice-covered waters. *Int. J. Nav. Archit. Ocean Eng.* **2016**, *8*, 66–72. [CrossRef]

26. Kim, E.; Amdahl, J. Discussion of assumptions behind rule-based ice loads due to crushing. *Ocean Eng.* **2016**, *119*, 249–261. [CrossRef]

27. Huang, Y. Experimental study on ice destruction mechanism. *Ship Build. China* **2003**, *44*, 380–386.

28. Suominen, M. Influence of load length on short-term ice load statistics in full-scale. *Mar. Struct.* **2017**, *52*, 153–172. [CrossRef]

29. Zhang, X. Steady Vibration of Ice-Induced Vertical Structure. Master's Thesis, Dalian University of Technology, Dalian, China, 2002.

30. Jones, K.; Eylander, J. Vertical variation of ice loads from freezing rain. *Cold Reg. Sci. Technol.* **2017**, *143*, 126–136. [CrossRef]

31. Hendrikse, H.; Metrikine, A. Ice-induced vibrations and ice buckling. *Cold Reg. Sci. Technol.* **2016**, *131*, 129–141. [CrossRef]

32. Aksenov, Y.; Hibler, W.D. Failure Propagation Effects in an Anisotropic Sea Ice Dynamics Model. In Proceedings of the IUTAM Symposium on Scaling Laws in Ice Mechanics and Ice Dynamics Fairbanks, Alaska, USA, 13–16 June 2000; Volume 94, pp. 363–372.

33. Gagnon, R.E. A numerical model of ice crushing using a foam analogue. *Cold Reg. Sci. Technol.* **2011**, *65*, 335–350. [CrossRef]

34. Sopper, R. The influence of water, snow and granular ice on ice failure processes, ice load magnitude and process pressure. *Cold Reg. Sci. Technol.* **2017**, *139*, 51–64. [CrossRef]

35. Qu, Y. Random Ice Load Analysis on Offshore Structures Based on Field Tests. Ph.D. Thesis, Dalian University of Technology, Dalian, China, 2006.

36. Yue, Q.J. Fatigue-Life Analysis of Ice Resistant Platforms. *Eng. Mech.* **2017**, *24*, 159–164.

37. Vaudrey. *Elastic Property Studies on Compressive and Flexural Sea Ice Specimens*; Civil Engeneering Lab: Port Hueneme, CA, USA, 1975; Volume 12, pp. 1–16.

38. Li, Z.J. Preliminary statistics on the elements involved in sea ice in Liaodong Bay. *Ocean Eng.* **1992**, *10*, 72–78.

39. *Regulations for Offshore Ice Condition and Application in China Sea, in Q/HSn 3000-2002*; The China National Offshore Oil Production Research Centre and the National Marine Environmental Forecasting Center: China, 2002; p. 34.

40. Qu, Y.; Bi, X. Ice-induced Jacket Structure Vibrationss in Bohai Sea. *J. Cold Reg. Eng.* **2017**, *14*, 81–92.

41. Bani-Hani, K.A.; Malkawi, A.I. A Multi-step approach to generate response-spectrum-compatible artificial earthquake accelerograms. *Soil Dyn. Earthq. Eng.* **2017**, *97*, 117–132. [CrossRef]

42. Leppäranta, M. A treatise on frequency spectrum of drift ice velocity. *Cold Reg. Sci. Technol.* **2012**, *76*, 83–91. [CrossRef]

43. Cacciola, P.; Amico, L.D.; Zentner, I. New insights in the analysis of the structural response to response-spectrum-compatible accelerograms. *Eng. Struct.* **2014**, *78*, 3–16. [CrossRef]

44. Cacciola, P.; Zentner, I. Generation of response-spectrum-compatible artificial earthquake accelerograms with random joint time-frequency distributions. *Probab. Eng. Mech.* **2012**, *28*, 52–58. [CrossRef]

45. Yaghmaei-Sabegh, S.; Jalali-Milani, N. Pounding force response spectrum for near-field and far-field earthquakes. *Sci. Iran.* **2012**, *19*, 1236–1250. [CrossRef]

46. Spanos, P.D.; Giaralis, A. Third-order statistical linearization-based approach to derive equivalent linear properties of bilinear hysteretic systems for seismic response spectrum analysis. *Struct. Saf.* **2013**, *44*, 59–69. [CrossRef]

47. Xiao, Y.G. Sectional least square fitting method for calibrating seismic design response spectrum. *World Earthq. Eng.* **2012**, *28*, 29–33.

Journal of
Marine Science and Engineering

Article

Model Tests on the Frequency Responses of Offshore Monopiles

Rui He [1,2,*] and Tao Zhu [1,2]

[1] Key Laboratory of Coastal Disaster and Defense, Hohai University, Ministry of Education, Nanjing 210024, China; zhutaohhu@hhu.edu.cn
[2] College of Harbor, Coastal and Offshore Engineering, Hohai University, Nanjing 210024, China
* Correspondence: herui@hhu.edu.cn

Received: 30 September 2019; Accepted: 31 October 2019; Published: 26 November 2019

Abstract: Monopiles are widely used to support offshore wind turbines as a result of the extensive development of offshore wind energy in coastal areas of China. An offshore wind turbine is a typical high-rise structure sensitive to dynamic loads in ocean environment such as winds, water waves, currents and seismic waves. Most of the existing researches focus on elastic vibration analysis, bearing capacity or cyclic degradation problems. There're very few studies on vibration of monopiles, especially considering the influence of static loads with different amplitudes, directions, and loading-unloading-reloading processes. In this paper, laboratory-scale 1 g model tests for a monopile in dry sands were carried out to investigate the frequency responses of the monopile under different loading conditions. The bearing capacities of the model monopile were obtained as references, and dynamic loads and static loads with different amplitudes were then applied to the monopile. It was found that (1) the first resonant frequency of the monopile decreases with the increase of dynamic load amplitudes; (2) the first resonant frequency of the monopile steadily increases under the lateral static load and loading-unloading-reloading processes; (3) the frequency responses of the monopile with static loads in different directions are also quite different; (4) damping of the monopile is influenced by the load amplitudes, load frequencies, load directions and soil conditions. Besides, all the tests were conducted in both loose sand and dense sand, and the results are almost consistent in general but more obvious in the dense sand case.

Keywords: monopile; horizontal vibration; combined static and dynamic loads; different loading directions; frequency response functions

---

1. Introduction

Monopiles have been widely used in offshore wind engineering for their low costs, short construction periods and small environmental constraints. So, they are recommended by DNV (DET NORSKE VERITAS) code to be the well suited foundation type in the offshore wind power industry for water depths below 25 m [1]. Richards and Byrne [2] pointed out that 87% of the built foundations of offshore wind turbines are monopile foundations with large diameters. In recent years, monopiles are also widely used in Jiangsu, Zhejiang and other coastal areas of China where the surface layer of the seabed is mostly soft soil, while the bottom layer of the seabed is mainly fine sand.

Offshore wind turbines are typical high-rise and flexible structures with low natural frequencies, which can be very close to the frequencies of offshore dynamic loads such as winds, waves, currents and seismic waves (Figure 1a). As many studies have mentioned [3–7], the natural frequencies of offshore wind turbines should be controlled in a very narrow interval (1p–3p) to avoid the harm of resonances. Accordingly, vibration characteristics or dynamic impedances of pile foundations are of great concern to designers. Numerous methods are available in the literature to obtain these vibration characteristics, including:

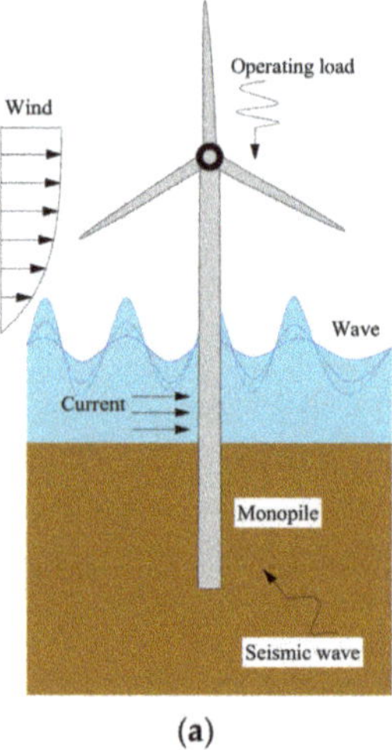

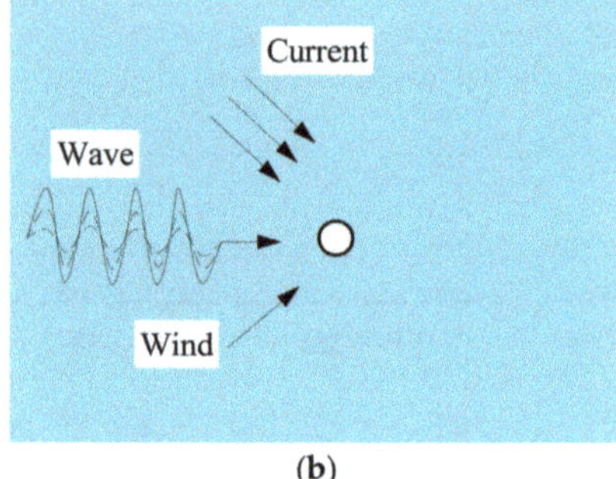

(a)

(b)

Figure 1. Loads on offshore wind turbines: (**a**) Front view; (**b**) Possible load direction from vertical view in water.

Analytic methods: Winkler type model—The soils around the pile are simplified into a series of independent 1D (one-dimensional) springs [8]; Plain strain model—The soils around the pile are treated as infinitely thin and independent 2D (two-dimensional) layers [9]; Virtual pile-soil model—The soil-pile system is decomposed into a fictitious pile and an extended 3D (three-dimensional) half-space [10]; Integral equation method, which is the most accurate analytical method to study 3D shell type foundations in offshore engineering—The theories of Cauchy singular and Fredholm integral equations are applied to solve the Green's function of pile and soil [11,12]; He et al. [13,14] obtained a rigorous analytical solution for coupled horizontal and rocking vibration of a monopile embedded in a porous seabed, and found that the effect of vertical shear stress on the monopile caused by horizontal loads and moments cannot be ignored.

Numerical methods: Zdravkovic et al. [15] presented a numerical study for laterally loaded monopiles; Sen et al. [16] presented a boundary element formulation for dynamic analysis of axially and laterally loaded single piles and pile groups; Latini and Zania [17] investigated the dynamic responses and the dynamic impedance of suction caissons by analyzing 3D finite element models in the frequency domain; Tao et al. [18] studied the influences of frequency, slenderness ratio, thickness–diameter ratio, soil–pile elastic modulus ratio and the existence of a scour hole on the dynamic impedances of monopile foundations; He et al. [19] analyzed the dynamic impedances and dynamic responses of large diameter rock-socketed monopiles under harmonic load based on a combined finite element–infinite element model; Ma et al. [20] presented a three-dimensional finite element model for analyzing the long-term performance of offshore wind turbines on monopiles in sand; Beuckelaers [21] brought insights into the behavior of monopile foundations and provided various modelling options to capture unloading, ratcheting and gapping effects for laterally loaded piles based on the Pile Soil Analysis (PISA) project.

Model experiments: Goit et al. [22] conducted dynamic experiments model soil–pile tests on a shaking table; Bhattacharya et al. [23] studied dynamic soil–pile interaction using a 1 g scale model test; Manna and Baidya [24] calculated the dynamic stiffness and damping of a slender pile and observed that the stiffness and damping decrease as the amplitude of load increases; Mohamed and Hesham [25] investigated the lateral vibration performance of two full-scale large-capacity helical piles and one driven pile installed in overconsolidated and structured clay, and observed a similar phenomenon to Manna and Baidya [24]; Lombardi et al. [5] investigated changes in pile natural frequency and damping after 32,000–172,000 cycles of horizontal loading; He et al. [26] studied the decreases of the pile's natural frequencies with the existence of a scour hole; Futai et al. [4] measured the natural frequency of piles in centrifuge tests by FFT (Fast Fourier Transform), and investigated the influences of the density of sand and the ratio of free length to embedded depth; Leblanc et al. [27] carried out a series of laboratory tests where a stiff pile in dry sand was subjected to 8000–60,000 cycles of combined moment and horizontal loading, and presented the accumulated rotation and changes in stiffness after long-term cyclic loading; Richards et al. [2] presented results from laboratory tests in dry sand, which explored pile rotation with multidirectional cyclic loadings.

Full scale tests: Shirzadeh et al. [28] identified the damping values of an offshore wind turbine on a monopile foundation using both field measurements and simulations; Damgaard et al. [29] presented field vibration tests and numerical studies of the cross-wind dynamic properties of offshore wind turbines. Full scale tests are very significant but too expensive, and there are very few full scale tests that can be referred to.

The analytic methods [8–16] and numerical methods [17–19] mostly depend on the assumption that the soil is linear elastic so only the dynamic responses under small stress/strain conditions can be obtained, and the nonlinearity of soil is ignored. When considering the nonlinearity of soil, bearing capacity analysis [30–35] and accumulated deformation [36–39] under static/cyclic loads are mostly studied. However, the influence of the static loads on the dynamic behavior of the soil-pile system is very scarce. Actually, the offshore wind turbines are exposed to lateral loads caused by winds, water waves, currents and seismic waves with different load directions, as shown in Figure 1. These types of loads can be decomposed to combinations of static loads and dynamic loads, while the directions of the static loads and dynamic loads can also be different, and the effect of static load amplitudes and directions on the dynamic responses of monopiles is very important but rarely understood. Besides, extreme conditions, such as typhoons on the southeast coast of China, have a very significant impact on offshore structures as they can force monopiles to reach ultimate bearing conditions, but whether the dynamic characteristics of monopiles under such conditions change during this loading-unloading-reloading process is still unclear. As the stress-strain state of the soil around the monopile is very complicated under coupled loads, a model test is conducted to study these topics preliminarily, which aims to investigate how the frequency responses of the monopile change under different lateral static load cases, including loading-unloading-reloading processes. Compared with the existing researches, the novelty of this paper is that it studies the evolution of vibration characteristics of monopiles under complex loading conditions, which is rarely investigated at present.

2. Similitude Relationships

Appropriate scaling laws constitute the first step to deduce the results of a prototype from an experimental study. Based on a perfect scaling law, the prediction of the prototype may be carried out from results obtained in model tests. However, a coupled wind-wave-structure-soil system is involved in the study of offshore wind turbines, which means that almost no physical model test technology can concurrently satisfy all the physical interactions. Specific analysis is needed for concerned research targets. In this model test, the following physical parameters are considered, as shown in Table 1:

(1) Geometric parameters: He et al. [13] pointed out that the dynamic impedances of a monopile are related to the embedded aspect ratio L/D, the elastic modulus ratio between pile and soil (E_p/E_s) and the thickness–diameter ratio (h/D). The length–diameter ratio (L/D) of modern monopiles used in offshore wind turbines is very small, from 3 to 8. The thickness–diameter ratio (h/D) is about 0.01. As a result, the L/D of the model pile in the test is chosen as 5, and h/D is 0.01.

(2) E_p/E_s: For cohesionless soil, when the shear strain amplitude is less than 10^{-4}, the shear modulus G_0 is mainly related to the void ratio e and the average effective principal stress σ_m' [40]. For round sand ($e < 0.8$), G_0 can be estimated as

$$G_0 = 6934 \times \frac{(2.17 - e)^2}{1 + e} (\sigma_m')^2 \text{ kPa} \tag{1}$$

where e is the void ratio related to the relative density D_r. For fixed relative density, σ_m' is proportional to the depth h_s. The average elastic modulus of soil E_s is simplified to the elastic modulus of sand at the depth of around 0.5–1.0 D, for simplicity. The pile is made of steel $E_p = 200$ GPa, so $(E_p/E_s)_{\text{model}} \approx 10{,}000$–$20{,}000$, while E_s in the prototype is about 10–250 MPa, with $(E_p/E_s)_{\text{prototype}} \approx 840$–$21{,}000$, which means monopiles in both the model and prototype act like rigid piles. However, tests in loose sand ($D_r = 10\%$, $E_p/E_s \approx 21{,}000$) and dense sand ($D_r = 88\%$, $E_p/E_s \approx 13{,}000$) are conducted to investigate the influence of the elastic modulus ratio.

(3) Dimensionless frequency (a_0): According to soil dynamics, the responses are frequency-dependent, and the nondimensional frequency $a_0 = r\omega / \sqrt{\mu_s/\rho_s}$ is especially useful when analyzing the obtained results, where r is the radius of the pile, ω is the circular frequency of the load, μ_s is the shear modulus of the soil, and ρ_s is the density of the soil. a_0 has been chosen including 0–0.5 in accordance with most pile dynamic analysis works.

Table 1. Dimensionless characters in the test.

Physical Parameters	Dimensionless Characters	Value ($D_r = 10\%$)	Value ($D_r = 88\%$)
Embedded depth L	L/D	5	5
Wall thickness of the pile h	h/D	0.01	0.01
Elastic modulus E	E_p/E_s	21,000	13,000
Static load H	H/H_u	0–1.08	0–1.86
Harmonic load Amplitude H_{amp}	H_{amp}/H_u	1.35–27.03%	0.65–3.25%
Frequency f	$r\omega / \sqrt{\mu_s/\rho_s}$	0–0.72	0–0.56

3. Experimental Formulation

3.1. Test Platform

The experimental investigation was carried out in the Offshore Geotechnical Engineering Laboratory in Hohai University. Tests were conducted in an iron tank with the size of 600 × 600 × 1500 mm, which was filled with dry Nanjing quartz sand. The box was elevated above the ground to install a valve to release the sand in the box conveniently when the test was finished. A coherence analysis between the tank and the pile was carried out at first to ensure that the vibration of pile foundations is not affected by tank vibration (Figure 2).

The monopile foundation was scaled at around 1:50–1:100 to an open-ended steel model pile with diameter $D = 60$ mm, length $L_0 = 600$ mm, and wall thickness $h = 0.6$ mm; both the free length and the embedded depth were $L = 300$ mm ($L/D = 5$). To simulate the mass of the nacelle, an iron block was welded on the top of the pile. In addition, the pile was placed in the center of the box by the hammer-driven method. The lateral dynamic load was applied by a vibration exciter, which can exert harmonic loads, sweeping loads and random loads (Figure 3).

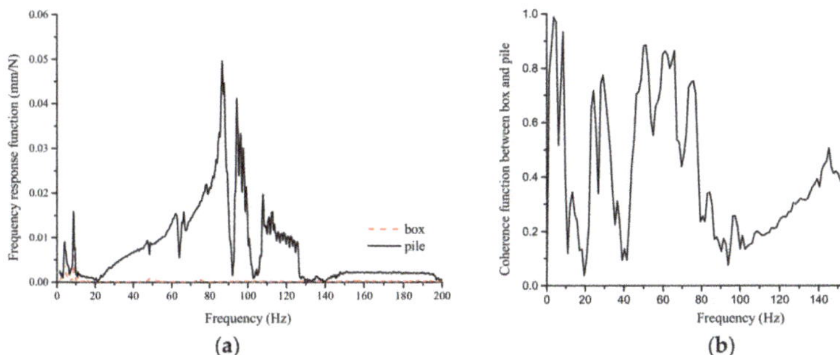

Figure 2. The influence of tank vibration: (**a**) Frequency response functions of the soil tank and the pile; (**b**) Coherence function between the soil tank and the pile.

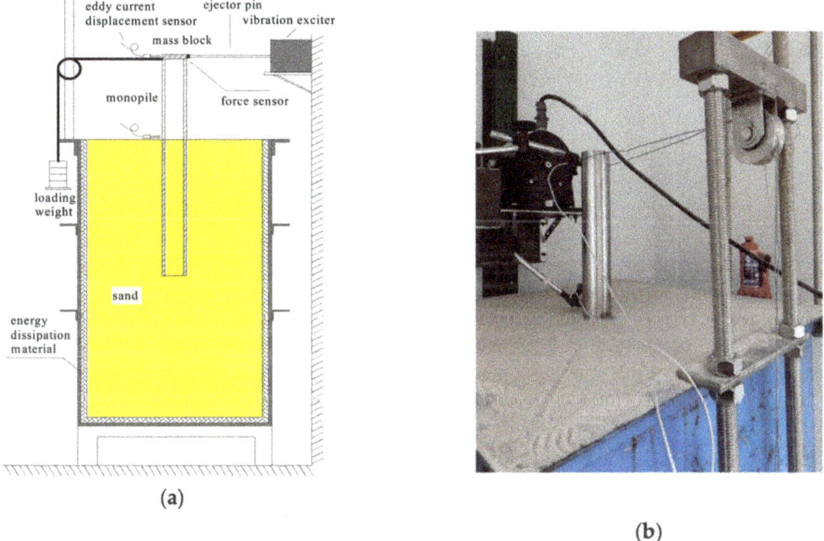

(**a**)

(**b**)

Figure 3. Experimental setup and sensor placement: (**a**) Sketch of the model; (**b**) Photo of the model.

3.2. Test Series

The following series of tests were included in the test process, as shown in Table 2:

(1) LBC: the lateral bearing capacity test in loose sand ($D_r = 10\%$, L.LBC) and dense sand ($D_r = 88\%$, D.LBC).

(2) HL: vibration characteristic test of monopile foundation under harmonic loads with different amplitudes. Harmonic loads with different frequencies and fixed amplitudes are applied to the pile top, and the displacements and the loads are measured. Then, change the amplitude of the loads from 1 N to 5 N in the dense sand case, and from 0.5 N to 10 N in the loose sand case.

(3) SL: the influence of the lateral static load on the vibration characteristics by hammer excitation with FRF method.

(4) L-U: three repeated loading-unloading processes to comprehend how extreme static loads influence the vibration characteristics of the pile.

(5) S-D: the influence of the static load on the vibration characteristics in different directions.

Table 2. Test series.

Test Name	D_r	f (Hz)	Dimensionless Frequency a_0	Amplitude H_{amp} (N)	H_{amp}/H_u	Static Load H (N)	H/H_u
L.LBC	10%	0	0	-	-	0–40	0–1.08
D.LBC	88%	0	0	-	-	0–165	0–1.06
L.HL	10%	40–200	0.14–0.72	1/2/3/4/5	0.65–3.25%	0	0
D.HL	88%	40–200	0.11–0.56	0.5/1/1.5/2/2.5	1.35–6.75%	0	0
	88%	40–50 100–200	0.11–0.14 0.28–0.56	4/6/8/10	10.81–27.03%	0	0
L.SL	10%	Random		Hammering	-	0–40	0–1.08
D.SL	88%	Random		Hammering	-	0–290	0–1.86
L.L-U	10%	Random		Hammering	-	0–20	0–0.54
D.L-U	88%	Random		Hammering	-	0–100	0–0.64
L.S-D1	10%					0	0
L.S-D2	10%					8	0.22
L.S-D3	10%	Random		Hammering	-	16	0.43
D.S-D1	88%					0	0
D.S-D2	88%					40	0.26
D.S-D3	88%					80	0.51

3.3. Preparation of the Sand

For the quartz sand used in the tests, the particle size is 0.2–0.6 mm, with the medium diameter 0.42 mm, minimum density 1.35 g/cm^3, and maximum density 1.57 g/cm^3. The gradation curve is shown in Figure 4. The friction angle of the sand is about 30°.

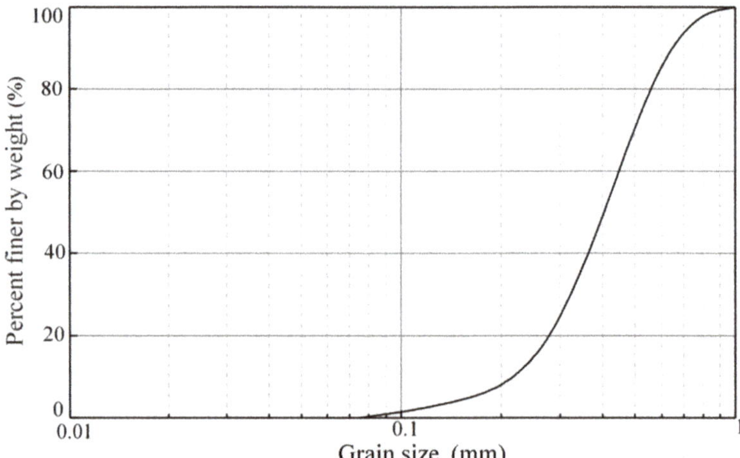

Figure 4. Gradation curve.

For sand, relative density D_r is of great importance in geotechnical tests [41]. The test will be more credible and repeatable if the relative density of the sand is well controlled. Various methods are used in the laboratory to reconstitute the sand, including pluviation, vibration and tamping. Pluviation through air is the most preferred method for its similarity to natural sand in the deposition mode [42]. According to the research now available, when other conditions remain unchanged, the relative density of sand is positively correlated with the falling distance [43].

In this paper, an experimental equipment of stationary pluviation was made referring to Chennarapu's method [43]. As shown in Figure 5, this device consists of three layers of sieves. The diameter of the sieve hole of the first layer (made of Acrylic plate) is 10 mm, and it has two layers (sheet1, sheet2) close together. When the sieve holes in each layer are staggered, the sieve can be filled with sand. When the sieve holes are aligned, the sand will flow out. The second layer (sheet3) and the third layer (sheet4) are made of stainless steel and the diameters of the sieve holes are 6 mm and 3 mm, respectively. To adjust the falling distance, the sieve is hung on a fixed pulley.

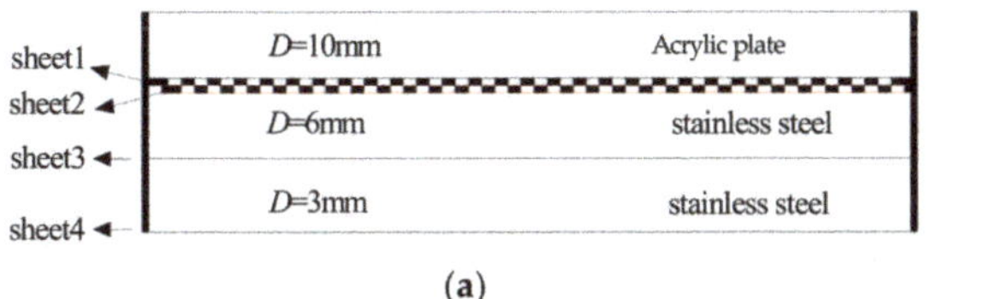

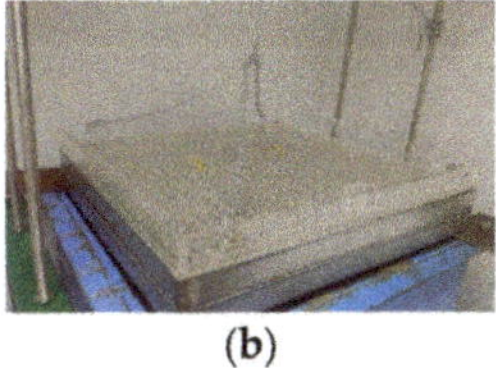

Figure 5. Pluviation equipment: (**a**) Sketch of the equipment; (**b**) Photo of the equipment.

The relationship between the falling distance h_f and the relative density D_r of the sand was observed before the experiment. As shown in Figure 6, nine sampling boxes are arrayed in the tank. After the boxes are overflowed by the soil, the relative density of the soil in the box is measured. To ensure that this device is reliable, 18 times' test results with the falling distance 30 cm were obtained, as shown in Figure 6.

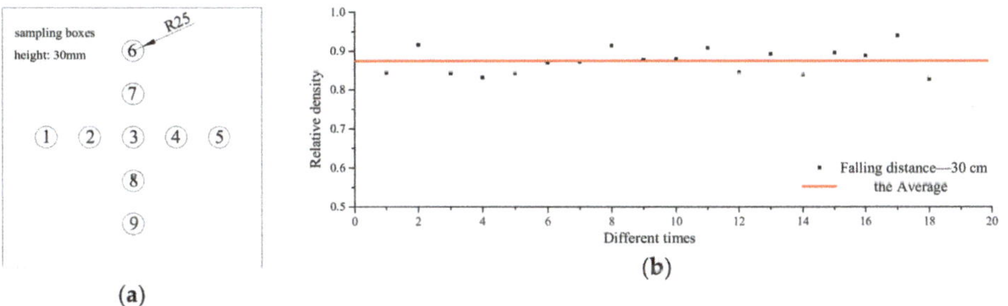

Figure 6. Relative density with the falling distance 30cm: (**a**) Sampling boxes; (**b**) Test reliability.

Finally, the relationship between h_f and D_r was mapped, and the falling distance can be found in Figure 7, where soil with relative density 0.75–0.9 is in need.

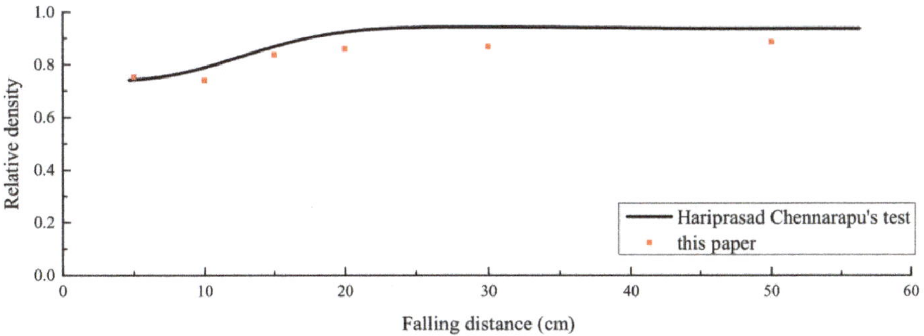

Figure 7. Relationship between the falling distance and the relative density.

On the other hand, loose sand was prepared by manual pouring from a very low height to achieve a density of $\rho_L = 1.39 \text{g/cm}^3$, and the corresponding relative density is 10%.

4. Test Results

4.1. Horizontal Bearing Capacity

In this paper, the horizontal bearing capacity of the monopile foundation (H_u) was obtained as a reference of the dynamic load amplitudes. According to Cuéllar et al. [44], the ultimate bearing capacity of a pile foundation under horizontal static load can take the corresponding load when the displacement of the pile top reaches 0.1 D. According to the Code for Pile Foundation in Port Engineering [45], the loading process is divided into 10 stages, and 0.1 Hu is loaded each time. When the displacement evolution rate at the loading point is less than 1×10^{-5} m/min in 30 minutes, start the next loading stage. Stop loading when total displacement reaches 0.1 D. From Figure 8, it can be found that the displacements are stable after each load. From Figure 9a, it can be found that the horizontal bearing capacity H_u is about 156 N and 37 N for dense sand and loose sand, respectively. The corresponding evolution of the rotation angle is also shown in Figure 9b.

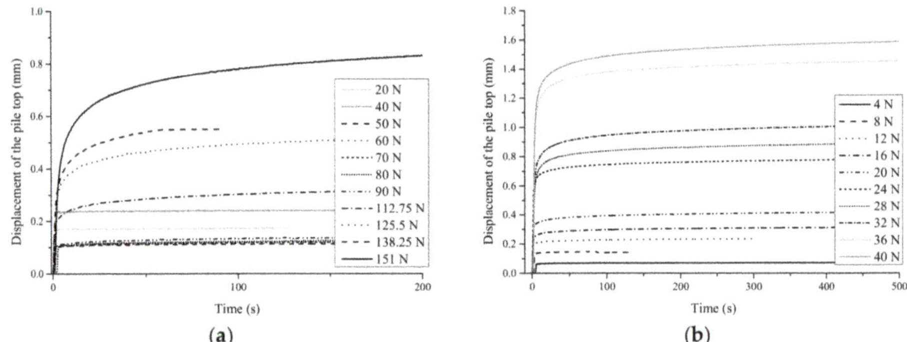

Figure 8. Y-T curves: (**a**) Dense sand case; (**b**) Loose sand case.

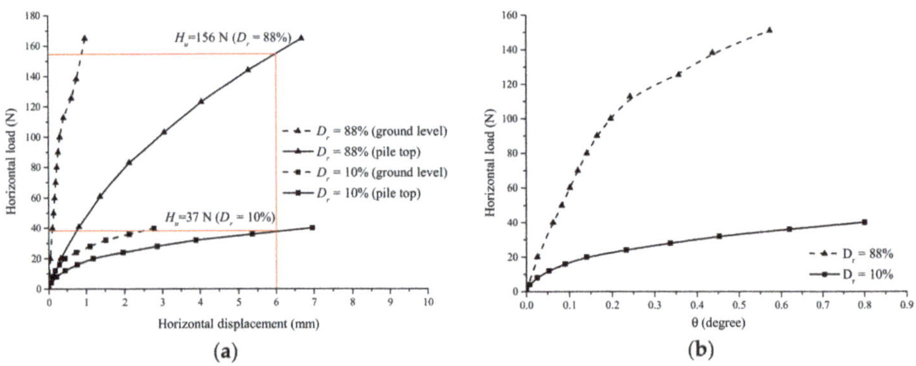

Figure 9. Horizontal bearing capacity: (**a**) Force - horizontal displacement curve; (**b**) Force- rotation angle curve.

4.2. Vibration Characteristics under Dynamic Loading with Different Amplitudes

Research on the influence of load amplitudes on the vibration characteristics of traditional slender piles has been well studied, but there are very few results for rigid monopiles. In this section, harmonic loads with different frequencies from 40 Hz to 200 Hz are applied on the top of the pile, and the test group is repeated with different amplitudes from 1 N to 5 N (dense sand, $0.65\%H_u$–$3.25\%H_u$), from 0.5 N to 2.5 N (loose sand, $1.35\%H_u$–$6.75\%H_u$). Harmonic loads with amplitudes of 4–10 N ($10.81\%H_u$–$27.03\%H_u$) are applied on the pile in loose sand to study the influences of soil nonlinearity.

Actually, it would be more realistic if the amplitude of the harmonic load in the test could reach 30% H_u or more because the cyclic load in the ocean can reach 30% of the ultimate bearing capacity. However, when the load amplitude was large enough and the frequency of the load was near the natural frequency, the sand around the pile vibrated violently and flowed toward the center so that the properties of the sand were different and the results cannot be put together for comparison. Time-domain sampling data under harmonic load with frequency 120 Hz and amplitude 5N are shown in Figure 10a. The displacement is obtained when it is stable (Figure 10b), and the loading time should be as short as possible to avoid the densification of the sand.

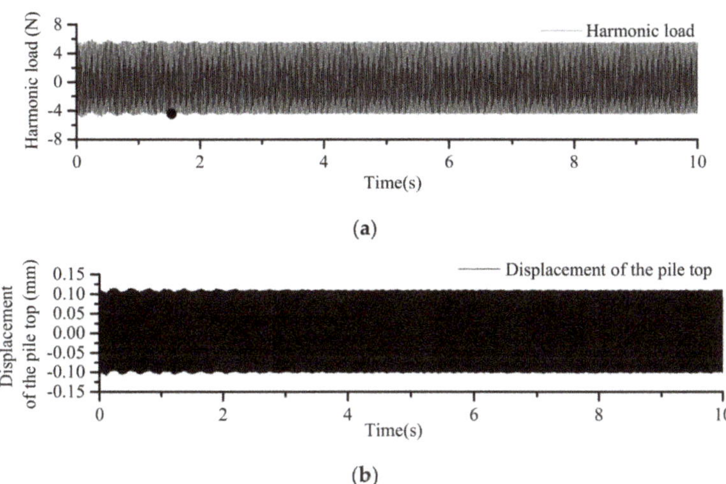

(a)

(b)

Figure 10. Time-domain sampling: (**a**) Harmonic load; (**b**) Displacement of the pile top.

Figures 11 and 12 show the frequency response curves of the monopile with different dynamic load amplitudes. The natural frequency decreases with the increase in the dynamic load amplitude H_{amp} in both dense sand and loose sand. In Figure 12b, H_{amp}/H_u is up to 27.03%, but the response is too large to be captured when the loading frequency is around the first natural frequency. According to the knowledge of structural dynamics, the natural frequency f_1 and stiffness K are positively correlated, and K is bound up with the shear modulus G of the soil. Hardin and Drnevich [46] obtained a negative correlation between the shear modulus and strain of sand by tests. Therefore, natural frequency decreases with the increase in the load amplitude, as Figure 13 shows. When H_{amp}/H_u is 0–4%, f_1 decreases with H_{amp}/H_u linearly; when H_{amp}/H_u is larger, the curve enters the nonlinear segment for the loose sand case. According to structural dynamics [47], the damping ratio of the pile can be obtained as $\xi = \frac{f_b - f_a}{2f_1}$ (Figure 14a), where f_1 is the frequency corresponding to the peak of the frequency response function curve (FRF_{max}), and f_a and f_b are the frequencies corresponding to $FRF_{max}/\sqrt{2}$. The obtained damping ratio of the monopile in the model test is given in Figure 14b, and it can be found that the damping ratio increases with the increase in load amplitudes during both dense sand and loose sand cases.

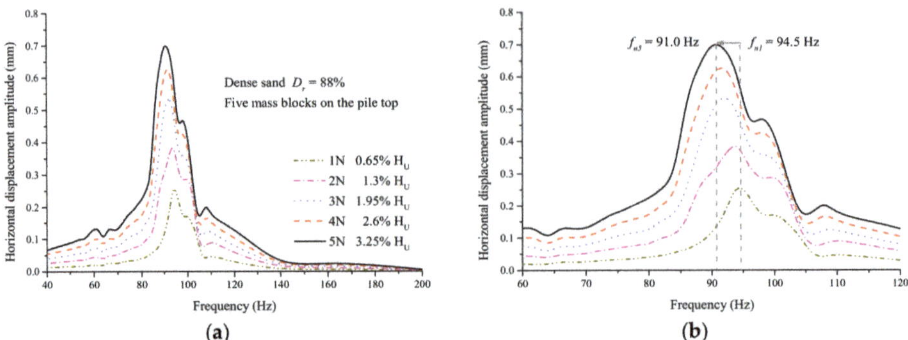

Figure 11. Horizontal displacement amplitudes with different load amplitudes and frequencies in dense sand: (**a**) 40–200 Hz; (**b**) 60–120 Hz.

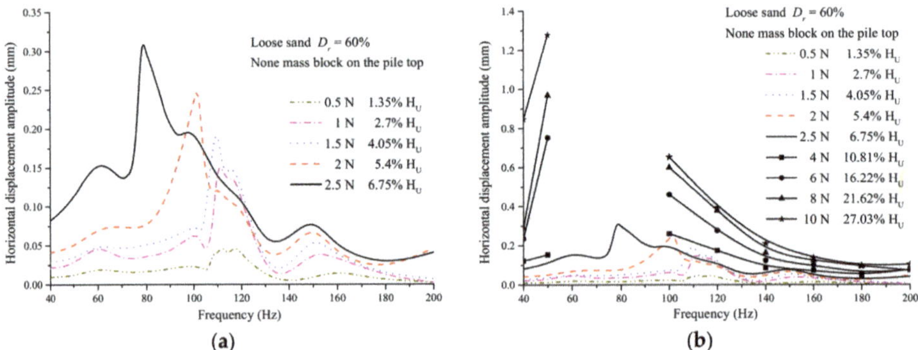

Figure 12. Horizontal displacement amplitudes with different load amplitudes and frequencies in loose sand: (**a**) 0.5–2.5 N; (**b**) 0.5–10 N.

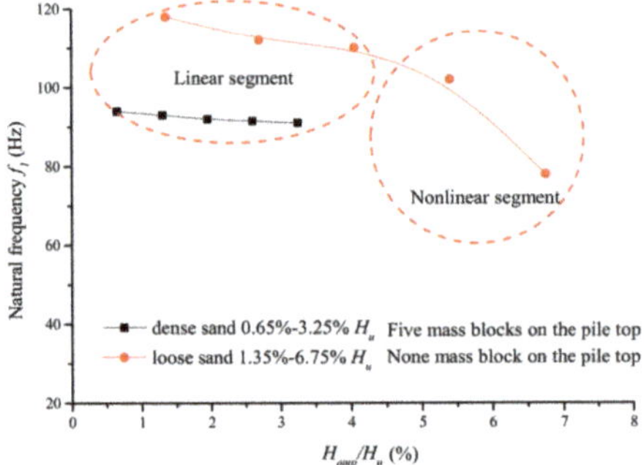

Figure 13. Influence of H_{amp}/H_u on natural frequency f_1.

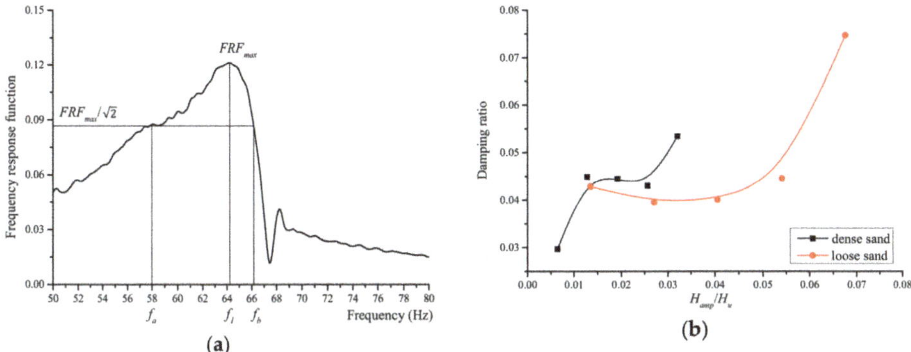

Figure 14. Calculation of damping ratio: (**a**) Method; (**b**) Model test results.

4.3. Vibration Characteristics under Different Static Loads

Few studies are concerned with the vibration characteristics under different lateral static loads; however, in the coastal areas of China where typhoons are frequent, it is important to study the vibration characteristics of monopile foundations under extreme loads.

In this section, hammer excitation was exerted on the pile top and frequency response analysis (FRA) was applied to obtain the vibration characteristics. In order to ensure the reliability of the hammering method, a comparison with random excitation was carried out, and the two results coincide well, as shown in Figure 15. Figure 16 shows the frequency response functions of the pile with different lateral static loads. Figure 17 shows the relationship between frequency and lateral static load, and the color of the pictures represents the ratio of *FRF* (frequency response function) to FRF_{max} (the maximum value of each frequency response curve), which helps us observe the peaks of the curves clearly, and the increasing of f_1 with the increase of the static load can be obtained.

Similar results occur in both the loose sand case and dense sand case, where f_1 increases under the loading process by and large. It can be obtained from Figure 17a that f_1 increases until static load is 220 N (much larger than the static bearing capacity 154 N) in the dense sand case. In Figure 17b, there are two distinct resonance frequencies, f_1 and f_2, in the loose sand case, which are caused by the horizontal vibration and rocking rotation of the monopile when the constraint supplied by the soil is insufficient. The first resonance frequency increases with the increase in static load, as in the dense sand case. However, the second resonance frequency almost keeps the same during the loading process.

The reason for the increase in the first natural frequency under static load is thought partly due to the increase in the shear modulus G of soils around the monopile. As Seed and Idriss [48] pointed out, the shear modulus G_0 of sand is positively correlated with confining pressure. As shown in Figure 18, the average stress around the monopile along the loading direction will be larger under static load than no static load case [49]. When the static load increases to a certain extent, the sand gradually reaches the limit state, and the resonance frequency decreases. On the other hand, it should be noted that the change of natural frequency may partly due to the constraint of the wire rope and the inertia of the added balance weights (the static force is exerted by the wire rope and balance weights). For monopiles in ocean environment, part of the forces are due to inertial effects (refer to Morison's equation for wave loading), which is similar to the case in this study, to some extent. However, further tests in a flume or in the field are needed to verify the results obtained here, and the contributions of the two parts should also be separated in future. This preliminary phenomenon states that the resonance frequency of the wind turbine may still be changing under horizontal static loads or even extreme loads. So, it is necessary to be concerned about the occurrence of resonance even during the ultimate bearing capacity design state.

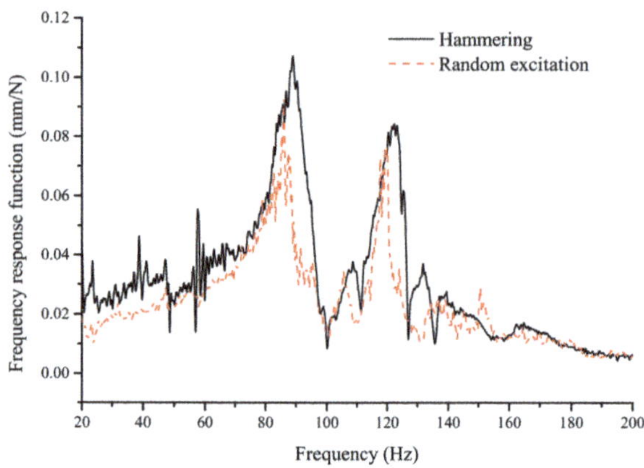

Figure 15. Frequency response functions of hammering and random excitation.

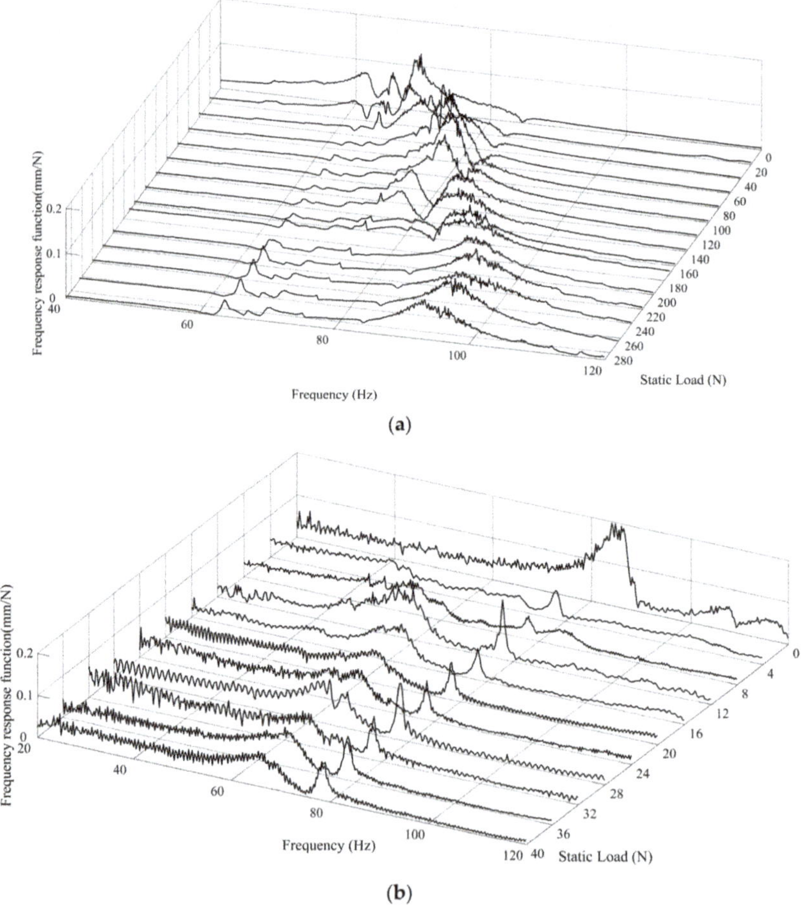

(**a**)

(**b**)

Figure 16. Frequency response functions of the pile with different lateral static loads: (**a**) Dense sand; (**b**) Loose sand.

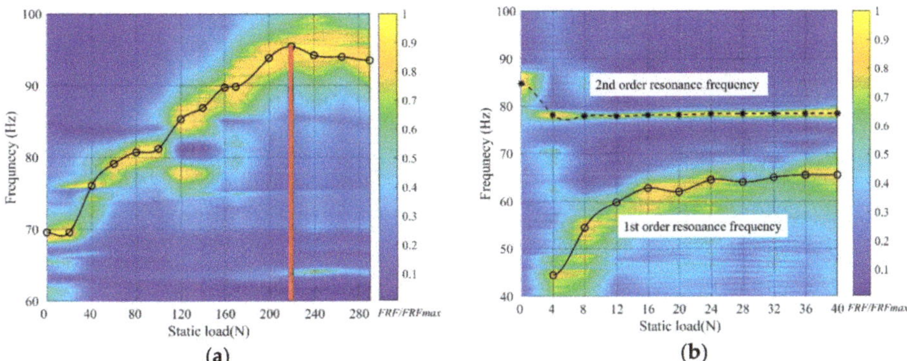

Figure 17. The relation between resonant frequencies and lateral static load: (**a**) Dense sand case; (**b**) Loose sand case.

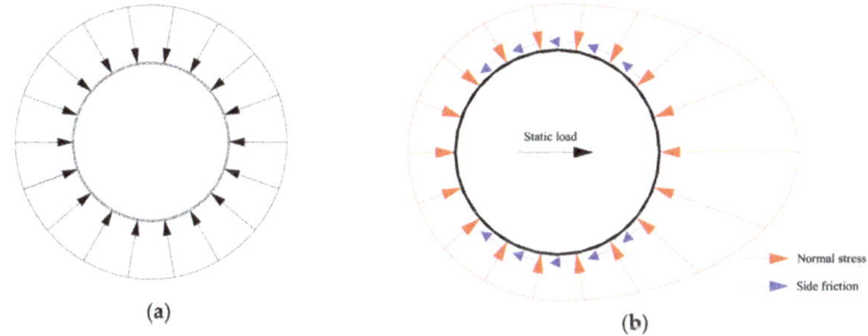

Figure 18. Stresses of soil around the pile: (**a**) No static load; (**b**) with static load.

Furthermore, the change in frequency responses during the loading-unloading-reloading process is shown in Figures 19 and 20. It can be observed from Figure 19 that the natural frequencies of the unloading process are larger than those of the loading process; the natural frequencies of the loading starting points and finishing points all increase after each loading and unloading process in dense sand case. However, this phenomenon is not obvious in loose sand case (Figure 20), but it proves the repeatability of the phenomena in Figure 17b.

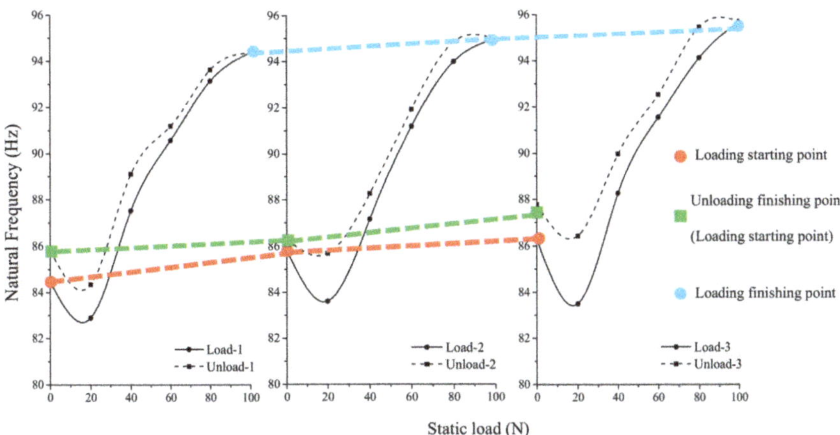

Figure 19. The influence of the loading and unloading process on natural frequency in dense sand.

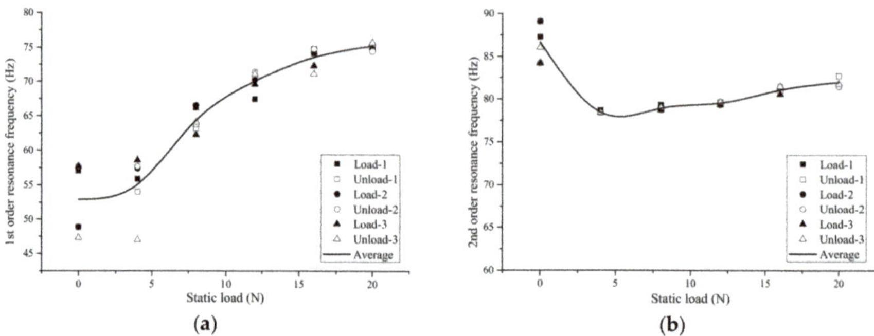

Figure 20. The influence of the loading and unloading process on natural frequency in loose sand: (**a**) First order resonance frequency f_1; (**b**) Second order resonance frequency f_2.

4.4. Vibration Characteristics in Different Directions under Different Static Loads

The directions of loads such as wind, waves and current are not always the same, and the foundation can vibrate in different directions. For simplicity, it is useful to study the vibration characteristics of the monopile in different directions under the main static loads. In this section, static load was applied to the top of the pile in DIR-1, and the hammer excitation was exerted in DIR-1, DIR-2, DIR-3, and DIR-4, respectively. The displacements were acquired in four directions by displacement sensors simultaneously, as shown in Figure 21. Based on the fundamental principle of modal analysis, the FRF curves of four directions were obtained under the condition that the pile is loaded in DIR-1 by static load (0 N, 40 N, and 80 N in dense sand case, and 0 N, 8 N, and 16 N in loose sand case, respectively).

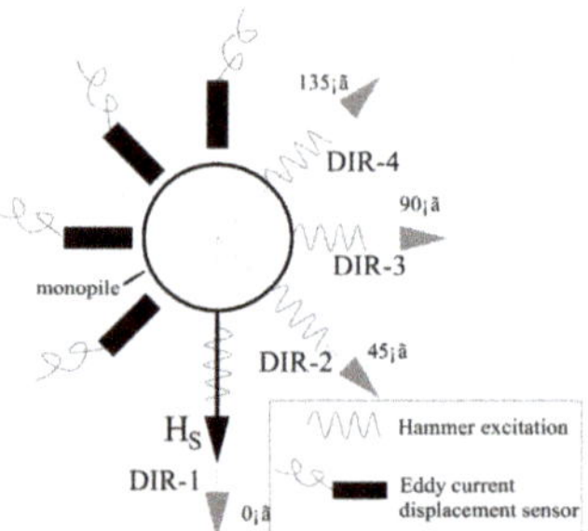

Figure 21. Layout of this experiment.

The results for dense sand case and loose sand case can be found in Figures 22 and 23, respectively. In this section, the results in dense sand are mainly analyzed, while the results in the loose sand case are similar, although not so obvious as in dense sand case.

As mentioned in the previous section, the first frequency f_1 of DIR-1 increases as expected with the increase of the static load in DIR-1. This proves once again that the stiffness provided by the soil-pile system increases with the increase in horizontal static load. However, it is different in DIR-3, which is perpendicular to DIR-1, where f_1 almost remains unchanged with the increase in the static load, but the amplitude of the FRF curve increases, which means a decrease in the damping ratio of the pile in DIR-3, as shown in Figure 24a. Damping ratio here is composed of two parts: the radiative damping, which increases with the increase in frequencies, and decreases with load amplitudes; and the material damping, which increases with the soil nonlinearity. The results for DIR-2 and DIR-4 lie between the results for DIR-1 and DIR-3. Theoretically, the results of DIR-2 and DIR-4 should be similar, and this is true from a general point of view.

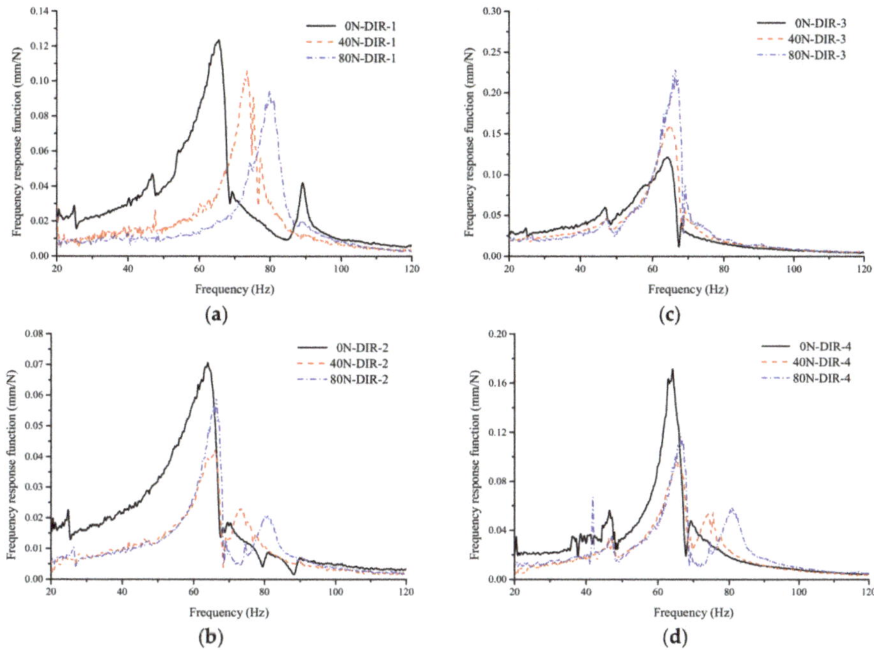

Figure 22. Frequency response functions of different directions under different lateral static loads:
(**a**) Direction-1; (**b**) Direction-2; (**c**) Direction-3; (**d**) Direction-4.

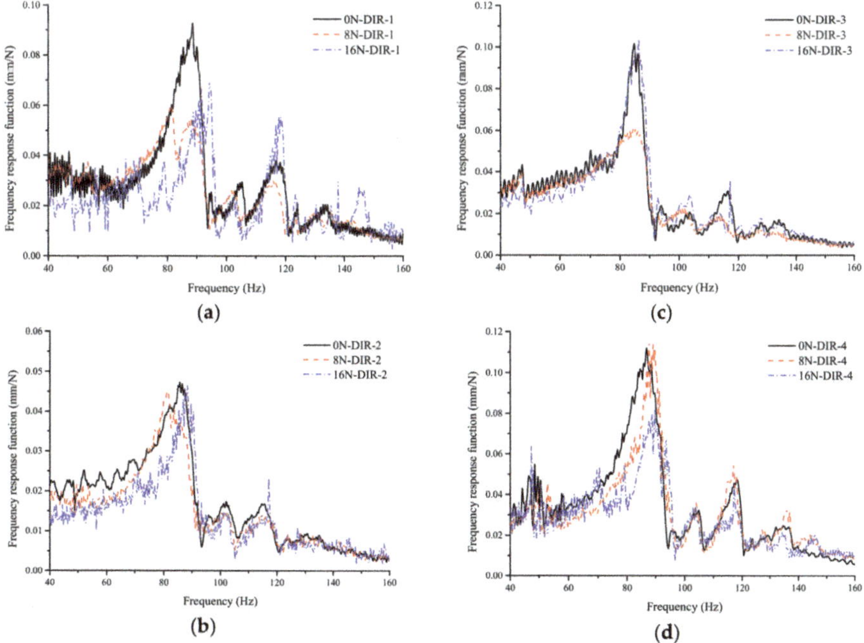

Figure 23. Frequency response functions of different directions under different lateral static loads in
loose sand: (**a**) Direction-1; (**b**) Direction-2; (**c**) Direction-3; (**d**) Direction-4.

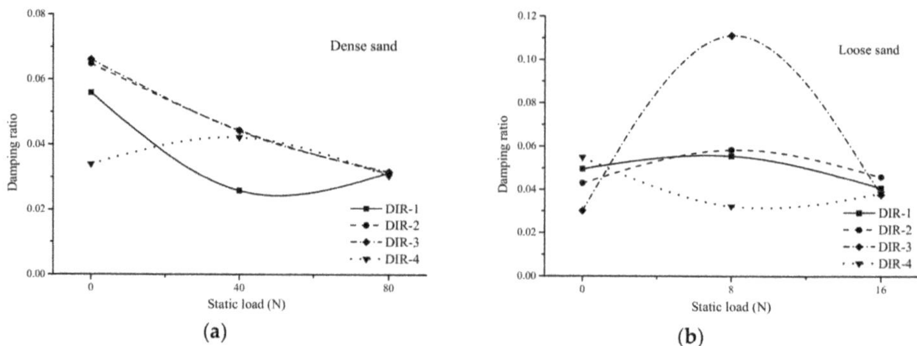

Figure 24. Total damping ratio of the model monopile: (**a**) Dense sand case; (**b**) Loose sand case.

Based on the results above, the differences of dynamic stiffness and damping in different directions should be focused on when the directions of the wind, waves or current are crossed.

5. Conclusions and Outlook

5.1. Conclusions

In this paper, the results of laboratory-scale 1 g model tests for monopiles in dry sands (D_r = 10% and 88%) are presented. The vibration characteristics of the model monopile under different lateral loading amplitudes, directions and loading-unloading cycles are analyzed by the FRF method. The main conclusions are:

(1) When there is no static load, the first natural frequency f_1 decreases with the increase of the amplitudes of the dynamic loads in both dense sand and loose sand cases;

(2) The first natural frequency f_1 increases with the increase in the lateral static load generally in both the dense and loose sand cases; Loading-unloading-reloading to the capacity process can increase the first resonance frequency of the monopile in dense sand, but this phenomenon is not observed in loose sand case;

(3) The frequency responses of the monopile in the direction perpendicular to the static loading are quite different from those in the static loading direction, as soils around the monopile are under different stress conditions, and this is more obvious in the dense sand case.

5.2. Outlook

In this study, several simplifications were used: the blades and nacelle on the top of the monopile are simplified as mass blocks; the stress level of soil is lower than the field case, and it is only taken as guaranteed that both the monopiles behave like rigid ones; the full drainage condition of saturated sand is modeled as dry sand. In order to further study the vibration characteristics of monopiles of offshore wind turbines, different seabed geological conditions, soil stress levels, drainage conditions, and actual wave, current and wind loads should be taken into account. Besides, theoretical or numerical models that can explain the observed results should be developed.

Author Contributions: Conceptualization, supervision, R.H.; model test, writing—original draft, T.Z.; data analysis, writing—review & editing, R.H. and T.Z.

Funding: The authors would like to acknowledge Grant No. 51879097 from the support of the National Natural Science Foundation of China, the Grant No. BK20190074 from the Natural Science Foundation of Jiangsu Province, and the Grant No. 2019B53114 and 2018B12714 from Fundamental Research Funds for the Central Universities.

Conflicts of Interest: The authors declare no conflict of interest. Rui He and Tao Zhu.

Nomenclature

L	embedded length of the monopile	H	horizontal static load
D	diameter	H_{amp}	harmonic load amplitude
E_p	elastic modulus of the monopile	H_u	ultimate bearing capacity of the pile
E_s	elastic modulus of soil	ρ_s	soil density
h	thickness of the monopile	L_0	length of the monopile
G_0	initial shear modulus of soil	f	load frequency
e	void ratio	h_f	falling distance
$\sigma_m{'}$	average effective principal stress	ρ_L	density of loose sand
h_s	soil depth	K	stiffness of structure
D_r	relative density	ξ	damping ratio
a_0	dimensionless frequency	$FRFmax$	peak of the FRF curve
r	radius of the monopile	f_1	first order resonance frequency
ω	circular frequency	f_a, f_b	the frequencies corresponding to $FRFmax/\sqrt{2}$
μ_s	shear modulus of soil	f_2	second order resonance frequency

References

1. DNV. *Offshore Standard DNV-OS-J101: Design of Offshore Wind Turbine Structures*; DNV: Oslo, Norway, 2010.
2. Richards, I.A.; Byrne, B.W.; Houlsby, G.T. Monopile rotation under complex cyclic lateral loading in sand. *Géotechnique* **2019**. [CrossRef]
3. Krolis, V.D.; Zwaag, G.L.V.D.; Vries, W.D. Determining the embedded pile length for large-diameter monopiles. *Mar. Technol. Soc. J.* **2010**, *44*, 24–31. [CrossRef]
4. Futai, M.M.; Dong, J.; Haigh, S.K.; Madabhushi, S.P.G. Dynamic response of monopiles in sand using centrifuge modelling. *Soil Dyn. Earthq. Eng.* **2018**, *115*, 90–103. [CrossRef]
5. Lombardi, D.; Bhattacharya, S.; Muir Wood, D. Dynamic soil-structure interaction of monopile supported wind turbines in cohesive soil. *Soil Dyn. Earthq. Eng.* **2013**, *49*, 165–180. [CrossRef]
6. Damgaard, M.; Bayat, M.; Andersen, L.V.; Ibsen, L.B. Assessment of the dynamic behaviour of saturated soil subjected to cyclic loading from offshore monopile wind turbine foundations. *Comput. Geotech.* **2014**, *61*, 116–126. [CrossRef]
7. Leblanc, C. Design of Offshore Wind Turbine Support Structures: Selected topics in the field of geotechnical engineering. Ph.D. Thesis, Aalborg University, Aalborg, Danmark, 2009.
8. Dobry, R.; Gazetas, G. Dynamic response of arbitrarily shaped foundations. *J. Geotech. Eng.* **1986**, *112*, 109–135. [CrossRef]
9. Novak, M.; Nogami, T. Soil-pile interaction in horizontal vibration. *Earthq. Eng. Struct. D* **1977**, *5*, 263–281. [CrossRef]
10. Muki, R.; Sternberg, E. On the diffusion of an axial load from an infinite cylindrical bar embedded in an elastic medium. *Int. J. Solids Struct.* **1969**, *5*, 587–605. [CrossRef]
11. Pak, R.Y.S.; Jennings, P.C. Elastodynamic response of pile under transverse excitations. *J. Eng. Mech.* **1987**, *113*, 1101–1116. [CrossRef]
12. He, R.; Pak, R.Y.S.; Wang, L.Z. Elastic lateral dynamic impedance functions for a rigid cylindrical shell type foundation. *Int. J. Numer. Anal. Met.* **2017**, *41*, 508–526. [CrossRef]
13. He, R.; Kaynia, A.M.; Zhang, J.S. A poroelastic solution for dynamics of laterally loaded offshore monopiles. *Ocean Eng.* **2019**, *179*, 337–350. [CrossRef]
14. He, R.; Kaynia, A.M.; Zhang, J.S. Influence of vertical shear stresses due to pile-soil interaction on lateral dynamic responses for offshore monopiles. *Mar. Struct.* **2019**, *64*, 341–359. [CrossRef]
15. Zdravkovic, L.; Taborda, D.M.G.; Potts, D.M.; Jardine, R.J.; Gretlund, J.S. Numerical Modelling of Large Diameter Piles under Lateral Loading for Offshore Wind Applications. In Proceedings of the International Symposium on Frontiers in Offshore Geotechnics, Oslo, Norway, 10–12 June 2015.
16. Sen, R.; Davies, T.G.; Banerjee, P.K. Dynamic analysis of piles and pile groups embedded in homogeneous soils. *Earthq. Eng. Struct. Dyn.* **1985**, *13*, 53–65. [CrossRef]
17. Latini, C.; Zania, V. Dynamic lateral response of suction caissons. *Soil Dyn. Earthq. Eng.* **2017**, *100*, 59–71. [CrossRef]

18. Tao, W.Y.; He, R.; Zheng, J.H. Analysis on horizontal-rocking vibrations of monopile supporting wind turbine. *J. Hohai Univ. Nat. Sci.* **2018**, *46*, 260–267.

19. He, R.; Ji, J.; Zhang, J.S.; Peng, W.; Sun, Z.F.; Guo, Z. Dynamic impedances of offshore rock-socketed monopiles. *J. Mar. Sci. Eng.* **2019**, *7*, 134. [CrossRef]

20. Ma, H.; Yang, J.; Chen, L. Numerical analysis of the long-term performance of offshore wind turbines supported by monopiles. *Ocean Eng.* **2017**, *136*, 94–105. [CrossRef]

21. Beuckelaers, W. Numerical Modelling of Laterally Loaded Piles for Offshore Wind Turbines. Ph.D. Thesis, University of Oxford, Oxford, UK, 2017.

22. Goit, C.S.; Saitoh, M.; Mylonakis, G.; Kawakami, H.; Oikawa, H. Model tests on horizontal pile-to-pile interaction incorporating local non-linearity and resonance effects. *Soil Dyn. Earthq. Eng.* **2013**, *48*, 175–192. [CrossRef]

23. Bhattacharya, S.; Nikitas, N.; Garnsey, J.; Alexander, N.A. Observed dynamic soil–structure interaction in scale testing of offshore wind turbine foundations. *J. Soil Dyn. Earthq. Eng.* **2013**, *54*, 47–60. [CrossRef]

24. Manna, B.; Baidya, D.K. Nonlinear dynamic response of piles under horizontal excitation. *J. Geotech. Geoenviron.* **2010**, *136*, 1600–1609. [CrossRef]

25. Elkasabgy, M.; El Naggar, M.H. Lateral vibration of helical and driven steel piles installed in clayey soil. *J. Geotech. Geoenviron.* **2018**, *144*, 6018009. [CrossRef]

26. He, R.; Zhu, T.; Ma, B.; Tao, W.Y. Dynamic Responses of Mono-piles in the Presence of Scour Holes. In Proceedings of the China-Europe Conference on Geotechnical Engineering, Vienna, Austria, 13–16 August 2018.

27. Leblanc, C.; Houlsby, G.T.; Byrne, B.W. Response of stiff piles in sand to long-term cyclic lateral loading. *Géotechnique* **2010**, *60*, 79–90. [CrossRef]

28. Shirzadeh, R.; Devriendt, C.; Bidakhvidi, M.A.; Guillaume, P. Experimental and computational damping estimation of an offshore wind turbine on a monopile foundation. *J. Wind Eng. Ind. Aerodyn.* **2013**, *120*, 96–106. [CrossRef]

29. Damgaard, M.; Ibsen, L.B.; Andersen, L.V.; Andersen, J.K.F. Cross-Wind modal properties of offshore wind turbines identified by full scale testing. *J. Wind Eng. Ind. Aerodyn.* **2013**, *116*, 94–108. [CrossRef]

30. American Petroleum Institute. *Recommended Practice for Planning, Designing and Constructing Fixed Offshore Platforms—Working Stress Design*, 21st ed.; API Recommended Practice 2A-WSD (RP2A-WSD); API: Dallas, TX, USA, 2000.

31. Burd, H.J.; Beuckelaers, W.J.A.P.; Byrne, B.W.; Gavin, K.; Houlsby, G.T.; Igoe, D.; Jardine, R.J.; Martin, C.M.; McAdam, R.A.; Muir Wood, A.; et al. New data analysis methods for instrumented medium-scale monopile field tests. *Géotechnique* **2019**, 1–28. [CrossRef]

32. McAdam, R.A.; Byrne, B.W.; Houlsby, G.T.; Beuckelaers, W.J.A.P.; Burd, H.J.; Gavin, K.; Igoe, D.; Jardine, R.J.; Martin, C.M.; Muir Wood, A.; et al. Monotonic laterally loaded pile testing in a dense marine sand at Dunkirk. *Géotechnique* **2019**, 1–34. [CrossRef]

33. Zdravković, L.; Taborda, D.M.G.; Potts, D.M.; Abadias, D.; Burd, H.J.; Byrne, B.W.; Gavin, K.; Houlsby, G.T.; Jardine, R.J.; Martin, C.M.; et al. Finite element modelling of laterally loaded piles in a stiff glacial clay till at Cowden. *Géotechnique* **2019**, 1–40. [CrossRef]

34. Taborda, D.M.G.; Zdravković, L.; Potts, D.M.; Burd, H.J.; Byrne, B.W.; Gavin, K.; Houlsby, G.T.; Jardine, R.J.; Liu, T.; Martin, C.M.; et al. Finite element modelling of laterally loaded piles in a dense marine sand at Dunkirk. *Géotechnique* **2019**, 1–47. [CrossRef]

35. Martin, C.M.; Randolph, M.F. Upper-Bound analysis of lateral pile capacity in cohesive soil. *Géotechnique* **2006**, *56*, 141–145. [CrossRef]

36. Achmus, M.; Kuo, Y.; Abdel-Rahman, K. Behavior of monopile foundations under cyclic lateral load. *Comput. Geotech.* **2009**, *36*, 725–735. [CrossRef]

37. Allotey, N.; El Naggar, M.H. A numerical study into lateral cyclic nonlinear soil-pile response. *Can. Geotech. J.* **2008**, *45*, 1268–1281. [CrossRef]

38. Gerolymos, N.; Escoffier, S.; Gazetas, G.; Garnier, J. Numerical modeling of centrifuge cyclic lateral pile load experiments. *Earthq. Eng. Eng. Vib.* **2009**, *8*, 61–76. [CrossRef]

39. Hong, Y.; He, B.; Wang, L.Z.; Wang, Z.; Ng, C.W.W.; Masin, D. Cyclic lateral response and failure mechanisms of semi-rigid pile in soft clay: Centrifuge tests and numerical modelling. *Can. Geotech. J.* **2017**, *54*, 2016–2356. [CrossRef]

40. Hardin, B.O.; Richart, F.E. Elastic wave velocities in granular soils. *J. Soil Mech. Found. Div.* **1963**, *89*, 35–65.

41. Been, K.; Jefferies, M.G. A state parameter for sands. *Géotechnique* **1985**, *35*, 99–112. [CrossRef]

42. Oda, M.; Koishikawa, I.; Higuchi, T. Experimental study of anisotropic shear strength of sand by plane strain test. *J. Jpn. Soc. Soil Mech. Found. Eng.* **1978**, *18*, 25–38. [CrossRef]

43. Hariprasad, C.; Rajashekhar, M.; Umashankar, B. Preparation of uniform sand specimens using stationary pluviation and vibratory methods. *Geotechn. Geol. Eng.* **2016**, *34*, 1909–1922. [CrossRef]

44. Cuéllar, P.; Baeßler, M.; Rücker, W. Ratcheting convective cells of sand grains around offshore piles under cyclic lateral loads. *Granul. Matter.* **2009**, *11*, 379–390. [CrossRef]

45. Ministry of Transport of the People's Republic of China. *Code for Pile Foundation in Port Engineering*; JTJ 254-1998; Ministry of Transport of the People's Republic of China: Beijing, China, 2001.

46. Hardin, B.O.; Drnevich, V.P. Shear modulus and damping in soils: Design equations and curves. *J. Soil Mech. Found. Div.* **1972**, *98*, 667–692.

47. Chopra, A.K. *Dynamics of Structures: Theory and Applications to Earthquake Engineering*; PEARSON: London, UK, 2014.

48. Seed, H.B.; Idriss, I.M. Soil moduli and damping factors for dynamic response analysis. *J. Terramech.* **1972**, *8*, 109.

49. Sørensen, S.P.H.; Brødbæk, K.T.; Møller, M.; Augustesen, A.H. *Review of Laterally Loaded Monopiles Employed as the Foundation for Offshore Wind Turbines*; Aalborg University: Aalborg, Denmark, 2012.

Journal of
Marine Science and Engineering

Article

Riprap Scour Protection for Monopiles in Offshore Wind Farms

M.Dolores. Esteban [1,2,*], José-Santos López-Gutiérrez [2,*], Vicente Negro [2] and Luciano Sanz [1]

[1]   Departamento de Ingeniería Civil, Universidad Europea, 28670 Madrid, Spain; sanzluciano@hotmail.com
[2]   Grupo de Investigación de Medio Marino, Costero y Portuario, y Otras Áreas Sensibles,
     Universidad Politécnica de Madrid, 28040 Madrid, Spain; vicente.negro@upm.es
*   Correspondence: mariadolores.esteban@universidadeuropea.es (M.D.E.);
     josesantos.lopez@upm.es (J.-S.L.-G.)

Received: 27 October 2019; Accepted: 26 November 2019; Published: 2 December 2019

Abstract: The scour phenomenon is critical for monopile structures in offshore wind farms. There are two possible strategies: allowing the development of scour holes around the monopile or avoiding it by placing scour protection. The last one is the most used up to now. This paper is focused on the determination of the weight of the stones forming the scour protection. There are some formulas for the design of these parameters, having a lot of uncertainties around them. Some of them were created for fluvial environment, with a different flow to the marine one. Other formulas were elaborated specifically for coastal structures, closer to the coast than offshore wind farms, and with dimensions completely different. This paper presents the analysis of three formulas: Isbash, corresponding to fluvial environment, and Soulsby, and De Vos, corresponding to marine environment. The results of the application of those formulas are compared with real data of scour protection systems showing good results in five offshore wind facilities in operation (Arklow Bank phase 1, Egmond aan Zee, Horns Rev phase 1, Princess Amalia, and Scroby Sands), giving conclusion about the uncertainties of the use of these formulas and recommendations for using them in offshore wind.

Keywords: scour phenomenon; weight; size; nominal diameter; armour; monitoring

---

1. Introduction

In recent years there has been a growing concern about the effects of climate change and the sustainable development of the economy, as well as the search for new and more sustainable energy sources. All these aspects, apart from other considerations, have increased the importance of using marine energies such as waves, currents, tides, etc. Offshore wind energy is one of the marine energies; although it can be considered that the resource is not exclusive to the marine environment, some documents consider it as a marine renewable energy because the facilities are installed in the sea. Not all marine energies are in the same state of development; for instance, wave energy converters are in general in initial stages of Technology Readiness Level (TRL), while the offshore wind sector can be considered at a commercial stage [1].

Figure 1 [2] represents, in Europe, the installed annual offshore wind power, with blue bars, and the cumulative power, with the red line, having reached 15,780 MW at the end of 2017. Including facilities with partial connection to the network, at the end of 2017 there were in Europe 92 offshore wind farms and 4,149 wind turbines connected to the network, located in 11 countries (Table 1) [2].

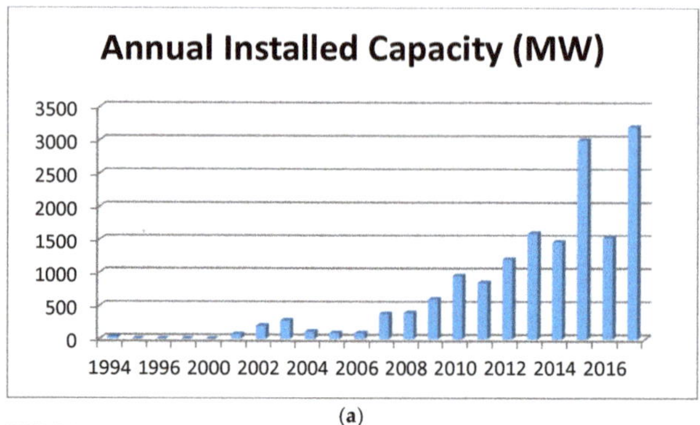

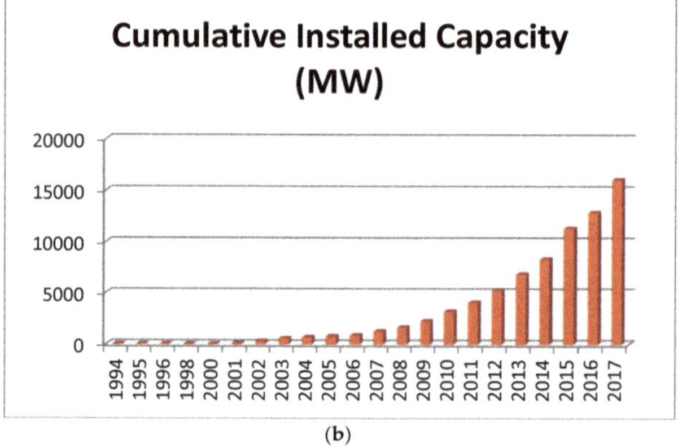

Figure 1. Offshore wind power (MW) in Europe at the end of 2017: annual new power installed (**a**) and cumulative power (**b**). Reproduced from WindEurope 2018 with permission.

Table 1. Number of offshore wind facilities with wind turbines connected to the network, number of turbines connected, and power connected (MW), in Europe at the end of 2017. Reproduced from WindEurope 2018 with permission.

Country	Number of Farms	Number of Turbines Connected	Capacity Installed	Capacity Installed/Decommissioned in 2017 (MW)
UK	31	1,753	6,835	1,679
Germany	23	1,169	5,355	1,247
Denmark	12	506	1,266	−5
Netherlands	7	365	1,118	0
Belgium	6	232	877	165
Sweden	5	86	202	0
Finland	3	28	92	60
Ireland	2	7	25	0
Spain	1	1	5	0
Norway	1	1	2	0
France	1	1	2	2
Total	92	4,149	15,780	3,148

Worldwide, the picture changes a little, with the appearance of China in the third position of power installed, as well as with the prominent positions of Vietnam, Japan, South Korea, the United States, and Taiwan, with a total installed capacity worldwide of 18,814 MW (Figure 2) [3].

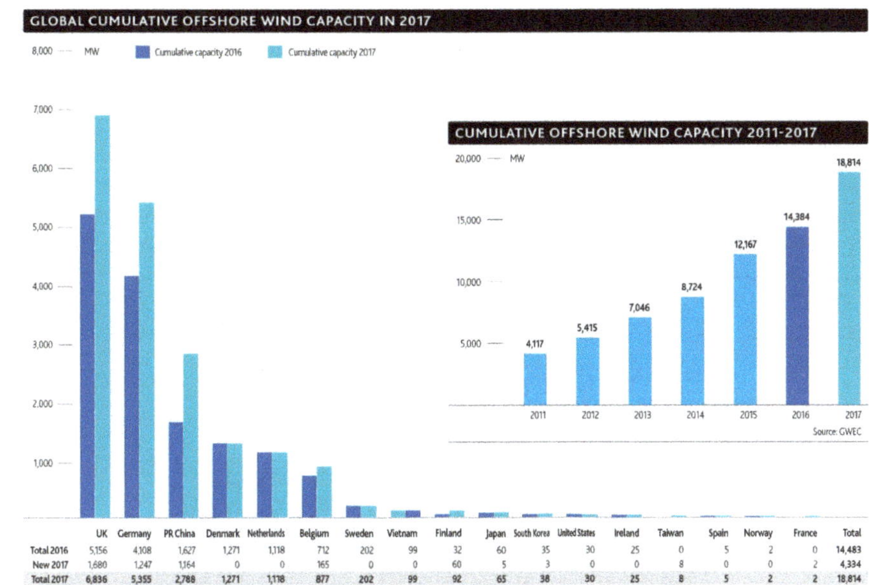

	UK	Germany	PR China	Denmark	Netherlands	Belgium	Sweden	Vietnam	Finland	Japan	South Korea	United States	Ireland	Taiwan	Spain	Norway	France	Total
Total 2016	5,156	4,108	1,627	1,271	1,118	712	202	99	32	60	35	30	25	0	5	2	0	14,483
New 2017	1,680	1,247	1,164	0	0	165	0	0	60	5	3	0	0	8	0	0	2	4,334
Total 2017	6,836	5,355	2,788	1,271	1,118	877	202	99	92	65	38	30	25	8	5	2	2	18,814

Figure 2. Total offshore wind power installed worldwide at the end of 2016 and 2017. Reproduced from Global Wind Energy Council (GWEC) 2018 with permission.

Figure 3 represents the basis of many discussions and decisions made in the offshore wind industry over the last decade. It can be seen that one or another type of foundation was recommended for wind turbines depending on the depth of the site. It is appreciated that until 20–25 m, the most interesting types were the gravity-based foundation or gravity-based structure (GBF or GBS) and the monopile; between 20–25 m and 40–50 m the most viable options were the tripod and the jacket; and for larger depths, floating supports appeared as an alternative, with important doubts about its future short-term implantation at the beginning of this decade [4,5].

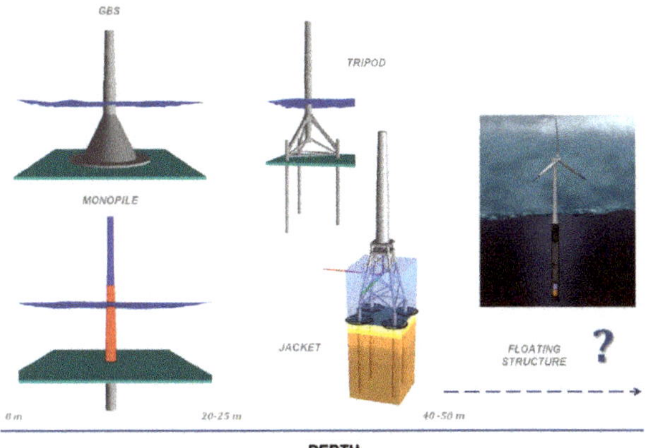

Figure 3. Main criteria for the selection of the foundation for offshore wind turbines depending on the depth. Reproduced from Esteban, M.D. 2009 with permission.

Although the main typologies have not changed very much, there have been certain modifications in the sector in relation to foundations, such as the development of new typologies or the use of previously known typologies in the offshore oil and gas engineering as bucket foundations. When locations for offshore wind facilities around 40 m appeared, a competition began to position the different typologies in the new depth range. At first it seemed that the competition was going to be between the jacket and the tripod, with a prominent role of the first of them. Anyway, some proposals have appeared at these depths of large monopiles (XXL monopiles) as well as new GBS designs [6,7].

According to WindEurope's report [2] of the statistics of offshore wind in Europe at the end of 2017, Figure 4 shows that the monopile is at the top of the classification, with 81.7%, followed by the jacket (6.9%), and the gravity foundation (6.2%), taking the floating platforms only seven representatives, being six of them SPARs and one of them semi-submersible.

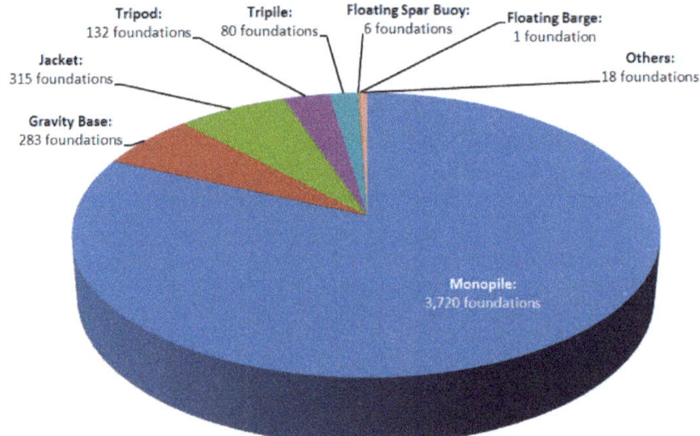

Figure 4. Share of foundation types in offshore wind turbines connected to the network in Europe, at the end of 2017. Reproduced from WindEurope 2018 with permission.

Although offshore wind sector is in a commercial stage, there are still a lot of uncertainties about the design, construction, maintenance, etc. [8]. Some of these uncertainties are related to the design of foundations; for instance, in the case of monopile foundations, it can be mentioned the scour, the hydrodynamic domains with current diameters and depths to which offshore wind farms are built, the difference in scale compared to conventional maritime piles, the liquefaction of the soil, the nonlinearity in the wave velocity field, etc. [9,10]. Among all the uncertainties, the scour around the pile can be emphasized [11]. It should be noted that scour has proved to be a key element in the design and maintenance of these facilities, as it has been observed given the evolution of the scour hole around the structure. In some cases, an important evolution of the hole has been detected only some years after the construction of the facility. Consequently, it is one of the main concerns in the offshore wind farms and they have to pay special attention to its evolution in the O&M phase of the facilities.

If this is not addressed in a suitable way and just in time, the scour phenomenon can cause operational unavailability, changes in the natural frequencies of the soil-structure system [12], fatigue life reduction [13], and even in extreme cases the collapse of the structure [14]. There is a significant need to advance in this field and therefore, it is very important to carry out some researches about scour.

There are two possible strategies to face this issue during the design of monopile foundations: (1) allowing the occurrence of the scour phenomenon, predicting the scour hole dimensions around the pile, it is the depth and length, when the scour has reached its equilibrium at medium-long term; so the design of the monopile is carried out considering the scour phenomenon is going to happen, and the existence of the hole is taken in to account in the design of the pile; (2) avoiding the occurrence of the scour phenomenon, by designing and installing an scour protection system around the monopile.

Both strategies must be accompanied by periodical monitoring campaigns of the scour around the structure, well only around the monopile in case of not having scour protection, or in the own scour protection, to validate its good working, and also around it. The strategy of using scour protection is the most common one up to now in the wind industry, having shown good results from the structural point of view, but driving some changes in local ecosystem [15,16]. Many studies consider expected scour [17,18] and some of them have demonstrated that it is more economical to increase the length of the piles to take into account the possible scour considering the strategy of allowing the scour phenomenon occurs. In addition, it is necessary to consider the scour in other structures of the installation as it is the case of the electrical cables in both the interarray and export cables.

There are different types of protection systems, being the most common ones: riprap, concrete mattress, geocontainers, or other more innovative as SSCC front mats [19–22] (Figure 5), being the most usual and even sanctioned by experience the riprap one. Many researches are currently being carried out in relation to the design of scour protection systems.

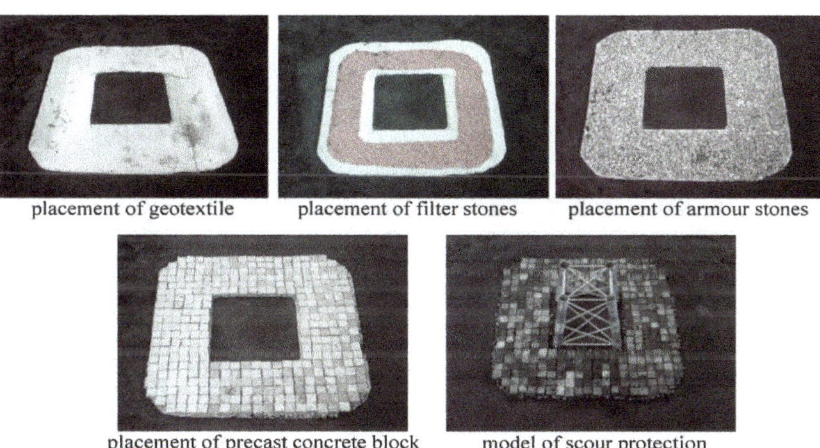

Figure 5. Different scour protection systems. Reproduced from Chen, H. et al. 2014 with permission.

Although the use of riprap as scour protection has proven to work well in maritime structures like breakwaters, quays, etc. there is not a clear formulation to be used for designing it in case of offshore wind farms. One of the key aspects in this design is the determination of the size and weight of the stones to be used in the upper part of the protection, it is the armour layer [23]. Below the armour layer, it is usual to place a filter to prevent seabed natural material from escaping through the gaps between armour units.

Among the uncertainties detected that affect the design of foundations, it is important to highlight the importance of the choice of the scale used in the physical models, because scour formulas are obtained in laboratory tests [10].

Regarding scour, there is much more experience in fluvial environment than in the marine one. However, with the development of the offshore wind industry, there has been a significant push in research about scour in recent years. Formulas used in fluvial engineering for the design of riprap weight and size are Isbash (1936) replace years with reference numbers, Shields (1936/77), MOP-54 (1936/75), Peterka (1958/66), Maynord (1970/88), Blaisell (1973), Breusers-Raudkivi (1977/91), Blogett-McConaughy (1981), HEC-11 (1989), Pilarczyk (1990/91), Escarameia/May (1992/98), etc., while formulas used for coastal structures are Cox/Campbell (1958/66), Soulsby (1997), Hoffmans-Verheij (1997), De Vos (2012), etc. [24–29].

So, there are a lot of existing formulas, none of them developed specifically for the design of the stones used as scour protection in offshore wind farms (flow type, depths, monopile dimensions, etc.) except De Vos formula [24]. There are two possible approaches for the design of the protection: static and dynamic, allowing the last one some movement of the armour layer stones, being the design more economical [28,29] as it happens in other maritime structures as breakwaters. Furthermore, there are more criteria to be considered such as the effect of the diameter value in the scour hole, etc., although it is very important to consider that scour in alone current flow is generally higher than in the case of wave plus current flow.

In some cases, it has been considered appropriate to use some of the fluvial engineering formulas, first due to the existence of a very small number of coastal engineering formulas and second due to the importance of currents in the scour phenomenon.

This paper presents a research with the main objective of analyzing the feasibility of different existing formulas for the design of armour units used in scour protection systems in offshore wind monopiles, emphasizing the uncertainties and giving some recommendations for their use. For that, the formulas are applied to five case studies of offshore wind farms in operation, and the results have been compared to the real data of the scour protection systems really constructed.

2. Methods

The main objective of the paper is to analyze, compare and verify the feasibility of the application of different equations for calculating the weight and size of the stones used as riprap for scour protection of monopiles in offshore wind facilities. It is important to emphasize that the scope of this paper is limited, on the one hand to monopile foundations because it is the type of substructure clearly majority, with 82% representation in European offshore wind farms. On the other hand, it is limited to riprap-type protections, also because it is the most used one so far [30], having shown its good performance so far.

The methodology followed is divided into four different steps. The first step is the review of the state of knowledge regarding the design formulations of riprap protections, both related to fluvial and marine environment. For this research, one fluvial environment formula has been used (Isbash) to be compared with two marine environment formulas (Soulsby, and De Vos).

The second step consists of the characterization of real case studies. For that, five offshore wind farms in operation have been selected, all of them with monopile foundations and riprap scour protections. For that, it has been necessary a review of different facilities throughout Europe, where more public information about these is available, selecting only those with all the information necessary for this research.

The information includes a general section about the facility (total power, turbine model, depth, distance to the coast, layout, etc.), another one about the foundation and the scour protection (monopile dimensions, soil type and scour protection dimensions), and another one about metocean conditions (wave design parameters associated to 50 years of return period and marine current design velocity). All this is explained in the next section, called case studies. During data collection, it is essential to analyze and verify all available information, discarding non consistent one, with the main objective of having consistent and complete information as input data.

The third step refers to the application of different design equation to calculate the weight and the size of the stones used as riprap scour protection to the five case studies identified and characterized previously. Three equations have been checked, with the objectives to analyze their possible use and the differences in the results. The input data for those equations are metocean climate identified in the second step, and the depth of the location to obtain the horizontal velocity at the seabed level due to waves, and the equations are selected during the state-of-the-art study.

Finally, in the fourth step a comparison is made between the results of riprap weight obtained in the practical application (third step) and the real data of the facilities, obtained by characterizing them (second step). The differences between the weights calculated with different formulas and the actual data documented in each case of study are valued, to end with a discussion and conclusions.

3. Description of the Formulas Used for Scour Protection

As mentioned in the introduction, there are numerous formulas to calculate the weight and size of the stones used for scour protection. While some of them are designed for fluvial environment, others are for the marine one. All equations have been formulated prior to the date of construction of the offshore wind farms except the De Vos formula. Three of these equations have been selected for this paper: Isbash (1936), corresponding to fluvial environment, and Soulsby (1997), and De Vos (2012), corresponding to marine environment.

Although the only formula that fits perfectly the type of structure to be protected against scour is the De Vos formula, two other formulas have been chosen because their wide use at the current time. The first is the Isbash equation for river environments and the second is the Soulsby equation for the marine environment. In the following paragraphs, each of the formulas used in the research is briefly described, following the list shown in Table 2 in which first is the one formula used in river engineering and later the two formulas used in maritime engineering.

The first formula was proposed by the Russian researcher Isbash in 1936, with the objective of dimensioning the average size of the scour protection material. This equation has been widely used in the margins of discharge channels and dam closures, where high turbulence flows act in a direct current regime. Subsequently, it was adopted by the ASCE-American Society of Civil Engineers in 1950, and its use and application are now widespread [27].

The next equation is the one proposed by the British researcher Soulsby in 1997 for marine environments. He proposed two equations, one as a function of the flow induced by the waves and the other as a function of the steady current. The formulations defined by Soulsby were based on several concepts, the most relevant ones the amplifying factor and the shear stress on the seabed. These formulations were contrasted by many tests carried out mainly in the laboratory.

The last of the formulas used was proposed by De Vos et al. in 2012. It is a dynamic design formula that determines the size of the stones used in scour protections for monopiles in combined wave and steady current conditions. It is an equation obtained from laboratory tests and it improves the results obtained by the static design formula that they proposed in a previous work.

Isbash formula has been considered for historical reasons. It was the first formula that appeared for the dimensioning of the average size of the protection material and it was widely used in areas with a very turbulent flow such as discharge channels. Soulsby equation was selected because it corresponds to cases where the flow generated by the waves at the base of the monopile is important, but without taking into account the current effect. That formula has been widely used in the marine environment. The fact of choosing these three equations is interesting to compare the behaviour of the three formulas applied to the five case studies.

Table 2. Equations of Isbash, Soulsby and De Vos.

Researcher (Year)	Equation
Isbash (1936)	$D_{50} = \dfrac{V_a^2}{g*N*(G_S-1)}$
Soulsby (1997)	$\text{Waves}: D_{n50} = \dfrac{97.9 *U_w^{3.08}}{T_P^{1.08}*[g*(s-1)]^{2.08}}$
De Vos (2012)	$\dfrac{S_{3D}}{N^{b0}} = a_0 \dfrac{u_m^3*T_{m-1,0}^2}{\sqrt{g*d}*(s-1)^{3/2}*D_{n50}^2} + a_1\left(a_2 + a_3 \dfrac{\left(\frac{U_c}{w_s}\right)^2*(U_c+a_4*U_m)^2*\sqrt{d}}{g*D_{n50}^{3/2}}\right)$

The meaning of the different symbols included in Isbash equation:

1. D_{50} is the median diameter of the stone, in feet.
2. V_a is the average velocity, in feet/s.

3.   g is the gravity acceleration, in feet/s^2.
4. G_s is the relative density between the stone and the water, dimensionless.
5.   N is the stability number of Isbash, dimensionless.

The meaning of the different symbols included in Soulsby equation:

1. D_{n50} is the nominal median diameter of the stone, in m.
2. U_w is the horizontal velocity due to waves at the seabed level, in m/s.
3. T_p is the wave peak period, in s.
4.   g is the gravity acceleration, in m/s^2.
5.   s is the relative density between the stone and the water, dimensionless.

The meaning of the different symbols included in De Vos equation:

1. S_{3D} is the three-dimensional damage of a scour protection, dimensionless.
2.   N is the number of waves, dimensionless.
3. a_0, a_1, a_2, a_3, a_4 and b_0 are parameters obtained to determine the equation.
4. U_m is the bottom orbital velocity, in m/s.
5. U_c is the average flow velocity.
6. $T_{m-1,0}$ is the spectral wave period order m−1,0, in s.
7.   g is the gravity acceleration, in m/s^2.
8.   d is the water depth, in m.
9.   s is the relative density between the stone and the water, dimensionless.
10. D_{n50} is the nominal median diameter of the stone, in m.
11. w_s is the fall velocity, in m/s.

In the De Vos formula, D_{n50} is determined from the median stone diameter D_{50} which is the stone size for which 50% of the stones is lighter by weight as $D_{n50}/D_{50} = 0.84$. If the same relationship is applied to W_{n50} and W_{50}, the coefficient is 0.593.

4. Case Studies

More than fifty offshore wind farms were identified, analyzed, and characterized. All of them satisfy the criteria of having monopile as foundation, and riprap as scour protection. Anyway, due to confidential issues, there is not a lot of public information of the offshore wind facilities. An essential part of the case study identification and characterization is to analyze and verify all available information, discarding non consistent data, with the main objective of having only consistent and complete information in the research.

Subsequently, and from this broad list, five case studies are selected, justifying their limited number due to the great difficulty encountered when identifying case studies with sufficient information, so as to be able to homogenize all the aspects and their main characteristics, reducing these to only the aspects of its location, the construction process, the foundation structure or those of the wind turbine, also incorporating maritime weather data, being necessary to perform in parallel an analysis of each of the documented data, contrasting and validating each one of them by different sources.

The selected wind farms are Arklow Bank phase 1 (Ireland), Edmond aan Zee (Netherlands), Horns Rev phase 1 (Denmark), Princess Amalia (Netherlands), and Scroby Sands (United Kingdom) (Figure 6), all of them located in the North Sea.

Figure 6. Case studies selected for the research: Arklow Bank phase 1 (Ireland), Egmond aan Zee (Netherlands), Horns Rev phase 1 (Denmark), Princess Amalia (Netherlands), and Scroby Sands (United Kingdom).

All information necessary for the different offshore wind farm case studies are summarized in Tables 3–5. Table 3 includes the general characteristics, Table 4 the characteristics of foundations and scour protections and Table 5 the metocean conditions. The information included in these three tables are taken, mainly, from the [29–45].

Table 3. General characteristics of offshore wind farm case studies.

General Characteristics	Offshore Wind Farms				
	Arklow Bank Phase 1	Egmond Aan Zee	Horns Rev Phase 1	Princess Amalia	Scroby Sands
Country	Ireland	Netherlands	Denmark	Netherlands	United Kingdom
Coordinate reference (datum)	WGS84	WGS84	WGS84	WGS84	WGS84
Coordinates (latitude and longitude)	52°47′22.4″ N–5°56′56.4″ W	52°36′21.6″ N–4°25′8.3″ E	55°31′47″ N–7°54′22″ E	52°35′24″ N–4°13′11.9″ E	53°38′56″ N–1°47′25″ E
Distance to the coast (km)	10	15	17	23	2.5
Area (km^2)	22	27	21	14	10
Mean sea level (MSL) (m)	8	20	14	24	12
Construction start (year)	2003	2005	2002	2006	2003
Comissioning (year)	2004	2007	2003	2008	2004
Turbine power (MW)	3.6	3	2	2	2
Turbine model	GE 3.6	V90	V80	V80	V80
Number of wind turbines	7	36	80	60	30
Total power of the wind farm (MW)	25.2	108	160	120	60
Turbine height (hub) (m)	149.8	105	100	100	100
Rotor diameter (m)	103	90	80	80	80
Layout rows	7	12	8	11	10
Layout columns	1	4	10	13	3
Average distance between turbines (m)	500	500	560	500	450

Table 4. Foundation and scour protection characteristics of offshore wind farm case studies.

Foundation &Scour Protection Characteristics	Offshore Wind Farms				
	Arklow Bank Phase 1	Egmond Aan Zee	Horns Rev Phase 1	Princess Amalia	Scroby Sands
Foundation type:	Monopile	Monopile	Monopile	Monopile	Monopile
Foundation diameter (m):	5	4.6	4.2	4	4.2
Driving length of the pile (m):	35	30	34	30	30
Scour protection type:	Riprap	Riprap	Riprap	Riprap	Riprap
Density of the material used as armour and filter (kg/m^3):	2,600.00	2,800.00	2,600.00	2,800.00	2,600.00
Soil type:	Sand	Sand	Sand	Sand	Sand
Soil median diameter (D_{50}) (mm):	0.20	0.20	0.15	0.45	0.40
Filter median diameter (D_{50}) (m):	0.05	0.05	0.20	0.17	0.15
Filter thickness (m):	0.6	0.4	1,00	0.90	1
Filter extension length (m):	20	24	20	24	25
Armour median diameter (D_{50}) (m):	0.42	0.40	0.55	0.50	0.45
Armour thickness (m):	1.20	1.80	1.80	1.50	1.30
Armour extension length (m):	15	18	15	18	15

Table 5. Metocean characteristics of offshore wind farm case studies.

Metocean Characteristics	Offshore Wind Farms				
	Arklow Bank Phase 1	Egmond Aan Zee	Horns Rev Phase 1	Princess Amalia	Scroby Sands
Significant wave height (H_s) (m)	5.6	3.6	5.2	7.7	3.17
Wave peak period (T_p) (s)	9	8	6.30	9.7	8.1
Marine current velocity (U_c) (m/s)	2.0	0.6	1.17	1.30	1.68
Bottom orbital velocity (U_m) (m/s)	2.70	0.73	1.15	1.61	1.08
Return period (T_R) (years)	50	50	50	50	50
Predominant wave direction	E–W to ENE	N–S	N–S	S–E	N–S
Predominant marine current direction	W–SW	S–E	N–S	S–E	N–S

Figures are prepared for the different five offshore wind farms, including the main dimensions of the monopile and scour protection: Arklow Bank phase 1 (Ireland) (Figure 7), Egmond aan Zee (Netherlands) (Figure 8), Horns Rev phase 1 (Denmark) (Figure 9), Princess Amalia (Netherlands) (Figure 10), and Scroby Sands (United Kingdom) (Figure 11).

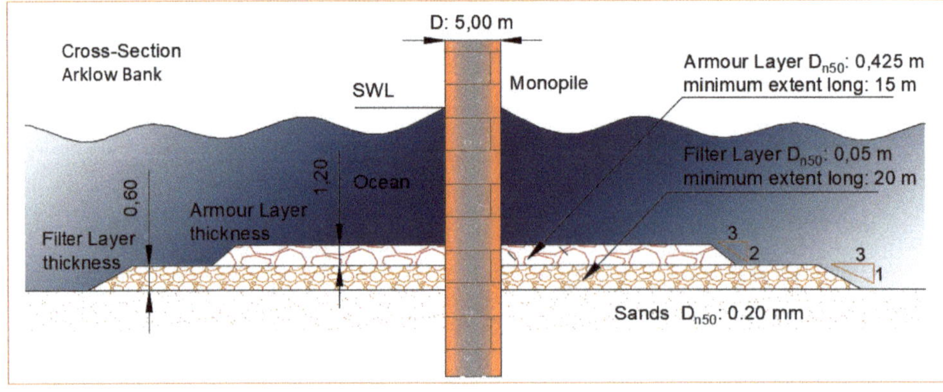

Figure 7. Main dimensions of monopile and scour protection of Arklow Bank phase 1.

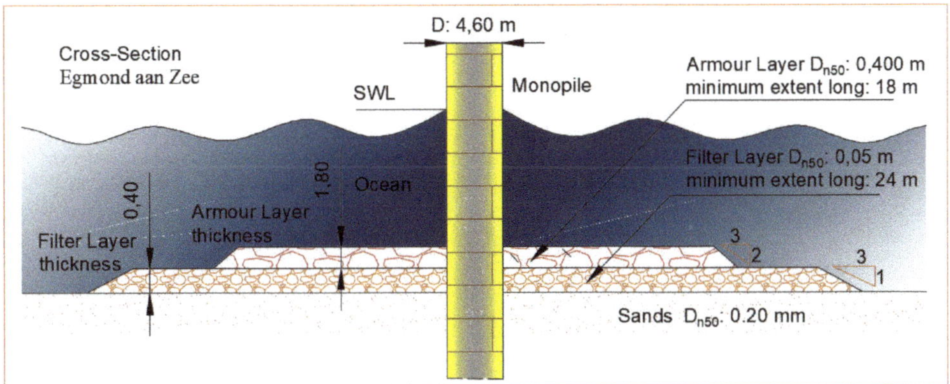

Figure 8. Main dimensions of monopile and scour protection of Egmond aan Zee.

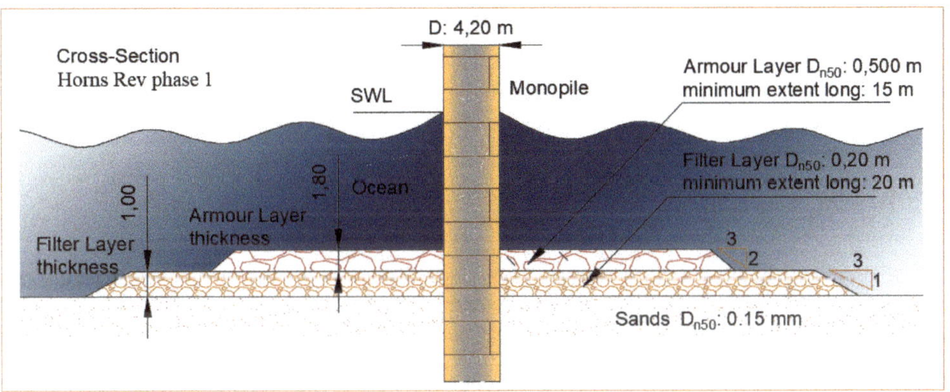

Figure 9. Main dimensions of monopile and scour protection of Horns Rev phase 1.

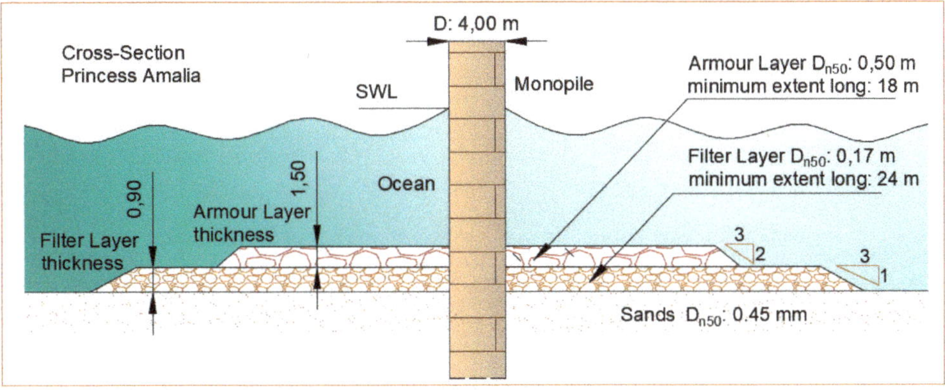

Figure 10. Main dimensions of monopile and scour protection of Princess Amalia.

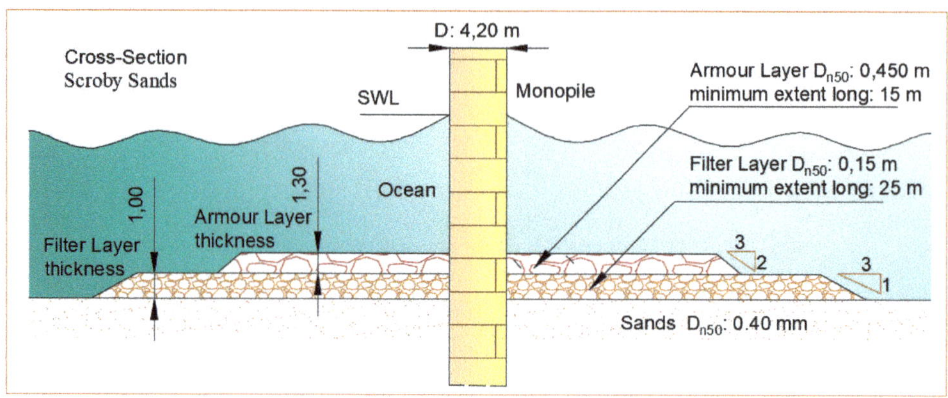

Figure 11. Main dimensions of monopile and scour protection of Scroby Sands.

5. Results

In order to check and validate each of the results obtained in the practical application, auxiliary calculations are performed to determine the orbital velocities at seabed level due to waves. Non-linear wave theories (e.g., stream function) are recommended to be applied in function of the zoning resulting from each of the study cases, for instance for the shallow water case study (Arklow Bank phase 1) high order Stokes' theory is inappropriate.

Only to note that wave velocity at seabed level has been calculated in all the cases according Airy wave theory as simplification, corresponding to the maximum horizontal velocity associated to the wave climate parameters related to 50 years return period. Table 6 includes the marine current, bottom orbital and design velocities.

Table 6. Marine current, bottom orbital, and design velocities of offshore wind farm case studies.

Metocean Characteristics	Offshore Wind Farms				
	Arklow Bank Phase 1	Egmond Aan Zee	Horns Rev Phase 1	Princess Amalia	Scroby Sands
Marine current velocity (U_c) (m/s):	2.00	0.60	1.17	1.30	1.68
Bottom orbital velocity (U_m) (m/s)	2.70	0.73	1.15	1.61	1.08
Design velocity	4.70	1.33	2.32	2.91	2.76

Based on the input data corresponding to the case studies, mainly metocean ones, the results of the application Isbash (1936), Soulsby (1997), and De Vos (2012) are included in Table 7. For De Vos, S_{3D} has taken as 1, and the number of waves as 2000.

Table 7. Results of the application of the three equations to the five case studies.

Case Study	W_{n50} (kg)			Real Weight (kg)
	Isbash	Soulsby	De Vos	
Arklow Bank 1	1,497.97	142,439.86	7,912.86	192.60
Egmond aan Zee	0.77	1.79	1.03	179.20
Horns Rev 1	21.44	662.54	11.27	432.60
Princess Amalia	84.34	1,330.58	78.95	350.00
Scroby Sands	61.43	1,464.54	50.44	236.90

Table 7 shows the results obtained after applying the three design formulations selected in the five case studies, with notable differences among them, which may be due to the fact that each researcher has proposed a different methodology in their application. The real average weight of the stones used for scour protection in each of the case studies has been included in the table in the right-hand column.

Likewise, it is highlighted as the most influential variable in each formulation the velocity of the flow at seabed level, defined as orbital velocity as it can be observed in the different equations, with which its results have an important sensitivity to any variation of the velocities of the flow, obtaining very different results. In order to calculate the design velocity of the current, it has been chosen to consider that waves follow the same direction as currents, this being the most unfavorable situation for the stability of the foundation against scour. It should be mentioned that there are other combinations of wave and currents actions such as waves opposite currents, alternate current such as those induced by tides, etc. However, these have not been the object of the research carried out. In all the cases, it has been considered mild slope and wave breaking coefficient in around of 0.7 referred to the mean sea level.

6. Discussion

The practical application allows to compare and determine differences existing between the two weights analyzed, on the one hand the result of applying the formulations and, on the other hand the one identified in the characterization of the case studies, the real projects, quantifying their difference by means of three different methodologies:

(a)　Ranges relationship represented by the value of the difference between the real and the calculated weights.

(b)　Relative error ratio represented by the absolute value of the difference between real and calculated weights divided by the real weight.

(c)　Step factor relationship defined as the square root of the division between the real and the calculated weights.

All this made possible to detect that the results are very different, specifically using Soulsby equation, designed for the flow produced by the waves on the seabed without taking into account the interaction with the current.

Furthermore, it was observed that the velocities in the case studies are between 1 and 5 m/s, the value of the riprap protection is very related to the velocity. In case of higher velocities, bigger riprap dimensions, being clearly the highest dimension for Arklow Bank, with 4.7 m/s of velocity at sea bottom. Lowest values correspond to Egmond aan Zee, with 1.33 m/s of velocity at sea bottom.

Next table (Table 8) shows the results obtained after comparing these calculated weights using the three methodologies mentioned for each of the case studies, Arklow Bank phase 1 (Ireland), Edmond aan Zee (Netherlands), Horns Rev phase 1 (Denmark), Princess Amalia (Netherlands) and Scroby Sands (United Kingdom).

In the case of Arklow Bank phase 1, the weights obtained are very conservative with values between 10 and 1000 times higher than those placed in reality, with dispersion ratios between 0.001 and 0.13; the values obtained by means of Soulsby equation are very striking 10 times higher than those obtained by means of the other two equations. This is because the value of the combined velocity (waves and current) is also the highest of all case studies with a high value (4.70 m/s).

Table 8. Comparison between calculated and real weights in offshore wind farm case studies.

Case Study		W_{n50} (kg)		
		Isbash	Soulsby	De Vos
Arklow Bank Phase 1	Calculated weight	1,497.97	142,439.86	7,912.86
	Ranges relationship	1,305.37	142,247.26	7,720.26
	Relative error	6.78	738,56	40.08
	Step factor relationship	0.36	0.03	0.14
Egmond aan Zee	Calculated weight	0.77	1.79	1.03
	Ranges relationship	178.43	177.41	178.17
	Relative error	1.00	0.99	0.99
	Step factor relationship	15.25	10.005	13.19
Horns Rev Phase 1	Calculated weight	21.44	662.54	11.27
	Ranges relationship	411.16	229.94	421.33
	Relative error	0.95	0.53	0.97
	Step factor relationship	4.49	0.81	6.19
Princess Amalia	Calculated weight	84.34	1,330.58	78.95
	Ranges relationship	265.66	980.58	271.05
	Relative error	0.76	2.80	0.77
	Step factor relationship	2.04	0.51	2.10
Scroby Sands	Calculated weight	61.43	1,464.54	50.44
	Ranges relationship	175.47	1,227.64	186.46
	Relative error	0.74	5.18	0.79
	Step factor relationship	1.96	0.40	2.18

In the case of Egmond aan Zee, the three equations give a result with the same order of magnitude, and representing between 0.4% and 1% of the weight actually placed. In this case, the value of the velocity considered in the calculation is the lowest of all (1.33 m/s).

In the rest of cases, the combined current-wave speed considered is between 2 and 3 m/s. Soulsby equation gives very high values of the weight of the stones between 1.5 and 6.2 times the weight of the stones placed, and with values of the ratios one order of magnitude lower than those obtained with the other formulas. On the other hand, the values of the ratios for the Isbash and De Vos formulas are similar. Differences can also be observed with respect to the weights placed. In the case of the Princess Amalia and Scroby Sands parks, the value obtained represents around 25% of the weight placed, while in Horns Rev, it is 3.5%.

It shows that in case of high velocity values at sea bottom, it can be interesting to carry out laboratory tests to try to save money, because in that offshore wind farm, there is important differences between calculated and real weight values using any of the three selected equations. Egmond aan Zee has some important differences between calculated and real weight. In fact, the comparison of the values in the case of the highest and lowest velocity value at sea bottom gives worst results than in case with intermediate values.

For De Vos formula, the expected value of parameter S_{3D} for the diameters of the existing stones in the scour protections has been analyzed. The results obtained are the following: for Arklow Bank phase 1 the value is 3.95, for Egmond aan Zee the value is 0, for Horns Rev phase 1 and for Scroby Sands it is 0.20, for Princess Amalia it was 0.26 is obtained. In view of the values obtained, it follows that all installations, except for Arklow Bank phase 1 and Egmond aan Zee, are in the range of values posed for the equation (0.2–1.0). Therefore, it can be deduced that in Horns Rev phase 1, Princess Amalia and Scroby Sands, the stone sizes placed would be a little smaller than those obtained with the static approximation. In Arklow Bank phase 1, the stone size placed is much smaller than the result of the static approach and in Egmond aan Zee it will be close to the result of the static approach.

Considering the relative error ratio, it is observed that the Isbash and De Vos formulae give very similar values, lower than 1.00, in all cases except for Arklow Bank phase 1. On the other hand, Soulsby's formula gives the best value of the relative error ratio for Horns Rev phase 1 (0.53), and a value of this ratio very similar to those of the other two formulas in Egmond aan Zee.

According Table 9 [46], the D_{50} can be obtained knowing the velocity at seabed level. For instance, in Princess Amalia, the design velocity is 2.91 m/s, and according the table the minimum D_{50} recommended is 0.40 m for the scour protection armour. $D_{50} = 0.40$ m corresponds to $W_{50} = 173$ kg, considering 2800 kg/m^3 of density, D_{n50} is 0.84 multiplied by D_{50}, and if the same relationship is applied to W_{n50} and W_{50}, the coefficient is 0.593. So, $W_{50} = 173$ kg is converted to $W_{n50} = 103$ kg. This value is in the magnitude order established in the Table 9 in the case of Isbash, and De Vos equations, although Table 9 is more conservative than the use of the equation. So, it can be used for estimate values.

Table 9. Stone size according the velocity of the flow. Adapted from [46].

Ub	D_{50} (m)
1	0.05
2	0.20
3	0.40
4	0.70
5	1.10
6	1.60

7. Conclusions

It is concluded that there is a very small number of formulas applicable for the design of the foundation protection systems of offshore wind turbines. Furthermore, it has been observed that the formulas are not very clear, not allowing a rapid determination, because most of them have not been applied or contrasted for offshore wind farms, having been only verified for coastal structures or in hydraulic channels. An additional uncertainty has to be taken into account, regarding the dimensions used in these structures; these dimensions are completely different to the usual ones included in coastal and offshore engineering.

A high degree of confidentiality is imposed in the offshore wind sector, as a result of which the utilities do not share the information, taking care of their know-how. Anyway, not sharing the information, mainly the failures, make the industry advances slower than in the case of looking for synergies to solve the problems and the uncertainties.

It is also verified that when applying the formulations of scour protections, both for continuous and oscillatory regime, their results presented important inconsistencies, because in general these are thought for depths of the order of 5 m, and not for usual depth of offshore wind facilities, more than 15–20 m depth.

With respect to the comparisons of the weights, in the case of Arklow Bank phase 1, the weights obtained are very conservative with values between 10 and 1000 times higher than those placed in reality, with step factor relationships between 0.03 and 0.36. It shows that in case of high velocity values at sea bottom, it can be interesting to carry our laboratory test to try to save money, because in that offshore wind farm, there is important differences between calculated and real weight values using any of the three selected equations

From the comparison of the results obtained with the different formulas selected, it can be deduced that the Soulsby formula is not applicable because it gives the highest values of the weights of the stones to be placed. Only in the case of the Egmond aan Zee offshore wind farm, the value obtained is close to those of the other two formulas, which can be interpreted as meaning that it can be used in locations with a low speed range, less than 2 m/s.

On the other hand, it can be observed that the values obtained by the Isbash, and De Vos equations are similar in parks whose bottom velocities are between 2 and 3 m/s. Differences can also be observed with respect to the weights placed. In the case of the Princess Amalia and Scroby Sands parks, the value obtained represents around 25% of the weight placed, while in Horns Rev, it is 3.5%.

For the De Vos formula, the expected value of parameter S_{3D} for the diameters of the existing stones in the scour protections has been analyzed. In view of the values obtained, it follows that all installations, except for Arklow Bank phase 1, and Egmond aan Zee, are in the range of values posed for the equation (0.2–1.0). Therefore, it can be deduced that in Horns Rev phase 1, Princess Amalia, and Scroby Sands, the stone sizes placed would be a little smaller than those obtained with the static approximation. In Arklow Bank phase 1, the stone size placed is much smaller than the result of the static approach and in Egmond aan Zee it will be close to the result of the static approach.

Considering the relative error ratio, it is observed that the Isbash and De Vos formulae give very similar values, lower than 1.00, in all cases except for Arklow Bank phase 1. On the other hand, Soulsby's formula gives the best value of the relative error ratio for Horns Rev phase 1 (0.53), and a value of this ratio very similar to those of the other two formulas in Egmond aan Zee.

Evidently, although with the Isbash formula similar results to the De Vos formula have been obtained in all the facilities analyzed except for Arklow Bank phase 1, this last formula has better possibilities of adjustment to the conditions of this type of foundations when considering more factors, not only the velocity of the current at the bottom, so that it can offer a better result of the average weight of the stones to be placed in the scour protection systems.

For a rapid estimation, Table 9 can be used because it gives conservative values ad it is very easy to use. Anyway, it is first estimate, and more calculations and possible laboratory test should be done to design the riprap protection.

This makes it essential to continue the research about the scour, for which it would be appropriate to have real data on offshore wind farms in operation, mainly based on monitoring the scour around the structures along the time with field campaign with multibeam echosounders.

Author Contributions: Conceptualization, V.N.; methodology, J.-S.L.-G.; investigation, L.S.; writing—original draft preparation, M.D.E.; writing—review and editing, J.-S.L.-G. and V.N.; supervision, M.D.E.

Funding: Authors give thanks the Agustín de Betancourt Foundation (FAB) for the support received over the past few years.

References

1. Esteban, M.D.; López-Gutiérrez, J.S.; Diez, J.J.; Negro, V. Methodology for the design of offshore wind farms. *J. Coast. Res.* **2011**, *64*, 496–500.
2. WindEurope. *Offshore Wind in Europe. Key Trends and Statistics 2017*; Research Report; Iván Pineda WindEurope: Brussels, Belgium, 2018.
3. Global Wind Energy Council Web Page. Available online: http://www.gwec.net (accessed on 7 August 2018).
4. Esteban, M.D. Propuesta de Una Metodología Para la Implantación de Parques Eólicos Offshore. Ph.D. Thesis, Universidad Politécnica de Madrid, Madrid, Spain, 2009.
5. Esteban, M.D.; Diez, J.J.; López-Gutiérrez, J.S.; Negro, V. Integral management applied to offshore wind farms. *J. Coast. Res.* **2009**, *56*, 1204–1208.
6. Esteban, M.D.; Couñago, B.; López-Gutiérrez, J.S.; Negro, V.; Vellisco, F. Gravity based structures for offshore wind turbine generators: Review of the installation process. *J. Ocean Eng.* **2015**, *110*, 281–291.
7. Negro, V.; López-Gutiérrez, J.S.; Esteban, M.D.; Alberdi, P.; Imaz, M.; Serraclara, J.M. Monopiles in offshore wind: Preliminary estimate of main dimensions. *J. Ocean Eng.* **2017**, *133*, 253–261. [CrossRef]

8.	Esteban, M.D.; López-Gutiérrez, J.S.; Negro, V.; Matutano, C.; García-Flores, F.M.; Millán, M.A. Offshore wind foundation design: Some key issues. *J. Energy Resour. Technol.* **2015**, *137*, 1–5.

9.	Negro, V.; López-Gutiérrez, J.S.; Esteban, M.D.; Matutano, C. Uncertainties in the design of support structures and foundations for offshore wind turbines. *J. Renew. Energy* **2014**, *63*, 125–132. [CrossRef]

10.	Luengo, J.; Negro, V.; García-Barba, J.; López-Gutiérrez, J.S.; Esteban, M.D. New detected uncertainties in the design of foundations for offshore wind turbines. *J. Renew. Energy* **2019**, *131*, 667–677. [CrossRef]

11.	Yamini, O.A. Numerical modeling of sediment scouring phenomenon around the offshore wind turbine pile in marine environment. *J. Environ. Earth Sci.* **2018**, *77*, 766.

12.	Li, H.; Ong, M.C.; Leira, B.J.; Myrhaug, D. Effects of soil profile variation and scour on structural response of an offshore monopile wind turbine. *J. Offshore Mech. Arct. Eng.* **2018**, *140*, 1–10. [CrossRef]

13.	Rezaei, R.; Duffour, P.; Fromme, P. Scour influence on the fatigue life or operational monopile-supported offshore wind turbines. *J. Wind Energy* **2018**, *21*, 683–696. [CrossRef]

14.	Matutano, C.; Negro, V.; López-Gutiérrez, J.S.; Esteban, M.D. Scour prediction and scour protections in offshore wind farms. *J. Renew. Energy* **2013**, *57*, 358–365. [CrossRef]

15.	Causon, P.D.; Gill, A.B. Linking ecosystem services with epibenthic biodiversity change following installation of offshore wind farms. *J. Environ. Sci. Policy* **2018**, *89*, 340–347. [CrossRef]

16.	Esteban, M.D.; López-Gutiérrez, J.S.; Diez, J.J.; Negro, V. Offshore Wind Farms: Foundations and influence on the Littoral Processes. *J. Coast. Res.* **2011**, *64*, 656–660.

17.	Stroescu, I.E.; Frigaard, P.; Fejerskov, M. Scour Development around Bucket Foundations. *Int. J. Offshore Polar Eng.* **2016**, *26*, 57–64. [CrossRef]

18.	Margheritini, L.; Martinelli, L.; Lamberti, A.; Frigaard, P. Scour around Monopile Foundation for Off-Shore Wind Turbine in Presence of Steady and Tidal Currents. In Proceedings of the 30th International Conference on Coastal Engineering (ICCE 2006), San Diego, CA, USA, 3–8 September 2006; Volume 3, pp. 2330–2342.

19.	Subsea World News Web Page. Available online: http://www.subseaworldnews.com (accessed on 7 April 2017).

20.	Offshore Technology Web Page. Available online: http://www.offshore-technology.com (accessed on 7 April 2017).

21.	Geosynthetica Web Page. Available online: http://www.geosynthetica.net (accessed on 7 April 2017).

22.	Chen, H.; Yang, R.; Hwung, H. Study of hard and soft countermeasures for scour protection of the jacket-type offshore wind turbine foundation. *J. Marine Sci. Eng.* **2014**, *2*, 551–567. [CrossRef]

23.	Fazeres-Ferradosa, T.; Taveira-Pinto, F.; Romão, X.; Vanem, E.; Reis, M.T.; das Nevez, L. Probabilistic design and reliability analysis of scour protections for offshore windfarms. *J. Eng. Fail. Anal.* **2018**, *91*, 291–305. [CrossRef]

24.	National Cooperative Highway Research Program. *Riprap Design Criteria, Recommended Specifications and Quality Control*; Research sponsored by the American Association of State Highway and Transportation Officials in Cooperation with the Federal Highway Administration: National Cooperative Highway Research Program, NCHRP REPORT 568; Transportation Research Board: Washington, DC, USA, 2006.

25.	Matutano, C.; Negro, V.; López Gutiérrez, J.S.; Esteban, M.D.; Campo, J.M. Dimensionless wave height parameter for preliminary design of scour protection in offshore wind farms. *J. Coast. Res.* **2013**, *65*, 1633–1638. [CrossRef]

26.	Whitehouse, R.J.S.; Harris, J.M.; Sutherland, J.; Rees, J. The nature of scour development and scour protection at offshore windfarm foundations. *Mar. Pollut. Bull.* **2011**, *62*, 73–88. [CrossRef] [PubMed]

27.	González-Ortega, J.M. Análisis de Procesos de Erosión Local en Márgenes de Cauces Fluviales con Curvatura en Planta. Ph.D. Thesis, Universidad Politécnica de Madrid, Madrid, Spain, 2004.

28.	De Vos, L.; De Rouck, J.; Troch, P.; Frigaard, P. Empirical design of scour protections around monopile foundations. Part 2: Dynamic approach. *Coast. Eng.* **2012**, *60*, 268–298. [CrossRef]

29.	Arboleda Chavez, C.E.; Stratigaki, V.; Wu, M.; Troch, P.; Schendel, A.; Welzel, M.; Villanueva, R.; Schlurmann, T.; de Vos, L.; Kisacik, D.; et al. Large-scale experiments to improve monopile scour protection design adapted to climate change—The PROTEUS project. *Energies* **2019**, *12*, 1709. [CrossRef]

30.	Wind Europe Web Page. Available online: http://windeurope.org (accessed on 3 February 2017).

31.	4C Offshore Web Page. Available online: http://www.4coffshore.com (accessed on 1 April 2017).

32.	Belwind offshore wind project Web Page. Available online: http://www.belwind.eu (accessed on 1 April 2017).

33.	Power Technology Web Page. Available online: http://www.power-technology.com (accessed on 3 April 2017).

34. Petersen, T.U. Scour around Offshore Wind Turbine Foundations. Ph.D. Thesis, Technical University of Denmark, Lyngby, Denmark, 2014.
35. The Wind Power Web Page. Available online: http://www.thewindpower.net (accessed on 7 April 2017).
36. Louwersheimer, W.F. Scour around an Offshore Wind Turbine. Ph.D. Thesis, Delft University of Technology, Delft, The Netherlands, 2007.
37. Noordzee Wind Web Page. Available online: http://www.noordzeewind.nl (accessed on 10 April 2017).
38. Carroll, B.; Cooper, B.; Dewey, N. *COWRIE Scour Sed-09: A Further Review of Sediment Monitoring Data*; Technical Report (COWRIE); 2010. Available online: https://tethys.pnnl.gov/publications/further-review-sediment-monitoring-data (accessed on 10 February 2017).
39. Vanagt, T.; Faasse, M. *Development of Hard Substratum Fauna in the Princess Amalia Wind Farm-Monitoring Six Years after Construction*. Technical Report. 2014. Available online: https://www.researchgate.net/publication/271963891_Development_of_hard_substratum_fauna_in_the_Princess_Amalia_Wind_Farm_Monitoring_six_years_after_construction_eCOAST_report_2013009 (accessed on 10 February 2017).
40. Official Princess Amalia (Q7) Wind Farm Web Page. Available online: http://www.q7wind.nl (accessed on 22 April 2017).
41. Jensen, M.S.; Juul Larsen, B.; Frigaard, P.; DeVos, L.; Christensen, E.D.; Asp Hansen, E.; Solberg, T.; Hjertager, B.H.; Bove, S. *Offshore Wind Turbines Situated in Areas with Strong Currents: Laboratory Tests*; Technical report of Offshore Center Danmark: Esbjerg, Denmark, 2006.
42. Gerdes, G.; Tiedemann, A.; Zeelenberg, S. *Case Study: European Offshore Wind Farms - A Survey for the Analysis of the Experience and Lessons Learns by Developers of Offshore Wind Farms -*; Technical Report; Deutsche WindGuard GmbH: Varel, Germany, 2005.
43. Horns Rev Offshore Wind Farm Web Page. Available online: http://www.hornsrev.dk (accessed on 1 May 2017).
44. GE Renewable Energy Web Page. Available online: http://www.gerenewableenergy.com (accessed on 5 May 2017).
45. Whitehouse, R.J.S.; Harris, J.M.; Rees, J. *Dynamics of Scour Pits and Scour Protection: Synthesis report and recommendations (milestones 2 and 3): A Report for the Research Advisory Group: Final Report*; Technical Report; Department of Energy and Climate Change: London, UK, 2008.
46. Spanish Ministry of Public Works. *Geotechnical Recommendation for Design of Maritime and Harbor Works (ROM 0.5-05)*; Technical Standards; Puertos del Estado: Madrid, Spain, 2005.

Journal of
*Marine Science
and Engineering*

Review

Foundations in Offshore Wind Farms: Evolution, Characteristics and Range of Use. Analysis of Main Dimensional Parameters in Monopile Foundations

Sergio Sánchez, José-Santos López-Gutiérrez, Vicente Negro and M. Dolores Esteban *

Research Group on Marine, Coastal and Port Environment and other Sensitive Areas, Universidad Politécnica de Madrid, E28040 Madrid, Spain; sergio.sanchez.lopez@alumnos.upm.es (S.S.); josesantos.lopez@upm.es (J.-S.L.-G.); vicente.negro@upm.es (V.N.)
* Correspondence: mariadolores.esteban@upm.es

Received: 21 October 2019; Accepted: 1 December 2019; Published: 2 December 2019

Abstract: Renewable energies are the future, and offshore wind is undoubtedly one of the renewable energy sources for the future. Foundations of offshore wind turbines are essential for its right development. There are several types: monopiles, gravity-based structures, jackets, tripods, floating support, etc., being the first ones that are most used up to now. This manuscript begins with a review of the offshore wind power installed around the world and the exposition of the different types of foundations in the industry. For that, a database has been created, and all the data are being processed to be exposed in clear graphic summarizing the current use of the different foundation types, considering mainly distance to the coast and water depth. Later, the paper includes an analysis of the evolution and parameters of the design of monopiles, including wind turbine and monopile characteristics. Some monomials are considered in this specific analysis and also the soil type. So, a general view of the current state of monopile foundations is achieved, based on a database with the offshore wind farms in operation.

Keywords: offshore wind farm; support structure; monopile; jacket; GBS; tripod; floating

---

1. Introduction

Wind energy can be considered nowadays a main participant in the energy market and power generation [1,2]. In 2016, it overtook coal as the second-largest form of generation capacity in Europe, only being overpassed by gas, and showing the highest development rate of every source considered [3]. Although the offshore wind energy sector is still far from the stage of maturity of its older brother (onshore wind) [4], the similarities between both have allowed offshore wind turbines [5] to become efficient and commercial-ready in a relatively short time [6]. During 2018, a total of 4496 MW were installed in six different countries [7]. China became the largest investor of this field (1800 MW), overtaking the United Kingdom for the first time (1312 MW), who had led the market during the past years. Germany (969 MW), Belgium (309 MW), Denmark (61 MW) and South Korea (35 MW) were the other participants.

Referring to the total installed capacity, the United Kingdom is still far ahead of any other country, covering 35.0% of the total power. Germany and China stay behind, accounting for 27.3% and 19.6%, respectively. Denmark, Sweden and the Netherlands, who started designing and testing these installations in the 90s, are still playing an important role as industry manufacturers and technology developers; however, the capacity installed in these countries has not gone up to that scale. Table 1 gives some details of the picture at the end of 2018.

Table 1. Capacity installed by country at the end of 2018 (UK: United Kingdom; GER: Germany; CHI: China; DEN: Denmark; BEL: Belgium; NET: Netherlands; SWE: Sweden; VIE: Vietnam; JAP: Japan; KOR: Korea; FIN: Finland; USA: United States of America; IRE: Ireland; SPA: Spain; NOR: Norway; FRA: France) (Own elaboration based on [3,7]).

Country	Wind Farms	Power Installed (MW)	% Total Power Installed	Turbines Installed	Average Wind Farm Power (MW)	Average Turbine Capacity (MW)
UK	37	8184	35.0%	1920	221.2	4.3
GER	23	6375	27.3%	1314	277.2	4.9
CHI	36	4581	19.6%	1224	127.3	3.7
DEN	14	1327	5.7%	514	94.8	2.6
BEL	8	1187	5.1%	274	148.4	4.3
NET	6	1118	4.8%	365	186.3	3.1
SWE	5	201	0.9%	86	40.2	2.3
VIE	2	99	0.4%	62	49.5	1.6
JAP	11	84	0.4%	35	7.6	2.4
KOR	3	71	0.3%	24	23.7	3.0
FIN	3	68	0.3%	24	22.7	2.8
USA	1	30	0.1%	5	30.0	6.0
IRE	1	25	0.1%	7	25.0	3.6
SPA	2	10	0.0%	2	5.0	5.0
NOR	1	2	0.0%	1	2.0	2.0
FRA	1	2	0.0%	1	2.0	2.0

According to [8–15], the main advantages of offshore wind farms over onshore ones can be briefly described as follows:

- The existing wind resource in the sea is higher than in nearby coasts.
- Due to its location offshore, the visual and acoustic impact is lower than wind farms on land, which allows better use of the existing wind resource, with larger turbines and the use of more efficient blade geometries. Likewise, the lower surface roughness in the sea favors the use of lower tower heights.
- It provides a bigger creation of employment in the phases of construction, assembly and maintenance, due to the greater complexity during installation and exploitation.
- Possibility of integration in mixed marine complexes.
- Spaciousness of the environment.
- More constant and stable energy generated.

However, these marine facilities also have significant disadvantages with respect to terrestrial ones, which are limiting their development: non-existence of electrical infrastructures close to the location; more severe environmental conditions; evaluation of wind resource more complex and expensive; and above all, its higher investment ratios and operating expenses, needing specific technologies for construction and foundations, transport and assembly at sea, laying of electrical networks underwater and operation and maintenance tasks [16,17]. Furthermore, there is some research emphasizing the advantages of harnessing wind and waves at the same facility [18–20].

In order to face the investment costs and become more productive and, therefore, more attractive to the market investors, offshore wind farms must generate as much power as possible with the lowest construction and installation costs [21–23]. Developing more and more efficient turbines has become the greatest challenge of the sector from the early years, while the costs of foundations and electric installations remain stable [24,25]. During the last 25 years, the average power capacity of the turbines has increased by 1500% (Figure 1), evolving from the first Vestas and Nordtrank used in 1995, with a capacity of 500 kW (Figure 2), to the 7–8 MW turbines that are already being used on some of the largest and newest projects these days [26]. The European Offshore Wind Deployment Centre in Scotland, and Horns Rev 3 in Denmark, are currently using Vestas V164-8.4 MW turbines, the most powerful turbines nowadays for commercial power generation. Currently, the industry is talking about exceeding 10 MW unit power [27,28].

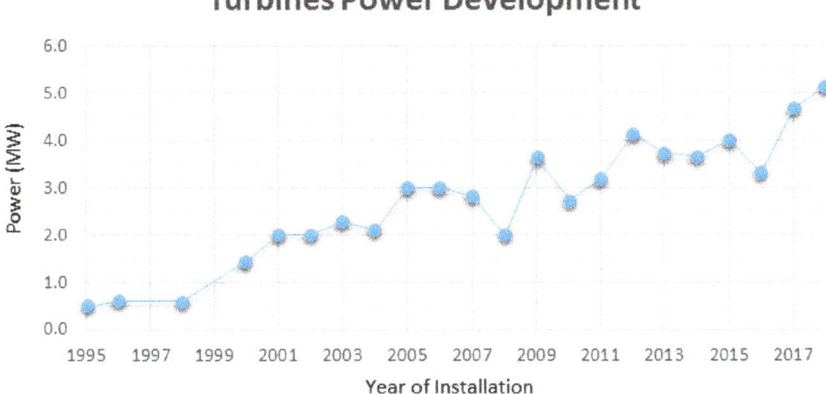

Figure 1. Average offshore wind turbine power installed every year.

2. Typologies of Foundations

Located in such a dynamic and extremely powerful element as the sea is, foundations become one of the main elements of these projects, receiving over one-third of the total cost [29–32]. As they must support the wind turbines, absorbing all the forces and loads and providing a safe and stable base, defining the right typology of foundation can have a huge impact both on the economic and technical sides, becoming especially vital these last years as wind farms are being located further from the coast, and every element must be designed and optimized in detail, to avoid performance problems and reduce maintenance works.

As was previously discussed in the article, offshore wind farms, even nowadays, are placed in a few selected locations. There are several conditions that limit the range of use of this type of installations [33–36]:

- The depth is vital when defining installation costs. Greater depths associates higher costs and the use of more complex and specialised technologies, something that is reflected in the final investment of the wind farm, reducing its profitability. Closed seas, located within continental platforms, have an average depth significantly lower than oceans and open seas, being more convenient for this type of project [37].

- Meteocean climate, in particular, wave height values [38–40]. Sea zones with high wave heights make installation and maintenance work more complicated, requiring more workload and economic resources, in addition to damaging the foundations and causing phenomena such as to scour the seabed [41–45], which weakens the structure and requires closer surveillance and maintenance work.

- The distance to coast [46]. In order to have a higher intensity wind and less impact on the landscape and the coast, in the location of this type of facilities, it is sought to distance them as far as possible from the coastline, since this also allows the use of turbines of greater nominal size and power. However, as it has been discussed before, this usually leads to an increase in depth, with the inconveniences that this entails. Therefore, the optimal location for a wind farm is the one that is placed as far from the coastline as possible while keeping a low depth.

- Spontaneous or unexpected phenomena such as earthquakes, tsunamis or extreme meteorological offshore events. Additional information from location, ground composition and soil-foundation interactions become much more relevant at regions where these events appear more frequently [47–49].

Other factors to consider are the typology of the stratum of the seabed on which engineers support the foundations [50], possible impacts to shipping routes or other offshore installations, and sea climate. Furthermore, it is important taking into account that there are still some uncertainties for the design of the foundations of these facilities [51].

As wave height is a dynamic value that may vary in a significant way for every wind turbine inside each farm, depending on several different factors, the paper is focused on the other main conditioners: the depth of the seabed and the distance from the coast. The development of new technologies and manufacturing and construction procedures [52], together with the increase in the size of wind farms and turbines and the increase in society's awareness of visual and environmental impacts [53,54], has led to the displacement of these installations further from the coastline, as it can be seen in Figure 2 (the size of the bubbles represents the capacity installed).

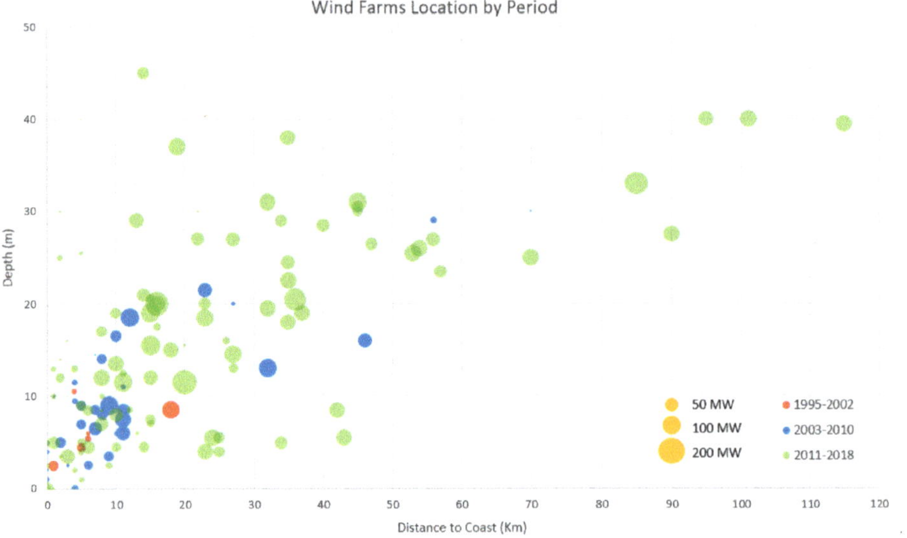

Figure 2. Offshore wind farms in operation classified by depth and distance from the coast at the end of 2018 (the size of the bubbles represents the capacity installed).

Table 2 shows the average depth and distance to coast in different locations and also the ratio depth/distance. In the case of lower depth and distance to coast, the more convenient and cost-efficient the construction project of the wind farm will be [55].

Regarding the wind farms operating these days, low-depth seas belonging to the continental platforms are the optimal location for these installations, where the ratio between the seabed depth and the distance is under or close to 1 m/km.

In Europe, the North Sea, Baltic Sea and Irish Sea stand as the best spots and mostly the first that have lead the sector since the beginning of the century. In the case of Asia, seas such as Eastern China, Southern China and the Yellow Sea provide an even better ratio, which can undoubtedly be a reason to explain the fast development and huge interest that this sector has brought to this continent.

Table 2. Average depth and distance conditions of installed farms by location at the end of 2018.

Location	Wind Farms	Power Installed (MW)	Power Installed (%)	Depth (m)	Distance to Coast (km)	Ratio Depth/Distance
North Sea	61	12,933	55.4%	23.4	30.0	0.78
Irish Sea	15	2938	12.6%	13.1	10.1	1.29
Eastern China Sea	16	2574	11.0%	5.3	16.7	0.32
Baltic Sea	20	2186	9.4%	12.4	10.9	1.14
Yellow Sea	13	1365	5.8%	7.5	11.3	0.66
South China Sea	10	777	3.3%	6.5	8.3	0.78
English Channel	1	400	1.7%	29.0	13.0	2.23
Japan Sea	6	66	0.3%	13.4	1.5	8.94
Philippines Sea	7	53	0.2%	39.9	6.3	6.34
Atlantic Ocean	4	42	0.2%	21.4	7.3	2.95
Lake Vanern	1	30	0.1%	9.5	4.0	2.38

Depending on the depth of the seabed, as well as the surrounding conditions and sea climate, different solutions have been used since the first offshore wind turbines were designed. Some of the most important typologies of foundations are defined as follows [55–59]:

- Gravity-based structure (GBS): concrete-based structure which can be constructed with or without small steel or concrete skirts. The base width can be adjusted to suit the actual soil conditions. The proposed design includes a central steel or concrete shaft for transition to the wind turbine tower [60,61].

- Monopile: the monopile support structure is a simple design by which the tower is supported by the monopile, either directly or through a transition piece, which is a transitional section between the tower and the monopile. The monopile continues down into the soil. The structure is made of cylindrical steel tubes [62,63].

- Tripod: three-leg structure made of cylindrical steel tubes. The central steel shaft of the tripod makes the transition to the wind turbine tower. The tripod can have either vertical or inclined pile sleeves. Inclined pile sleeves are used when the structure is to be installed with a jack-up drilling rig. The base width and pile penetration depth can be adjusted to suit the actual environmental and soil conditions [64].

- Jacket: similar to the tripods described above, with the difference of having four piles instead of three. These metal piles are linked together thanks to a lattice that provides strength and stability to the whole structure. They have dimensions similar to the tripods but given their greater adaptability to diverse conditions and stability, they are more widespread than these, until being the second most used typology only behind the monopiles [65,66].

- Floating: The floating support structure consists of a floating platform and a platform anchoring system. The platform has a transition piece to install the tower on top of that. The platform can have several typologies: spar, semisubmersible and tension leg platform (TLP) [67,68].

Besides the previously mentioned types of foundations, there are others in development to be used in the offshore wind industry. One of the most known is the suction caisson being used in oil and gas with very good results [69–71]. This typology provides a stable and light base, whose strength comes directly, not from its weight or depth, but from the compaction of the supporting soil through the suction of the water. Suction caissons work very well in soft soils such as clays [72,73].

The seabed depth has often been used as the main conditioner to choose the most convenient typology for a project. Various articles and reports have attempted to settle the range of employment, establishing approximate parameters based on the behavior that each one of them experiences when facing sea climates. Table 3 includes some of the most accepted criteria [74,75].

Table 3. Range of use of typology by seabed depth.

Foundation	Ashuri and Zaaijer, 2007	DNV, 2013	Iberdrola, 2017
GBS	0–10 m	0–25 m	0–30 m
Monopile	0–30 m	0–25 m	0–15 m
Tripod/Jacket	>20 m	20–50 m	>30 m
Floating	>50 m	>50 m	>50 m

However, the depth of the seabed is not the only condition to be considered. Other factors such as the availability of resources [76] and technology and experience in a certain type over the others make different markets bet on different solutions in similar environments.

Figures 3 and 4 show the comparison of the typologies in Europe and in Asia according to the parameters of depth and distance to the coast.

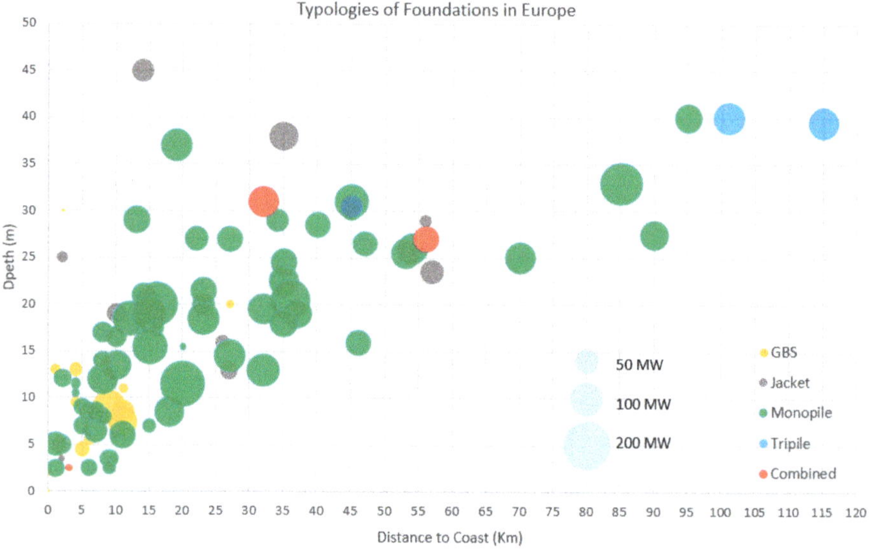

Figure 3. Typologies of foundations in Europe at the end of 2018 according to the parameters of depth and distance to the coast.

The maturity of the sector in Europe has allowed the installation of wind farms in remote areas not yet considered in Asia, both in distance and depth. The best example is the Global Tech I Park, in Germany (400 MW, 115 Km from the coast, 80 x 5 MW turbines, AREVA M5000 wind turbine, situated in the German Exclusive Economic Zone (EEZ), 180km away from the Bremerhaven Emden in the north-west, Germany). In Europe, monopile foundations after decades of experience and acquired knowledge, present a high percentage of use compared to the other typologies in Europe (77%). The specialisation in their design and manufacturing makes them, due to cost and range of aptitude, the most appropriate alternative in most wind farms located in shallow and intermediate depths.

Meanwhile in Asia, the arrival of offshore wind energy in a continent without tradition in facilities of this type has led to the use of various solutions. On the other hand, the optimal conditions of some Asian seas with shallow depths at great distances, explains the use of simple and inexpensive typologies such as pile cap concrete foundations. Wind farms with several different typologies are also common, mainly pile cap foundations and monopile or jacket structures.

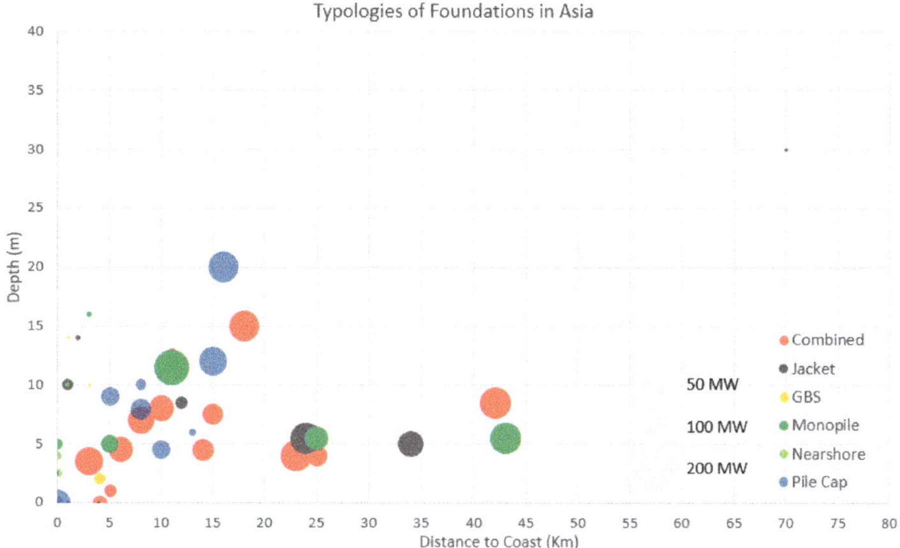

Figure 4. Typologies of foundations in Asia at the end of 2018 according to the parameters of depth and distance to the coast.

Although floating foundations have not been represented in the previous graphs, to facilitate the visualisation of the data, they will be analysed separately. As it can be extracted from previous text, all the conventional typologies present problems or a high degree of uncertainty when depths are greater than 50 meters. However, it is in these regions where the wind resource is the strongest and most stable, and the distance to the coastline minimises the impact on the environment and the landscape. Due to these reasons, many market participants started to investigate and develop new solutions that allow the installation and operation of wind farms at greater depths and distances, without compromising the safety of wind turbines or requiring huge installation and maintenance costs. Some of the advantages that floating platforms can offer compared to the other types are:

- Allowing the installation of wind farms in regions with great potential in the production of wind energy but which cannot be supported with current typologies.
- They offer a competitive alternative in medium depths (30–50 m), since they do not require a preparation of the seabed nor is it necessary to carry out works in the large-scale environment. Therefore, the impact on the ecosystem is lower.
- They allow countries without a continental platform and great depths at a short distance from the coast (America, Southern Europe, among others), to step forward in the market to which they barely had access to a few years ago.

The biggest disadvantage of this type of project is the need for research and development until it is viable and used in real installations, together with the enormous associated cost involved. Despite all this, several demonstrations and pilot projects have already been carried out, and the behaviour and profitability of this type of proposal are currently being analysed. Table 4 includes the floating projects already operating.

Table 4. Floating projects in operation at the end of 2018 (JAP: Japan; NOR: Norway; UK: United Kingdom; FRA: France).

Wind Farm	Country	Power (MW)	Turbines	Depth (m)	Distance (Km)	Year
Fukushima Floating -2	JAP	12.0	2	122.5	20	2018
Fukushima Floating -1	JAP	2.0	1	122.5	20	2013
Sakiyama Wind Turbine	JAP	2.0	1	40.0	5	2016
Hywind	NOR	2.0	1	220.0	10	2009
Hywing Scotland	UK	30.0	5	103.0	25	2018
Floatgen Project Demo	FRA	2.0	1	30.0	22	2018

With a total of 154 offshore wind farms classified and analysed using all the data set created for this research, it has been limited to the range of use of each one of the considered typologies, establishing both the areas of frequent use and the singular projects that are outside these limits, and which is interesting to mention since they establish the limits of these foundations in the future (Table 5, Figure 5).

Table 5. Typologies of foundations by range of use (1995–2018).

Typology	Depth (m)	Distance (Km)	Max. Depth (m)	Max Distance (Km)
Pile Cap	0–15	0–15	15	15
Combined	0–15	0–30	30	35
GBS	0–20	0–15	30	30
Monopile	0–30	0–60	40	100
Jacket	5–50	5–60	50	70
Tripod	25–50	40–120	40	120
Floating	>50	5–25	220	25

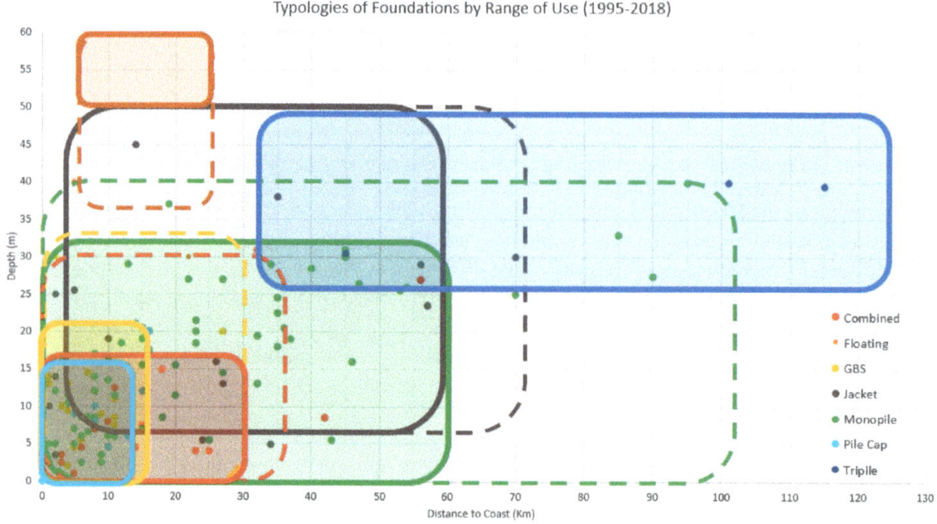

Figure 5. Typologies of foundations by depth and distance to coast (1995–2018).

In Figure 5, the areas inside the solid lines represent the regions where each typology has been used various times, while the areas outside the solid lines but inside the dashed ones represent the regions where these typologies have been used for single projects. Those dashed-line areas can be of interest, as they set the possibilities of these different solutions in the short and medium-term.

As it can be seen from the Table 5 and Figure 5, there are three differentiated zones for which different solutions are used, depending mainly on the depth of the seabed but also on the distance to the coast and location of the wind farm:

- In shallow depths (0–20 m), the typologies used differ depending on the continent. In Europe, GBS and monopile foundations have been the most frequent option, with the former being used at distances less than 15–20 km from the coast, and the latter at greater distances. In Asia, on the other hand, the pile cap foundations and the combined wind farms (mainly pile cap and monopiles) have been chosen for practically all the foundations.
- At medium depths (20–40 m), the most commonly used foundations are monopile and jacket typologies. The main difference between the two is the use of the second ones in depths from 35–40 m. It is in these depths when the monopiles begin to experience buckling and instability phenomena, being necessary to resort to more stable structures with better behavior under severe marine conditions. Bac Lieu wind farm, the first one in Vietnam, stands as one of the examples of pile cap foundations [77].
- On the other hand, it is important to highlight the use of tripod foundations. Although there are currently only three wind farms with this typology, their excellent behavior in areas far from the coast makes them a competitive alternative at medium-high depths and large distances, these regions being potentially the most demanded for installation of wind farms in Europe during the coming years.
- At greater depths (>50 m) floating foundations are the only typology that has been used to date. It can, therefore, be assumed that above this depth these typologies are those that a priori can be considered as the only possible alternative, although in particular conditions other typologies (usually tripod or jacket) may be used as well.

3. Monopiles: Evolution and Parameters of Design

As has been explained previously, if there is a typology that stands out above the others, both in frequency and range of use, it is the monopile typology. Considered an alternative since the first years of development of this sector, it has several characteristics that explain its popularity and widespread use:

- Simplicity: a simple and easily standardised design allows it to be manufactured in series without the need for cutting-edge technology, shortening construction and installation times and, therefore, reducing costs without compromising the safety and operability of the installation.
- Adaptability: linked to the previous point. Its simple design makes it possible to adapt to different dimensions and characteristics, without excessive complications to external conditions, avoiding the need for a large amount of field data, providing competitive solutions in shallow depths as well as in larger ones.
- Behavior: possibly the main strength of this typology. Able to stand favourably against external forces without increasing installation and maintenance costs. Mainly in high distances and average depths (25–35 m) where typologies such as GBS and HRPC (High Rise Pile Cap) are not competitive, and the use of jacket and tripod foundations can greatly increase the cost of the project, monopiles are considered the best alternative, even nowadays.

Monopiles began to be used mainly in 2002 when offshore wind was considered for the first time a competitive alternative in some parts of Europe, and the first relevant projects came up (Horns Rev 1, 160 MW, Denmark, Vestas V80-2.0 MW, North Sea, Denmark). These large wind parks needed to move away from the coast to minimise environmental and social impact, in addition to seeking a more powerful and stable wind resource. GBS foundations could not compete at intermediate depths and jacket structures were not economically viable.

During the next years, this typology experienced a great development (marked in green in Figure 6), as more resources were invested to reduce manufacturing costs while improving the building processes. Some companies, collecting the knowledge obtained from each project, soon became solid specialised brands who developed new solutions and techniques, spreading the use of monopiles out of their initial range. At the top of its popularity in 2014, monopile foundations were used in 76.7% of the total wind turbines operating in the world by that time (Figure 6).

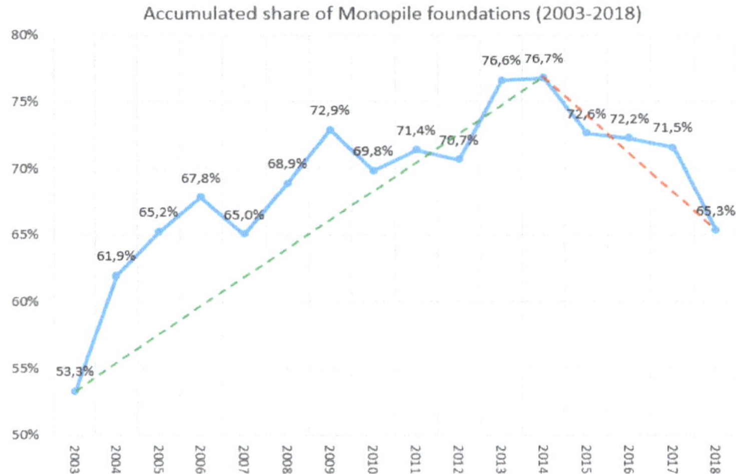

Figure 6. Accumulated share of monopile foundations (2003–2018).

The decreasing period takes place as of 2014 (highlighted in red), and presents a decreasing trend from the 77% existing in that year to the current 65%. To explain this fact, two facts must be considered.

On the one hand, the tendency to move wind farms away from the coast towards greater depths favours the use of more solid and stable typologies (jacket, tripod), with a better performance in the face of more severe marine climates.

On the other hand, the development of the sector in the Asian countries with different environmental conditions and technical solutions, has favoured the decrease of this quota. However, as it has been already seen, turbines have barely been installed at intermediate depths in Asia and when they begin to do so, monopiles will be considered again.

In order to adapt to the needs of each wind farm, monopiles have evolved over the years. Its main dimensions, diameter and length have increased to be able to support increasingly high and heavy turbines and greater depths. If it is analysed, these changes during the last 10 years in which 90% of all existing monopiles have been installed, the results are exposed in Table 6.

Table 6. Monopiles characteristics (2009–2018).

Year	Power Installed (MW)	Turbines Installed	Average Turbine Power, Pn (MW)	Average Diameter, ø (m)	Average Length, L (m)	Average Depth, d (m)	Average Distance to Coast, DC (Km)
2009	457	127	3.6	4.85	37.8	7.6	5.7
2010	862	311	2.8	4.26	40.4	11.7	20.2
2011	232	72	3.2	5.00	43.0	19.3	15.0
2012	184	51	3.6	6.00	55.0	27.0	22.0
2013	2149	599	3.6	5.13	47.8	13.1	15.7
2014	1063	317	3.4	5.48	58.8	19.2	24.4
2015	2072	564	3.7	5.47	55.4	17.9	27.8
2016	698	191	3.7	5.68	62.2	9.5	16.5
2017	3144	609	5.2	6.83	69.0	23.6	46.1
2018	3598	608	5.9	7.26	69.1	22.4	27.4

As can be seen, both the length and the diameter of the piles have evolved proportionally to the power of the turbines, maintaining constant ratios. However, in order to analyse the relationship of these dimensions with the average depth and distance to the coast, it is seen that the ratio is not constant and there are relevant deviations.

In order to study these correlations more carefully, some monomials from the previous table are shown in Table 7.

Table 7. Monopiles monomials (2009–2018).

Year	ø/L (m/m)	ø/Pn (m/KW)	L/Pn (m/KW)	ø/d (m/m)	L/d (m/m)	ø/DC (m/Km)	L/DC (m/Km)	d/DC (m/Km)
2009	0.13	1.35	10.5	0.64	5.0	0.85	6.6	1.33
2010	0.11	1.54	14.6	0.36	3.5	0.21	2.0	0.58
2011	0.12	1.55	13.3	0.26	2.2	0.33	2.9	1.29
2012	0.11	1.66	15.2	0.22	2.0	0.27	2.5	1.23
2013	0.11	1.43	13.3	0.39	3.6	0.33	3.0	0.83
2014	0.09	1.63	17.5	0.29	3.1	0.22	2.4	0.79
2015	0.10	1.49	15.1	0.31	3.1	0.20	2.0	0.64
2016	0.09	1.55	17.0	0.60	6.5	0.34	3.8	0.58
2017	0.10	1.32	13.4	0.29	2.9	0.15	1.5	0.51
2018	0.11	1.23	11.7	0.32	3.1	0.26	2.5	0.82
Average	0.11	1.48	14.2	0.37	3.5	0.32	2.92	0.86

The monomial with the lowest variation is the ratio Diameter/Length of the monopile, staying close to 0.10–0.11 during the last years. This parameter can be helpful at the time of designing the right length of the pile based on the diameter of the wind turbine.

Although the other coefficients show a higher variability, in order to obtain a valid correlation for wind farms with the current characteristics and conditions, those obtained from the data corresponding to the last years will be more useful for this study. With these parameters, engineers can make estimations of the dimensions of the foundations, considering both the conditions of the environment and the power of the chosen turbines.

Other parameters that may be of interest are the hub height and blade length of the turbines, which directly affect the electrical production capacity. In Table 8 and Figure 7, it can be seen how they have evolved their average values over the years, and their relationship with the length of the monopile foundations.

Table 8. Hub and blade heights of monopile wind farms by year (1996–2018).

Year	PN (MW)	Hub Height (m)	Total Height (m)	Blade Length (m)	Monopile Length (m)	Ratio Blade/Hub	Monopile/Hub	Monopile/Blade
1996	0.61	50.0	71.5	21.5	19.0	0.43	0.27	0.88
1998	0.60	41.5	60.0	18.5	21.0	0.45	0.35	1.14
2000	1.43	65.0	100.5	35.5	32.5	0.55	0.32	0.92
2002	2.00	70.0	110.0	40.0	30.0	0.57	0.27	0.75
2003	2.30	61.0	102.2	41.2	45.0	0.68	0.44	1.09
2004	2.16	67.0	110.3	43.3	40.0	0.65	0.36	0.92
2005	3.00	70.0	115.0	45.0	41.0	0.64	0.36	0.91
2006	3.00	75.0	120.0	45.0	49.0	0.60	0.41	1.09
2007	3.25	76.8	126.0	49.3	48.5	0.64	0.38	0.98
2008	2.00	59.0	99.0	40.0	54.0	0.68	0.55	1.35
2009	3.60	78.8	132.3	53.5	37.8	0.68	0.29	0.71
2010	2.77	70.0	114.3	44.3	40.4	0.63	0.35	0.91
2011	3.22	75.3	125.3	50.0	43.0	0.66	0.34	0.86
2012	3.61	90.2	150.2	60.1	55.0	0.67	0.37	0.92
2013	3.59	79.6	134.9	55.3	47.8	0.69	0.35	0.86
2014	3.35	83.2	139.7	56.5	58.8	0.68	0.42	1.04
2015	3.67	87.0	145.8	58.8	55.4	0.68	0.38	0.94
2016	3.65	91.3	152.3	61.0	62.3	0.67	0.41	1.02
2017	5.16	97.5	167.7	70.2	69.0	0.72	0.41	0.98
2018	5.92	99.8	174.2	74.4	69.1	0.75	0.40	0.93

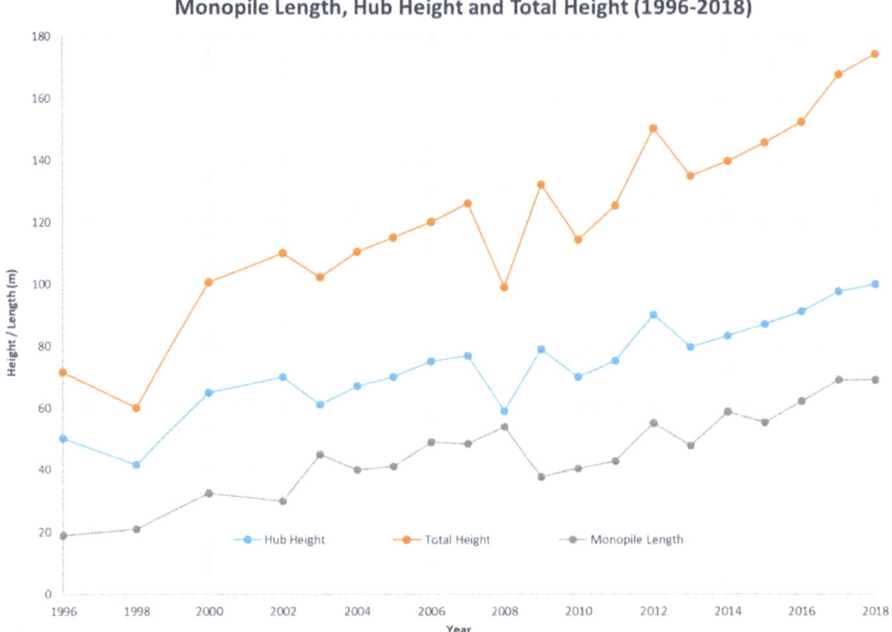

Figure 7. Evolution of Hub Height, Blade Length and Monopile Length (1996–2018).

Some interesting monomials can be extracted, like the ratios of Monopile Length/Blade Height, and Monopile Length/Hub Height, which have remained constant through the years (0.3–0.4 for the first one, and 0.9–1.0 for the second one). However, the monomial that shows a higher variation is the ratio Blade Length/Hub Height, from 0.43 in 1996 to 0.75 in 2018. Bigger blades lead to more efficient turbines and higher productions, and the location of offshore wind farms far from any affect on humans, and the lower surface roughness of the sea allows this ratio to keep growing.

Finally, some data on the predominant soil composition in each of these wind farms have also been included. In the same way as in previous sections, several monomials have been obtained. Table 9 includes some important figures related to the soil type, including the percentage considering the power installed and some of the monomials previously used.

Table 9. Soil composition on monopile wind farms.

Soil Type	Power (MW)	% Type of Soil (Total Power)	Pn (MW)	Wind Farms	Average Depth (m)	ø (m)	Lenght (m)	Ratio ø/L	Ratio Length/Depth
Sand/Clay	1582	10.3%	3.5	8	11.8	5.0	52.0	0.10	4.40
Sand/Bedrock	135	0.9%	2.0	4	9.3	3.6	32.1	0.11	3.47
Sand	9199	60.1%	4.2	39	17.9	5.6	57.8	0.10	3.23
Clay	2191	14.3%	4.1	10	14.8	5.4	48.5	0.11	3.28
Sand/Gravel	775	5.1%	3.5	3	12.0	5.1	40.7	0.13	3.39
Sand/Chalk	97	0.6%	3.6	1	9.0	-	34.0		3.78
Clay/Gravel	165	1.1%	3.0	1	16.0	4.5	55.0	0.08	3.44
Clay/Bedrock	576	3.8%	3.6	1	20.0	5.0	55.0	0.09	2.75
Chalk	595	3.9%	6.3	2	20.8	7.3	70.5	0.10	3.40

Placing our data on a graphic similar to Figure 5, it is shown in Figure 8 how monopile wind farms are located depending on average depth and distance to the coast for every different soil composition.

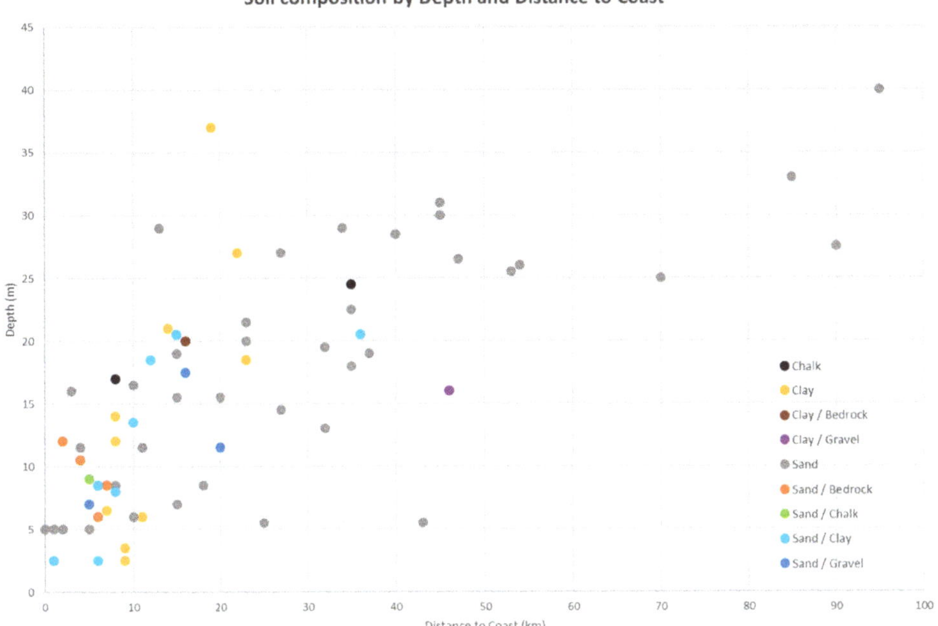

Figure 8. Soil composition of monopile windfarms by depth and distance to the coast.

4. Discussion of Results

The offshore wind sector awaits great potential both in the short and long term. The development of technology, the increase of competitiveness, reduction in costs and the interest by companies and institutions have motivated the development of a growing number of new projects year after year. In those areas where these wind facilities were already consolidated, that is, the regions surrounding the North Sea in Europe (Germany, the United Kingdom, the Netherlands, Denmark and Belgium), and China in Asia, this development has favoured the arrival of the first "megaprojects", installations of high power capacity carried out with the objective of not only supplying a significant quota of energy demand but with the ambitious objective of leading the production of electricity of their respective countries in not too long a space of time.

Regarding the countries that had not taken part in this sector yet, the consolidation of these projects together with the increase in the profitability of these facilities and the good results demonstrated, has led to the entry of new participants. A large number of European countries with access to the coast have already begun to analyze the viability and consent to the drafting of projects of this type, or in other cases, to carry out experimentation and analysis facilities that may favour the development of new wind farms in the future. Similarly, both the United States and various Asian countries have started to develop projects of this type over the past few years, some of which are already operating.

It must be highlighted the extensive use of monopiles, used in mostly all types of depths and environments. About this typology, it will be spoken of in more detail in the next lines. In addition, it can be observed the different trends and ranges of use for each of the foundations analysed. In shallow depths (0–15 m), it can be mainly found with foundations by gravity or GBS and pile cap foundations in Asia. These typologies have been progressively abandoned as wind turbines moved away from the coast and other typologies that needed less competent strata appeared. In the case of pile caps, they are still used in Asia in wind farms that are close to the coast. GBS foundations have been included in recent projects where the depth and sea conditions are a priori not convenient for them. At medium depths, it can be found the most used typologies today, the monopiles, and the jacket and tripod structures. Monopile foundations are currently used in more than 60% of wind turbines operating worldwide.

With regards to the analysis of the technical characteristics, the appearance of a constant value throughout all the foundations carried out from the beginning of these projects to the present, is the Diameter/Length ratio, which has remained unchanged for 20 years, always close to 0.10–0.11 and that seems to be the limit from which the phenomena of instability, vibration and deflection may appear.

Regarding the blade and hub heights, once more of this typology has evolved to be able to support bigger structures and turbines and maintain a constant ratio with both blade length and hub heights. If we look at the soil composition, we observe than this typology can adapt to different environments while keeping similar parameters and dimensions, proving it as a convenient and cost-effective solution even on heterogeneous locations.

For the jacket structures, their adaptability and excellent behavior in adverse conditions make them the best option in medium-high depths (20–40 m), especially in those areas where the monopiles can be instable and show deflection and bulking. On the contrary, its design and manufacturing complexity, unique for each wind turbine, and the cost associated with it has prevented its usual use until just a few years ago (2014–2015). However, since then its use quota has been increasing year after year, and as the technology on this typology becomes generalised and its own costs decrease, so will its expansion. As for the tripod foundations, although theoretically their features and benefits are at an intermediate point between monopiles and jacket structures, analysing the operational wind farms at present, it can be seen that their use is concentrated in the same area and in very similar conditions: high depths (30–40 m) and great distances to the coast (50–100 km).

5. Conclusions

The main conclusions of this research are:

- Offshore wind energy is experiencing constant growth in recent years, consolidating itself as one of the fields with more potential. For this form of energy generation to be competitive, several years of research and development have been necessary.
- Monopiles are the most used foundations in shallow (0–15 m) and intermediate depths (15–30 m). The simplicity of designing and manufacturing, together with a vast knowledge inherited from experience, especially in Europe, has led this typology to be the first option in more than 60% of the world's offshore wind foundations nowadays.
- Jacket and tripod structures remain as strong competitors to monopiles when the seabed depth is higher than 30 m. Floating solutions, which are beginning to be used for real energy production, are supposed to take the market by storm in the near future, bringing this sector to places that remain unavailable nowadays.
- Monopiles evolution has been studied according the following monomials: diameter/length, diameter/turbine power, length/turbine power, diameter/water depth, length/diameter, dimeter/distance to the coast, length/distance to the coast, and water depth/ distance to the coast, with the following average values: 0.11; 1.48; 14.2; 0.37; 3.5; 0.32; 2.92; 0.86.
- The Diameter/Length monomial has the lowest variation, staying close to 0.10–0.11 during the last years, so it can be used for a first and fast figure of the length of the pile.
- The other coefficient with a higher variability can be used for the first estimate in case of using the monomials value obtained in last years, the most current ones.

Author Contributions: Conceptualization, J.-S.L.-G.; methodology, V.N.; investigation, S.S.; writing—original draft preparation, S.S.; writing—review and editing, J.-S.L.-G., V.N. and M.D.E.; supervision, M.D.E.

Funding: Authors give thanks to the Agustín de Betancourt Foundation (FAB) for the support received over the past few years.

Acknowledgments: Authors give thanks to the Agustín de Betancourt Foundation (FAB) for the support received over the past few years.

Conflicts of Interest: The authors declare no conflict of interest. The funders had no role in the design of the study; in the collection, analyses, or interpretation of data; in the writing of the manuscript, or in the decision to publish the results.

References

1. Luengo, J.; Negro, V.; García-Barba, J.; Martín-Antón, M.; López-Gutiérrez, J.S.; Esteban, M.D.; Moreno, L.J. Preliminary Design for Wave Run-Up in Offshore Wind Farms: Comparison between Theoretical Models and Physical Model Tests. *Energies* **2019**, *12*, 1–17.

2. Mahdy, M.; Bajah, A. Multi criteria decision analysis for offshore wind energy potential in Egypt. *Renew. Energy* **2018**, *118*, 278–289. [CrossRef]

3. WindEurope. *Offshore Wind in Europe. Key Trends Statistics 2017*; Technical Report; WindEurope: Brussels, Belgium, 2018.

4. Esteban, M.D.; López-Gutiérrez, J.S.; Diez, J.J.; Negro, V. Methodology for the design of offshore wind farms. *J. Coast. Res.* **2011**, *SI64*, 496–500.

5. Hevia-Koch, P.; Jacobsen, H.K. Comparing offshore and onshore wind development considering acceptance costs. *Energy Policy* **2019**, *125*, 9–19. [CrossRef]

6. Esteban, M.D.; Espada, J.M.; Ortega, J.M.; López-Gutiérrez, J.S.; Negro, V. What about marine renewable energies in Spain? *J. Mar. Sci. Eng.* **2019**, *7*, 249. [CrossRef]

7. Global wind Energy Council. *GWEC Global Wind Report 2018*; Technical Report; Global wind Energy Council: Brussels, Belgium, 2019.

8. Spanish Institute for Diversification and Energy Saving (IDEA). *Offshore Wind Energy*; Technical Report; IDEA: Gandhinagar, India, 2015.

9. Li, J.; Yu, X.B. Onshore and offshore wind energy potential assessment near Lake Erie shoreline: A spatial and temporal analysis. *Energy* **2018**, *147*, 1092–1107. [CrossRef]

10. Enevoldsen, P.; Valentine, S.V. Do onshore and offshore wind farm development patterns differ? *Energy Sustain. Dev.* **2016**, *35*, 41–51. [CrossRef]

11. Kaldellis, J.K.; Apostolou, D.; Kapsali, M.; Kondili, E. Environmental and social footprint of offshore wind energy. Comparison with onshore counterpart. *Renew. Energy* **2016**, *92*, 543–556. [CrossRef]

12. Wüstemeyer, C.; Madlener, R.; Bunn, D.W. A stakeholder analysis of divergent supply-chain trends for the European onshore and offshore wind installations. *Energy Policy* **2015**, *80*, 36–44. [CrossRef]

13. Kumar, Y.; Ringernberg, J.; Depuru, S.S.; Devabhaktuni, V.K.; Lee, J.W.; Nikolaidis, E.; Andersen, B.; Afjeh, A. Wind energy: Trends and enabling technologies. *Renew. Sustain. Energy Rev.* **2016**, *53*, 209–224. [CrossRef]

14. Yang, J.; Chang, Y.; Zhang, L.; Hao, Y.; Yan, Q.; Wang, C. The life-cycle energy and environmental emissions of a typical offshore wind farm in China. *J. Clean. Prod.* **2018**, *180*, 316–324. [CrossRef]

15. Gil-García, I.C.; García-Cascales, M.S.; Fernández-Guillamón, A.; Molina-García, A. Categorization and analysis of relevant factors for optimal locations in onshore and offshore wind power plants: A taxonomic review. *J. Mar. Sci. Eng.* **2019**, *7*, 391. [CrossRef]

16. Esteban, M.D.; Diez, J.J.; López-Gutiérrez, J.S.; Negro, V. Integral management applied to offshore wind farms. *J. Coast. Res.* **2009**, *SI56*, 1204–1208.

17. Esteban, M.D.; López-Gutiérrez, J.S.; Diez, J.J.; Negro, V. Offshore wind farms: Foundations and influence on the littoral processes. *J. Coast. Res.* **2011**, *SI64*, 656–660.

18. Pérez-Collazo, C.; Greaves, D.; Iglesias, G. A review of combined wave and offshore wind energy. *Renew. Sustain. Energy Rev.* **2015**, *42*, 141–153. [CrossRef]

19. Veigas, M.; Iglesias, G. A Hybrid Wave-Wind Offshore Farm for an Island. *Int. J. Green Energy* **2015**, *12*, 570–576. [CrossRef]

20. Astariz, S.; Vazquez, A.; Sánchez, M.; Carballo, R.; Iglesias, G. Co-located wave-wind farms for improved O&M efficiency. *Ocean Coast. Manag.* **2018**, *163*, 66–71.

21. Lacal-Arántegui, R.; Yusta, J.M.; Domínguez-Navarro, J.A. Offshore wind installation: Analysing the evidence behind improvements in installation time. *Renew. Sustain. Energy Rev.* **2018**, *92*, 133–145. [CrossRef]

22. Schwanitz, V.J.; Wierling, A. Offshore wind investments—Realism about cost developments is necessary. *Energy* **2016**, *106*, 170–181. [CrossRef]

23. Dismukes, D.E.; Upton, G.B. Economies of scale, learning effects and offshore wind development costs. *Renew. Energy* **2015**, *83*, 61–66. [CrossRef]

24. Fathabadi, H. Novel high efficient speed sensorless controller for maximum power extraction from wind energy conversion systems. *Energy Convers. Manag.* **2016**, *123*, 392–401. [CrossRef]

25. Novaes, E.J.; Araújo, A.M.; da Silva, N.S.B. A review on wind turbine control and its associated methods. *J. Clean. Prod.* **2018**, *174*, 945–953. [CrossRef]

26. MHI Vestas. *MHI Vestas Achieves Final Turbine Installation at Horns Reef 3*; Technical Report; Vestas: Aarhus N, Aarhus, Denmark, 2019.

27. Velarde, J.; Bachynski, E. Design and fatigue analysis of monopile foundations to support the DTU 10 MW offshore wind turbine. *Energy Procedia* **2017**, *137*, 3–13. [CrossRef]

28. Bottasso, C.L.; Croce, A.; Gualdoni, F.; Montinari, P. Load mitigation for wind turbines by a passive aeroelastic device. *J. Wind Eng. Ind. Aerodyn.* **2016**, *148*, 57–69. [CrossRef]

29. Escobar, A.; Negro, V.; López-Gutiérrez, J.S.; Esteban, M.D. Software for predicting hydrodynamic pressures on offshore pile foundations: The next step in ocean energy development. *J. Coast. Res.* **2016**, *SI75*, 841–845. [CrossRef]

30. Oh, K.Y.; Nam, W.; Ryu, M.S.; Kim, J.K.; Epureanu, B.I. A review of foundations of offshore wind energy convertors: Current status and future perspectives. *Renew. Sustain. Energy Rev.* **2018**, *88*, 16–36. [CrossRef]

31. Matutano, C.; Negro, V.; López-Gutiérrez, J.S.; Esteban, M.D. Hydrodynamic regimes in offshore wind farms. *J. Coast. Res.* **2016**, *75*, 892–897. [CrossRef]

32. Wang, X.; Zeng, X.; Yang, X.; Li, J. Feasibility study of offshore wind turbines with hybrid monopile foundation based on centrifuge modeling. *Appl. Energy* **2018**, *209*, 127–139. [CrossRef]

33. Bosch, J.; Staffell, I.; Hawkes, A.D. Global levelised cost of electricity from offshore wind. *Energy* **2019**, 116357. [CrossRef]

34. Ma, H.; Yang, J.; Chen, L. Effect of scour on the structural response of an offshore wind turbine supported on tripod foundation. *Appl. Ocean Res.* **2018**, *73*, 179–189. [CrossRef]

35. Kin, H.G.; Kim, B.J. Feasibility study of new hybrid piled concrete foundation for offshore wind turbine. *Appl. Ocean Res.* **2018**, *76*, 11–21.

36. Esteban, M.D.; López-Gutiérrez, J.S.; Negro, V. Gravity-Based Foundations in the Offshore Wind Sector. *J. Mar. Sci. Eng.* **2019**, *7*, 64. [CrossRef]

37. Esteban, M.D.; López-Gutiérrez, J.S.; Negro, V.; Matutano, C.; García-Flores, F.M.; Millán, M.A. Offshore wind foundation design: Some key issues. *J. Energy Resour. Technol.* **2014**, *137*, 1–6.

38. Guzmán, R.F.; Müller, K.; Vita, L.; Cheng, P.W. Simulation requirements and relevant load conditions in the design of floating offshore wind turbines. In Proceedings of the ASME 2018, 37th International conference on Ocean, Offshore Arctic Engineering, Madrid, Spain, 17–22 June 2018.

39. Bredmose, H.; Jacobsen, N.G. Breaking wave impacts on offshore wind turbine foundations: Focused wave groups and CFD. In Proceedings of the ASME 2010, 29th International conference on Ocean, Offshore Arctic Engineering, Shanghai, China, 6–11 June 2010.

40. Passon, P.; Branner, K. Load calculation methods for offshore wind turbine foundations. *J. Ships Offshore Struct.* **2014**, *9*, 433–449. [CrossRef]

41. Matutano, C.; Negro, V.; López-Gutiérrez, J.S.; Esteban, M.D.; Hernández, A. The effect of scour protections in offshore wind farms. *J. Coast. Res.* **2014**, *SI70*, 12–17. [CrossRef]

42. Hyldal Sørensen, S.P.; Ibsen, L.B. Assessment of foundation design for offshore monopiles unprotected against scour. *Ocean Eng.* **2013**, *63*, 17–25. [CrossRef]

43. Predengast, L.J.; Reale, C.; Gavin, K. Probabilistic examination of the change in eigenfrequencies of an offshore wind turbine under progressive scour incorporating soil spatial variability. *Mar. Struct.* **2018**, *57*, 87–104. [CrossRef]

44. Matutano, C.; Negro, V.; López-Gutiérrez, J.S.; Esteban, M.D.; del Campo, J.M. Dimensionless wave height parameter for preliminary design of scour protections in offshore wind farms. *J. Coast. Res.* **2013**, *65*, 1633–1639. [CrossRef]

45. Escobar, A.; Negro, V.; López-Gutiérrez, J.S.; Esteban, M.D. Assessment of the influence of the acceleration field on scour phenomenon in offshore wind farms. *Renew. Energy* **2019**, *136*, 1036–1043. [CrossRef]

46. Jacobsen, H.K.; Hevia-Koch, P.; Wolter, C. Nearshore and offshore wind development: Costs and competitive advantage exemplified by nearshore wind in Denmark. *Energy Sustain. Dev.* **2019**, *50*, 91–100. [CrossRef]
47. Kim, D.H.; Lee, S.G.; Lee, I.K. Seismic Fragility of 5 a MW offshore wind turbine. *Renew. Energy* **2014**, *65*, 250–256. [CrossRef]
48. Andersen, L.V.; Andersen, T.L.; Manuel, L. Model Uncertainties for Soil-Structure Interaction in Offshore Wind Turbine Monopile Foundations. *J. Mar. Sci. Eng.* **2018**, *6*, 87. [CrossRef]
49. Mo, R.; Kang, H.; Li, M.; Zhao, X. Seismic Fragility Analysis of Monopile Offshore Wind Turbines uncer Different Operational Conditions. *Energies* **2017**, *10*, 1037. [CrossRef]
50. Carswell, W.; Arwade, S.R.; DeGroot, D.J.; Lackner, M.A. Soil–structure reliability of offshore wind turbine monopile foundations. *Wind Energy* **2014**, *18*, 483–498. [CrossRef]
51. Luengo, J.; Negro, V.; García-Barba, J.; López-Gutiérrez, J.S.; Esteban, M.D. New detected uncertainties in the design of foundations for offshore wind turbines. *Renew. Energy* **2019**, *131*, 667–677. [CrossRef]
52. MacKinnon, D.; Dawley, S.; Steen, M.; Menzel, M.P.; Karlsen, A.; Sommer, P.; Hansen, G.H.; Normann, H.E. Path creation, global production networks and regional development: A comparative international analysis of the offshore wind sector. *Prog. Plan.* **2019**, *130*, 1–32. [CrossRef]
53. Takacs, B.; Goulden, M.C. Accuracy of wind farm visualisations: The effect of focal length on perceived accuracy. *Environ. Impact Assess. Rev.* **2019**, *76*, 1–9. [CrossRef]
54. Bishop, I.D. The implications for visual simulation and analysis of temporal variation in the visibility of wind turbines. *Landsc. Urban Plan.* **2019**, *184*, 59–68. [CrossRef]
55. Rodriguez, S.; Restrepo, C.; Kontos, E.; Teixeira Pinto, R.; Bauer, P. Trends of offshore Wind projects. *Renew. Sustain. Energy Rev.* **2015**, *49*, 1114–1135. [CrossRef]
56. Wu, X.; Hu, Y.; Li, Y.; Yang, J.; Duan, L.; Wang, T.; Adcock, T.; Juang, Z.; Gao, Z.; Lin, Z.; et al. Foundations of offshore wind turbines: A review. *Renew. Sustain. Energy Rev.* **2019**, *104*, 379–393. [CrossRef]
57. Det Norske Veritas (DNV). *Offshore Standard, DNV-OS-J101, Design of Offshore Wind Turbine Structures*; Technical Report; DNV GL: Oslo, Norway, 2014.
58. Det Norske Veritas—Germanischer Lloyd (DNV-GL). *Offshore Standard, DNVGL-ST-0126, Support Structures for Wind Turbines*; Technical Report; DNV GL: Oslo, Norway, 2018.
59. Det Norske Veritas—Germanischer Lloyd (DNV-GL). *Offshore Standard, DNVGL-ST-0119, Floating Wind Turbine Structures*; Technical Report; DNV GL: Oslo, Norway, 2018.
60. Esteban, M.D.; Couñago, B.; López-Gutiérrez, J.S.; Negro, V.; Vellisco, F. Gravity based support structures for offshore wind turbine generators: Review of the installation process. *Ocean Eng.* **2015**, *110*, 281–291. [CrossRef]
61. Li, Y.; Ong, M.C.; Tang, T. Numerical analysis of wave-induced poro-elastic seabed response around a hexagonal gravity-based offshore foundation. *Coast. Eng.* **2018**, *136*, 81–95. [CrossRef]
62. Negro, V.; López-Gutiérrez, J.S.; Esteban, M.D.; Alberdi, P.; Imaz, M.; Serraclara, J.M. Monopiles in offshore wind: Preliminary estimate of main dimensions. *Ocean Eng.* **2017**, *133*, 253–261. [CrossRef]
63. Wang, P.; Zhao, M.; Du, X.; Liu, J.; Xu, C. Wind, wave and earthquake responses of offshore wind turbine on monopile foundation in clay. *Soil Dyn. Eathquake Eng.* **2018**, *113*, 47–57. [CrossRef]
64. Yeter, B.; Garbatov, Y.; Guedes Soares, C. Evaluation of fatigue damage model prediction for fixed offshore wind turbine support structure. *Int. J. Fatigue* **2016**, *87*, 71–80. [CrossRef]
65. Muskulus, M. Pareto-Optimal Evaluation of Ultimate Limit States in Offshore Wind Turbine Structural Analysis. *Energies* **2015**, *8*, 14026–14039. [CrossRef]
66. Zhu, B.; Wen, K.; Kong, D.; Zhu, Z.; Wang, L. A numerical study on the lateral loading behaviour of offshore tetrapod piled jacket foundations in clay. *Appl. Ocean Res.* **2018**, *75*, 165–177. [CrossRef]
67. Bento, N.; Fontes, M. Emergence of floating offshore wind energy: Technology and industry. *Renew. Sustain. Energy Rev.* **2019**, *99*, 66–82. [CrossRef]
68. Kausche, M.; Adam, F.; Dahlhaus, F.; Grobmann, J. Floating offshore wind-economic and ecological challenges of a TLP solution. *Renew. Energy* **2018**, *126*, 270–280. [CrossRef]
69. Wang, X.; Zeng, X.; Li, J.; Yang, X.; Wang, H. A review on recent advancements of substructures for offshore wind turbines. *Renew. Sustain. Energy Rev.* **2018**, *158*, 103–119. [CrossRef]
70. Skau, K.S.; Grimstad, G.; Page, A.M.; Eiksund, G.R.; Jostad, H.P. A macro-element for integrated time domain analyses representing bucket foundations for offshore wind turbines. *Mar. Struct.* **2018**, *59*, 158–178. [CrossRef]

71. Zhang, P.; Wei, W.; Jia, N.; Ding, H.; Liu, R. Effect of seepage on the penetration resistance of bucket foundations with bulkheads for offshore wind turbines in sand. *Ocean Eng.* **2018**, *156*, 82–92. [CrossRef]

72. Byrne, B.; Houlsby, G.; Martin, C.; Fish, P. Suction caissons foundations for offshore wind turbines. *Wind Eng.* **2002**, *26*, 145–155. [CrossRef]

73. Houlsby, G.; Byrne, B. Suction caisson foundations for offshore wind turbines and anemometer masts. *Wind Eng.* **2000**, *24*, 249–255. [CrossRef]

74. Ashuri, T.; Zaayer, M.B. Review of design concepts, methods and considerations of offshore wind turbines. In Proceedings of the 2007 European Offshore Wind Conference and Exhibition, Berlin, Germany, 4–6 December 2007.

75. Iberdrola. *Foundations in Offshore Wind Farms*; Technical report; Iberdrola: Bilbao, Spain, 2017.

76. Zheng, C.W.; Xiao, Z.N.; Peng, Y.H.; Li, C.Y.; Du, Z.B. Rezoning global offshore wind energy resources. *Renew. Energy* **2018**, *129*, 1–11. [CrossRef]

77. Van-Tan, T.; Tsai-Hsiang, C. Assessing the Wind Energy for Rural Areas of Vietnam. *Int. J. Renew. Energy Res.* **2013**, *3*, 523–528.

Journal of
Marine Science and Engineering

Article

Performance and Effect of Load Mitigation of a Trailing-Edge Flap in a Large-Scale Offshore Wind Turbine

Xin Cai [1], Yazhou Wang [1], Bofeng Xu [1,2,*] and Junheng Feng [1]

[1] College of Mechanics and Materials, Hohai University, Nanjing 211100, China; xcai@hhu.edu.cn (X.C.); wangyazhou@ctnei.com (Y.W.); fengjunheng0@163.com (J.F.)
[2] College of Energy and Electrical Engineering, Hohai University, Nanjing 211100, China
* Correspondence: bfxu1985@hhu.edu.cn

Received: 8 January 2020; Accepted: 21 January 2020; Published: 23 January 2020

Abstract: As a result of the large-scale trend of offshore wind turbines, wind shear and turbulent wind conditions cause significant fluctuations of the wind turbine's torque and thrust, which significantly affect the service life of the wind turbine gearbox and the power output stability. The use of a trailing-edge flap is proposed as a supplement to the pitch control to mitigate the load fluctuations of large-scale offshore wind turbines. A wind turbine rotor model with a trailing-edge flap is established by using the free vortex wake (FVW) model. The effects of the deflection angle of the trailing-edge flap on the load distribution of the blades and wake flow field of the offshore wind turbine are analyzed. The wind turbine load response under the control of the trailing-edge flap is obtained by simulating shear wind and turbulent wind conditions. The results show that a better control effect can be achieved in the high wind speed condition because the average angle of attack of the blade profile is small. The trailing-edge flap significantly changes the load distribution of the blade and the wake field and mitigates the low-frequency torque and thrust fluctuations of the turbine rotor under the action of wind shear and turbulent wind.

Keywords: offshore wind turbine; trailing-edge flap; load mitigation; free vortex wake

---

1. Introduction

Wind energy systems, especially those offshore, face difficult competition from traditional carbon-based energy sources with respect to cost competitiveness per kilowatt hour. To counter this problem, many energy systems have increased in size and power in order to achieve utility-scale production and to access higher winds aloft [1,2]. Recently, the large-scale offshore wind turbine has had more than 5 MW of the power and more than 80 m of the blade length. However, unsteady factors such as wind shear and turbulent wind have more negative effects on the stable operation of large wind turbines. The stability of the load and the output power has become significant issues in large-scale offshore wind turbines. There are many cases of fatigue failure of bearings and gearboxes prior to the end of the design life, which indicates the necessity of load mitigation control [3]. As the blade inertia of large-scale offshore machines is very large, traditional pitch control methods are unable to handle the fast-changing aerodynamic load fluctuations [4]. A new load control system has to be developed.

Investigations into the load mitigation of large wind turbines has mainly focused on two aspects. One aspect is research on advanced transmission systems [5]. The flexible coupled tower and blade [6] was investigated for the absorption of the instantaneous change in the torque and the reduction of the impact load of the gearbox and generator. An advanced hydraulic torque converter [7,8] is also an efficient transmission control structure. It absorbs the impact load caused by turbulent wind, and accurately adjusts the speed of the output shaft; this results in efficient speed control of the permanent magnet synchronous generator unit and even cancels the frequency converter. The VESTAS Company has successfully applied hydraulic torque converters to its wind turbines. The other aspect is research on smart rotors [4]. Smart rotors control the amount of wind energy absorption by the wind turbine using flow control technology, including passive or active flow control devices, to reduce the load fluctuation of the wind turbine at the source. Smart rotors not only reduce load fluctuations but also the swing amplitude of the blade and the noise level. Active flow control equipment is an efficient control mode and with quick response to airload can be achieved in complex and unsteady conditions. Passive flow control devices are simple and stable, and can be implemented by performing only minor modifications to the existing blade structure [9]. Hansen and Madsen [10] reviewed both types of devices, including deformable trailing-edge, microtabs, morphing, active twist, synthetic jets, active vortex generators, and plasma actuators. In these flow control devices, deformable trailing-edge flaps have been investigated by many scholars due to their simple structure and considerable adjustability [11,12]. Although, trailing-edge flaps are still in the research stage, this technology is the most likely approach to be put into practical application of the large-scale blade first.

Bak et al. [13] conducted a wind tunnel test of the wind turbine airfoil Risø-B1-18 equipped with an active trailing edge flap. Steady-state and dynamic tests were performed with certain deflections of the active trailing edge flap. The steady-state tests showed that deflecting the flap towards the pressure side resulted in higher lift values and deflecting the flap towards the suction side provided lower lift values. Lee and Su [14] analyzed joint trailing-edge flaps and obtained the basic aerodynamic characteristics of two-dimensional airfoils with flaps. Lu et al. [15] examined and optimized the flexible variable camber trailing-edge flap. Lackner and Kuik [16] investigated the load reduction capabilities of trailing edge flaps of a 5 MW wind turbine. The results showed that the use of trailing-edge flaps and the proposed feedback control approach were effective in reducing the fatigue loads of the blades relative to the baseline. Xu et al. [17] studied the trailing-edge flap control of large-scale floating wind turbines and found that the trailing-edge flap exhibited excellent power fluctuation mitigation of large floating wind turbines. Recent publications indicate that there are relatively few studies on the effect of the flap motion on the load fluctuation of wind turbine blades under unsteady conditions.

In this study, the free vortex wake (FVW) method [18] is used to analyze the influence of the deflection angle of the flaps on the wind turbine blade aerodynamic load and wake flow field and is described and validated firstly. Subsequently, the aerodynamic performance of the airfoil, with the trailing-edge flap, as well as the influence of the trailing-edge flap on the blade aerodynamic load and the wake flow field, are analyzed in detail. Finally we elaborate on the trailing-edge flap control strategy, which is proposed by Xu et al. [17], used for an offshore wind turbine and the control performance under different wind conditions.

2. FVW Model and Validation

Figure 1 shows the structural model of the wind turbine. The red parts in the figure are the trailing-edge flaps. The influence of the tower on the aerodynamic performance, which is much smaller than that of unsteady wind conditions due to the upwind structure [19], is neglected in the calculation. In this study, the FVW method is used to simulate the aerodynamic performance of the wind turbine with trailing flaps. The FVW method simulates the aerodynamic characteristics of the blades by attaching vortices on 1/4 chord lines. Because the gradient of the attachment of the vortices is non-uniform, the blades are discretized into a finite number of micro-segments by using the arc-cosine method. Finally, the whole blade is simulated as a Weissinger-L model [20]. The load distribution of the blade is obtained by calculating the velocities induced by the vortices in the wake.

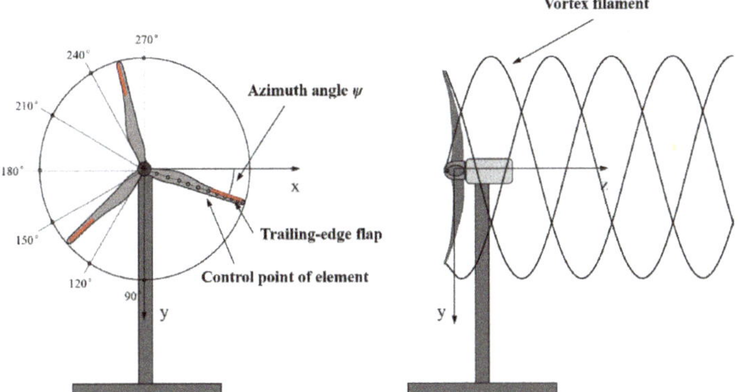

Figure 1. Wind turbine model with trailing-edge flap used in the free vortex wake (FVW) method.

The boundary of the blade element is defined by the following relationship,

$$(\bar{r})_i = \frac{(r_b)_i}{R} = \frac{2}{\pi}\cos^{-1}\left(1 - \frac{i-1}{N_E}\right) \tag{1}$$

where N_E is the number of blade elements ($N_E = 30$ in this study) and i is the element boundary number ($I = 2, \ldots, N_E + 1$). Consequently, there are N_E element control points and ($N_F + 1$) boundary points. The details of the FVW method for wind turbine aerodynamic calculations can be found in Ref. [18] and Ref. [21].

In order to verify the accuracy of the FVW method, we use it to model the NREL 5-MW wind turbine [22] and calculate the power and thrust of the rotor under stable wind conditions of 6 m/s to 18 m/s. The results are shown in Figure 2 and indicate that the calculated values (RotPwr result, RotThust result) are close to the calculated values obtained with the FAST software (RotPwr, RotThust) [22] at almost all wind speeds. Furthermore, some more validations of the FVW model comparing with the experimental results under the unsteady conditions including pitching case and yawed case can be found in Ref. [18]. Therefore, it is evident that the FVW model can be used to calculate the power and thrust of the wind turbine and that the calculation accuracy meets the research requirements.

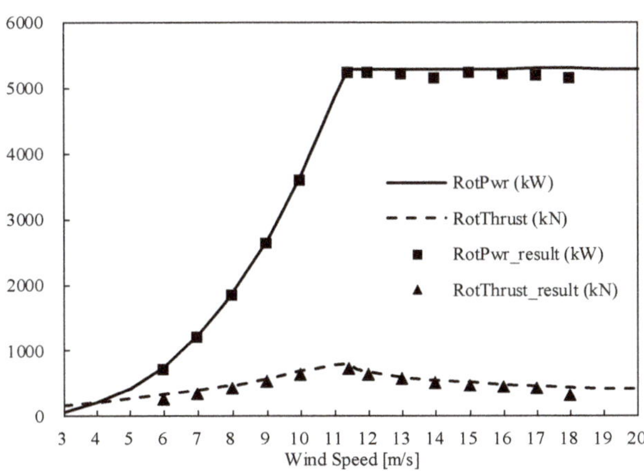

Figure 2. Power and thrust outputs of the NREL 5-MW wind turbine as a function of wind speed.

3. Aerodynamic Performance of Trailing-Edge Flap

The radial location of the trailing-edge flap should be close to the blade tip because of longer arm of force and smaller angle of attack at the outer part of the blade. Moreover, this layout can provide good control efficiency. The flap width should be appropriate to avoid damaging the blade structure. The size of the trailing-edge flaps of the NREL 5-MW wind turbine has been investigated in Ref. [17]. Here we also use the NREL 5-MW wind turbine as an example and the same size trailing-edge flaps as were described in Ref. [17], as shown in Figure 3. The flap width is 20% of the chord length; the radial flap length is 14 m in the radial direction of the blade, which is shown in red in the figure. The outboard location of the flap is 1.2 m from the blade tip. The thickness baseline of the profile with the flap was 18% and the NACA64-218 airfoil was used. The lift and drag coefficients of the NACA64-218 airfoil are calculated by the CFD (computational fluid dynamics) method [23]. The lift coefficient and lift-drag ratio of the NACA64-218 airfoil are shown in Figure 4. It is evident that the lift coefficient of the airfoil increases with the increase in the flap deflection angle at the same angle of attack. The larger the angle of attack, the larger the lift coefficient at the same flap deflection angle, but the rate of increase in the lift coefficient decreases with increasing angle of attack. It is noteworthy that when the angle of attack is greater than 8°, the lift coefficient of the airfoil does not increase or even decreases when the flap deflection angle is greater than 15°. According to the aerodynamic analysis of the trailing-edge flap by Zhang et al. [24], a stall of the trailing-edge will occur when the flap deflection angle is too large. This will result in a drop in the lift coefficient and an increase in drag. The lift-to-drag ratio of the airfoil increases first and then decreases with the increase in the flap deflection angle. The larger the angle of attack of the airfoil, the smaller the rate of increase is and the smaller the flap deflection angle of the maximum lift-drag ratio is. When the flap deflection angle is greater than 10°, the lift-to-drag ratios cease to increase or even decrease.

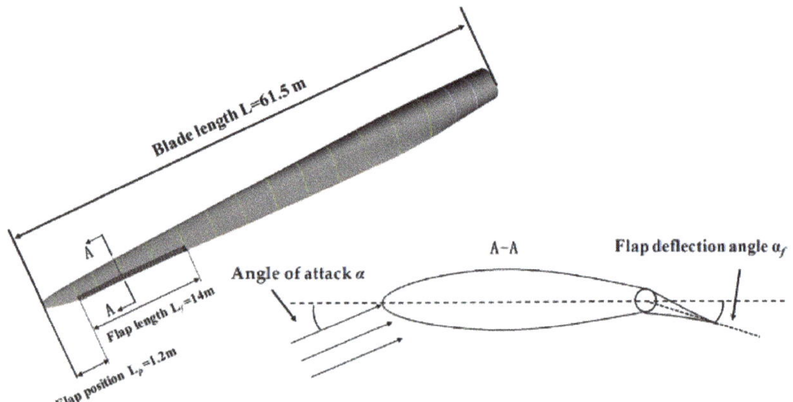

Figure 3. Blade structure with trailing-edge flap.

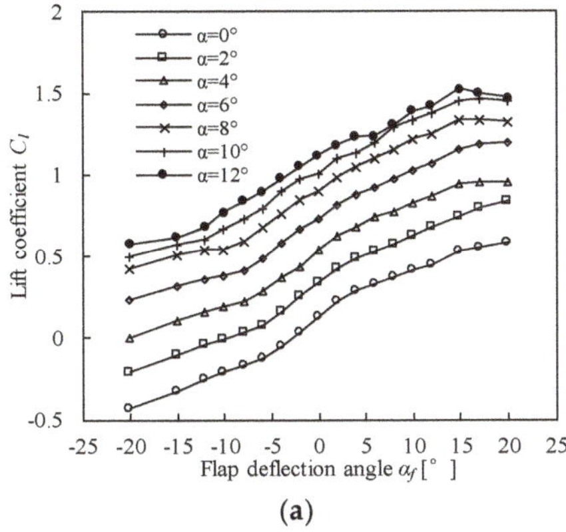 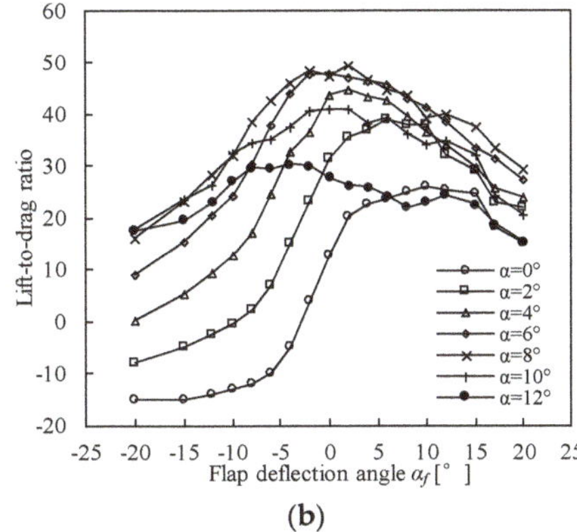

Figure 4. Aerodynamic performance of the airfoil NACA 64-218 with a trailing-edge flap. (a) Lift coefficient; (b) Lift-to-drag ratio.

4. Aerodynamic Characteristics of Trailing Flaps at Different Deflection Angles and Wind Speeds

Generally, the angle of attack has a large influence on the lift coefficient and lift-to-drag ratio of the airfoil with a flap. Therefore, it is necessary to know the angle of attack of the blade profile with the trailing-edge flap and its influence. Table 1 shows the rotor speed, blade pitch angle, and average angle of attack of the profile equipped with the trailing-edge flap of the NREL 5 MW wind turbine at different wind speeds. When the wind speed is less than the rated wind speed, the pitch angle is 0 and the rotor speed varies with the change in the wind speed. The average angle of attack of the selected profile increases slowly from 7.27° to 8.39°. When the wind speed is higher than the rated wind speed, the pitch angle increases, whereas the speed of the rotor does not change and the average angle of attack of the selected profile decreases gradually. As seen in Figure 4, the effect of the flap deflection angle on the aerodynamic performance differs for different angles of attack. Therefore, the effect of the flap deflection angle on the aerodynamic performance of the blade has to be determined.

Table 1. Rotor speed, blade pitch angle, and average angle of attack of the profile with the trailing-edge flap at different wind speeds.

Wind Speed (m/s)	Rotor Speed (rpm)	Pitch Angle (°)	Average Angle of Attack (°)
8	9.16	0.00	7.27
9	10.37	0.00	7.60
10	11.48	0.00	7.69
11.4	12.1	0.00	8.39
12	12.1	3.83	5.07
14	12.1	8.70	1.91
16	12.1	12.06	0.28
18	12.1	14.92	−0.85

The wind turbine torque and thrust for different flap deflection angles at three stable wind speeds of 8 m/s, 11.4 m/s, and 16 m/s are shown in Figure 5. The trends of the torque curves are similar at 8 m/s and 11.4 m/s. In the flap deflection angle range of −20°–0°, the torque increases with the increase in the flap deflection angle but it only increases slightly in the range of 0°–5° and even decreases in the range of 5°–10°. The torque values at 16 m/s increase in the range of −20°–10° and the slope of the curve decreases only when approaching 10°. The thrust curves are also similar at 8 m/s and 11.4 m/s. When the flap deflection angle is greater than 0°, the rate of increase in the thrust values is relatively small at these wind speeds, whereas the rate of increase in the thrust values at 16 m/s is greater. The control performance of the flap is related to the angle of attack of the profile. At high wind speeds, the smaller the angle of attack of the profile, the better the control performance is; this result is consistent with the results shown in Figure 4.

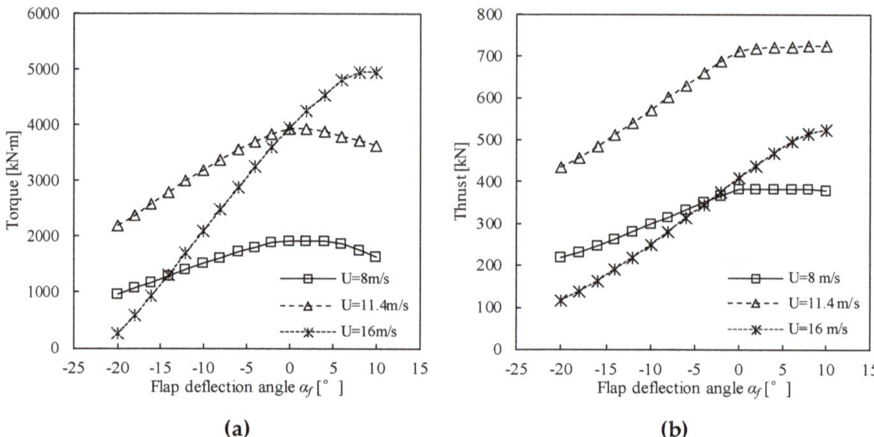

Figure 5. Effect of the trailing-edge flap deflection angle on the torque and thrust of the NREL 5-MW wind turbine rotor. (**a**) Low speed shaft torque, (**b**) Rotor thrust.

The FVW model simplifies the blade surface load to a series of centralized loads at the blade element control points. By analyzing the changes in the centralized loads, the influence of the flap deflection on the blade aerodynamic load distribution can be determined. Figures 6–8 show the aerodynamic load distributions of the blade control points at three wind speeds of 8 m/s, 11.4 m/s, and 16 m/s, respectively. It is evident that the curves exhibit a similar trend at 8 m/s and 11.4 m/s although the values in the two figures are different. At wind speeds of 8 m/s and 11.4 m/s, the tangential and normal forces of the tip flaps increase with the increase in the flap deflection angle. The deflection of flap only has a significant effect especially on the loads that are generated in the radial distribution of the flaps, but has little effect on the other position of the blade. It is worth noting that when the deflection angle of the flaps is equal to 5°, the tangential force of the flaps increases slightly and the normal force of the flaps increases considerably. However, the tangential force and the normal force of the other parts decrease slightly. When the wind speed is 16 m/s, the flap deflection also causes significant changes in the tangential force and normal force of the blade tip flaps and also has a considerable impact on the forces at the position near the flaps. The influence range is more than 20% of the blade length. A comparison of Figures 6–8 indicates that the smaller the angle of attack of the profile, the greater the change in the blade's aerodynamic force is when the flap deflection angle changes.

These results indicate that the adjustability of the trailing-edge flap is directly related to the angle of attack of the profile and when the wind speed is less than the rated wind speed of 11.4 m/s, the angle of attack of the profile is larger and there is little change. Therefore, in order to analyze the effect of the flap deflection on the operation of the wind turbine, only the wind speeds of 11.4 m/s and 16 m/s wind speeds (high angle of attack and small angle of attack) need to be considered.

Figures 9 and 10 show the distribution of the axial wind speed in the front and the back of the rotor of the wind turbine with a trailing-edge flap at 11.4 m/s, and 16 m/s, respectively, as determined by the FVW method. The deflection of the flaps has had little effect on the structure of the wake flow field, but still changes the axial velocity distribution near the blade. At a wind speed of 11.4 m/s, the axial velocity of the front and rear blades increases significantly with the decrease in the flap deflection angle, especially at the tip flaps. In contrast, the wind speed in the low wind speed region of the tip vortex is increasing. At a wind speed of 16 m/s, the distance between the tip vortices is larger and the decrease in the axial wind speed at the tip vortices does not change significantly with a change in the deflection angle of the flaps. As the flap deflection angle decreases, the change in the distribution of the axial wind speed at the tip flaps becomes more apparent but the change in the other parts of the blades is not significant.

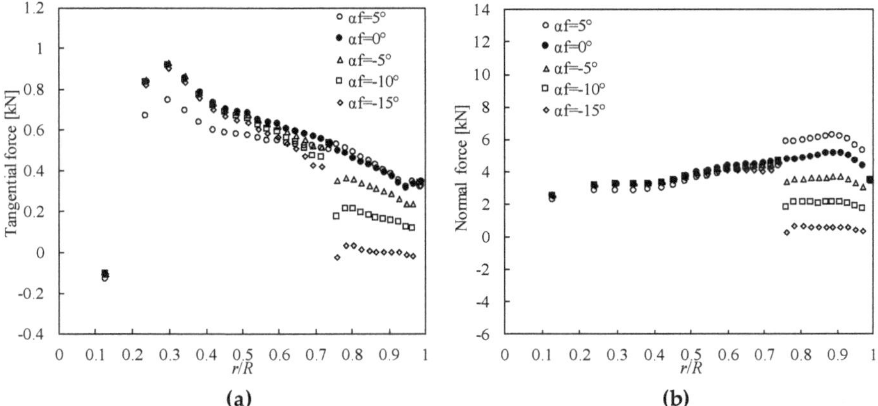

Figure 6. Aerodynamic force distribution of the blade control points for $U = 8$ m/s. (**a**) Tangential force to the rotor disc, (**b**) Normal force to the rotor disc.

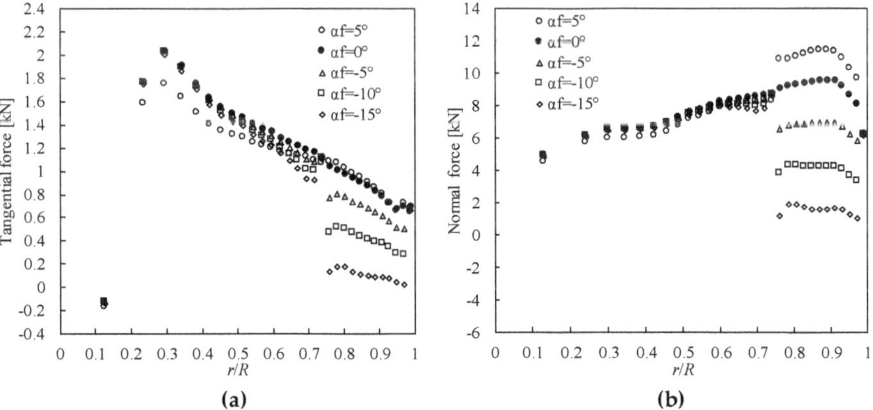

Figure 7. Aerodynamic force distribution of the blade control points for $U = 11.4$ m/s. (**a**) Tangential force to the rotor disc, (**b**) Normal force to the rotor disc.

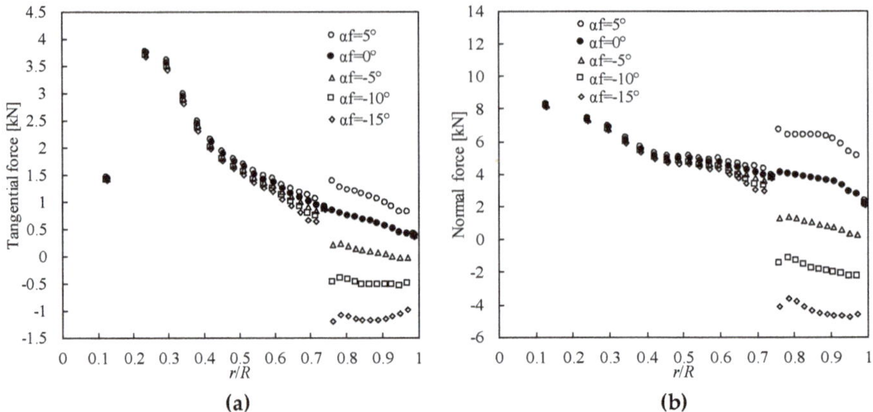

Figure 8. Aerodynamic force distribution of the blade control points for $U = 16$ m/s. (**a**) Tangential force to the rotor disc, (**b**) Normal force to the rotor disc.

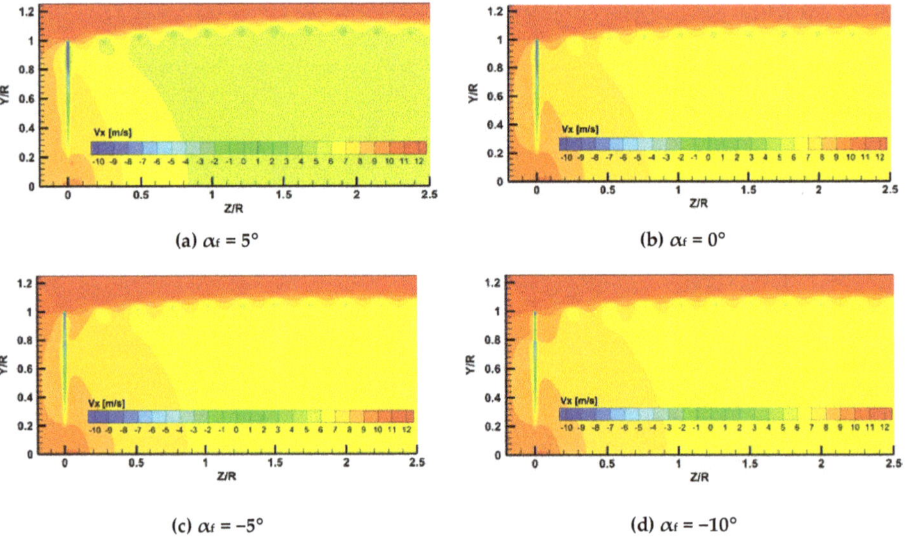

Figure 9. Axial velocity distribution on the plane with a $0°$ wake angle for $U = 11.4$ m/s. (Position: Axial direction from $-0.2\,R$ to $3\,R$, radial direction from 0 to $1.25\,R$). (**a**) $\alpha_f = 5°$, (**b**) $\alpha_f = 0°$, (**c**) $\alpha_f = -5°$, (**d**) $\alpha_f = -10°$.

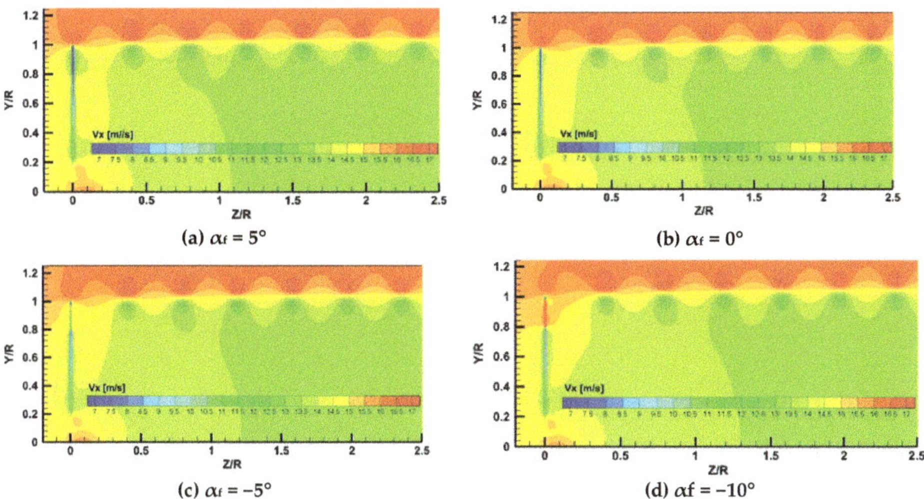

Figure 10. Axial velocity distribution on the plane with a 0° wake angle for U = 16 m/s. (Position: Axial direction from −0.2 R to 3 R, radial direction from 0 to 1.25 R). (**a**) $\alpha_f = 5°$, (**b**) $\alpha_f = 0°$, (**c**) $\alpha_f = −5°$, (**d**) $\alpha_f = −10°$.

5. Unsteady Wind Conditions

The trailing-edge flap is used to control the load fluctuation caused by unsteady wind conditions. The unsteady wind field of the wind turbines near the ground mainly includes wind shear and turbulent wind.

5.1. Wind Shear

Wind shear exists in the atmosphere near the ground and is affected by the thickness of the surface boundary layer. Common wind shear models are the exponential model and logarithmic model [25]. Here we choose the exponential model because the prediction of the exponential model agrees better with the measured value than that of the logarithmic model. The boundary layer wind speed is defined as,

$$U(h) = U(h_0)\left(\frac{h}{h_{hub}}\right)^{\alpha} \tag{2}$$

where $U(h)$ is the wind speed at a height of h and $U(h_0)$ is the wind speed at the reference height h_{hub}. The power law exponent α is usually in the range of 0.1–0.25. Figure 11 shows the wind shear distribution near the ground where the wind turbine is located. Generally, 0.2 is used on land and 0.1 is used over the ocean [26]. In this study, 0.1 is used.

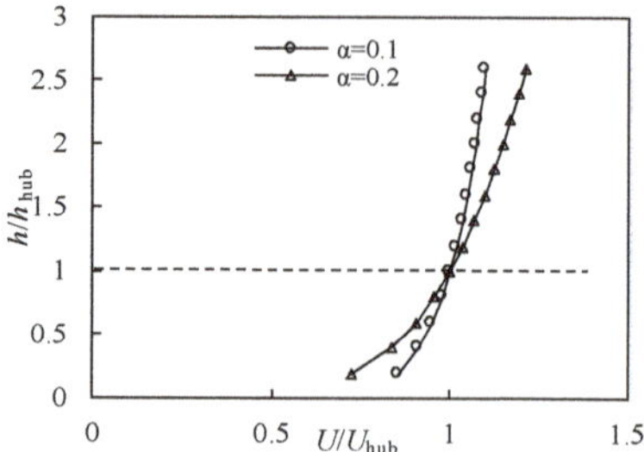

Figure 11. Atmospheric boundary layer profiles.

5.2. Turbulent Wind

Turbulent wind has a complex spatial distribution and in actual wind data, the turbulent intensity is rarely uniform. Therefore, we use the Blade 4.3 software [27] to generate turbulent wind data. The wavelet inverse transformation method [28] is used to create the turbulent wind field according to the advanced von Karman power density spectrum [27]. The surface roughness is 0.01 and the turbulent intensity is 9.58%. Figure 12a,b shows the data of the axial turbulent wind speed at the hub of the wind turbine at wind speeds of 11.4 m/s, and 16 m/s, respectively. Turbulent wind is unevenly distributed in the plane of the wind turbine but it is difficult to quickly measure the wind speed at different coordinate points and analyze the data using existing wind turbine measuring equipment. Therefore, the wind speed data at the hub should be simplified to determine the wind speed change in the entire plane of the wind turbine.

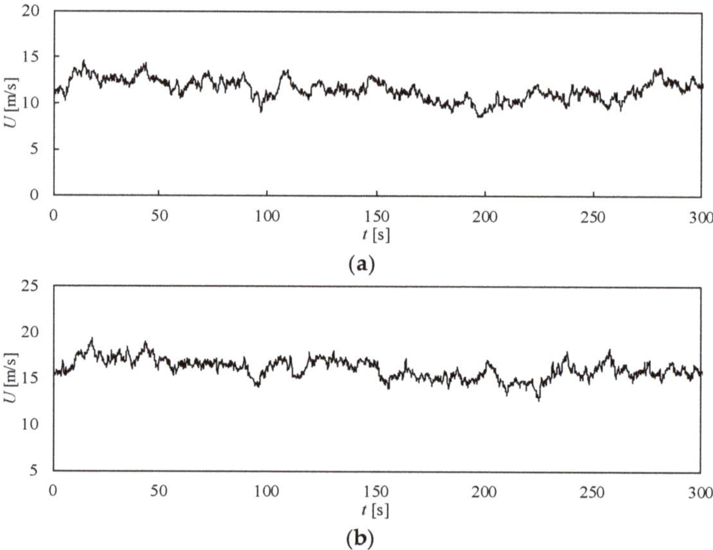

Figure 12. Axial velocities in turbulent conditions with a turbulence intensity of 9.58%. (**a**) The average wind velocity equals 11.4 m/s; (**b**) the average wind velocity equals 16 m/s.

6. The Trailing-Edge Flap Control Strategy

In Sections 3 and 4, the influence of the flaps on the aerodynamic load and wake flow field of the wind turbine was analyzed. The control performance of the flaps is related not only to the aerodynamic characteristics of the flaps but also to the control of the flaps. Here we adopt the flap control strategy proposed by Xu et al. [17]. This method is simple and efficient.

Under wind shear, the minimum wind speed occurs at the lowest point of the wind turbine ($\psi = 90°$) and the maximum wind speed occurs at the highest point of the wind turbine ($\psi = 270°$). The control of wind shear is based on the azimuth angle of the blades. The three blades of the wind turbine are controlled by three separate control units to deal with the load deviation at their respective locations. The control strategy of a single blade is defined as,

$$\alpha_s(t) = a \cdot \left(\frac{U_{hub} - U_{tip}}{|U_{hub} - U_{min}|}\right) \cdot \left|\frac{U_{hub} - U_{tip}}{|U_{hub} - U_{min}|}\right| \tag{3}$$

where a is the flap control factor of the wind shear, U_{hub} is the wind speed at the hub, U_{min} is the minimum value of U_{tip}, and U_{tip} is the wind speed at the blade tip, which is defined as,

$$U_{tip} = U_{hub}\left(\frac{h_{hub} - R\sin\psi}{h_{hub}}\right)^{\alpha} \tag{4}$$

where h_{hub} is the height of the hub, R is the blade length, ψ is the blade azimuth angle, and the power law exponent α is the same as in Equation (2).

The control of turbulent wind is based on the average wind speed within one second. Since the sampling period of the turbulent wind speed data in this study is 0.25 s, the control strategy of turbulent wind can be described as follows,

$$\alpha_t(t) = b \cdot (U - U_t) \tag{5}$$

where b is the control factor of the turbulence, U is the instantaneous wind speed, U_t is the average wind speed within one second prior to the time, which is expressed as.

$$U_t = \frac{u_t + u_{t-0.25} + u_{t-0.5} + u_{t-0.75} + u_{t-1}}{5} \tag{6}$$

where u_t is the wind speed at current time, $u_{t-0.25}$, $u_{t-0.5}$, $u_{t-0.75}$ and $u_{t-0.1}$ are the wind speed at times of 0.25 s, 0.5 s, 0.75 s, and 1 s before u_t.

The control factors a and b are essential to the effect of load mitigation and are dependent on the design of the wind turbine [17]. The control strategy engineers need to go through lots of debugging to obtain the appropriate values. In the following, the influence of value changes of factors a and b will be analyzed and the specific values for the NREL 5-MW wind turbine will be proposed.

7. Result and Discussion

7.1. Calculation Results of Wind Shear

Figure 11 shows that there are considerable differences in the wind speed in the vertical direction when the wind turbine tower is high and the blades are long. Figure 13 shows the tangential and normal forces at the control points for different blade azimuths on the NREL 5-MW wind turbine blades under wind shear with a hub wind speed of 11.4 m/s. It is evident that the aerodynamic load is considerably different for different blade azimuths due to the presence of wind shear. For every rotation cycle of the blade, this load change causes fluctuations in the blade's torque and thrust. The force change with the azimuth angle at the outer part of the blade, except the blade tip, which is mainly focused on noise abatement, appears more obvious than at the inner part.

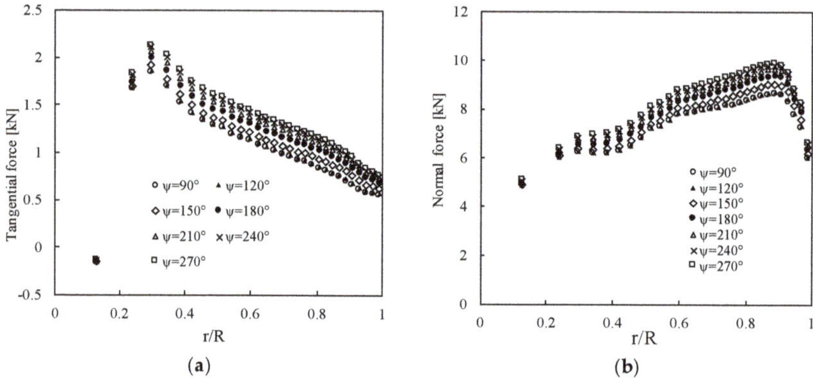

Figure 13. Tangential force and normal force at the control points in wind shear conditions. (**a**) Tangential force; (**b**) Normal force.

Figure 14 shows the results of different control factors under wind shear for U_{hub} = 11.4 m/s. The results of $a = 0$ are these which were conducted without using control strategies but at a constant flap angle of 0°. Although, the analysis of the blade's aerodynamic load distribution shows that wind shear has a large influence on the load distribution of the blades, it is observed in Figure 14 that the fluctuations of the torque and thrust ($a = 0$) of the rotor are not very large and are mainly attributed to the superposition of the three blades of the rotor. The torque and thrust ($a = 0$) fluctuations of a single blade are relatively large as shown in Figure 15.

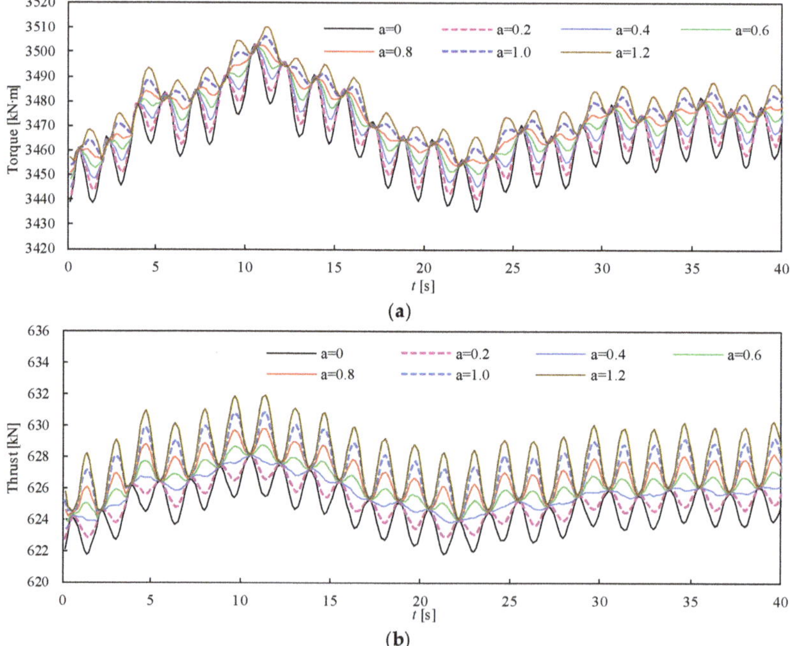

Figure 14. Torque and thrust response of the wind turbine rotor under wind shear conditions for U_{hub} = 11.4 m/s. (**a**) Torque response; (**b**) thrust response.

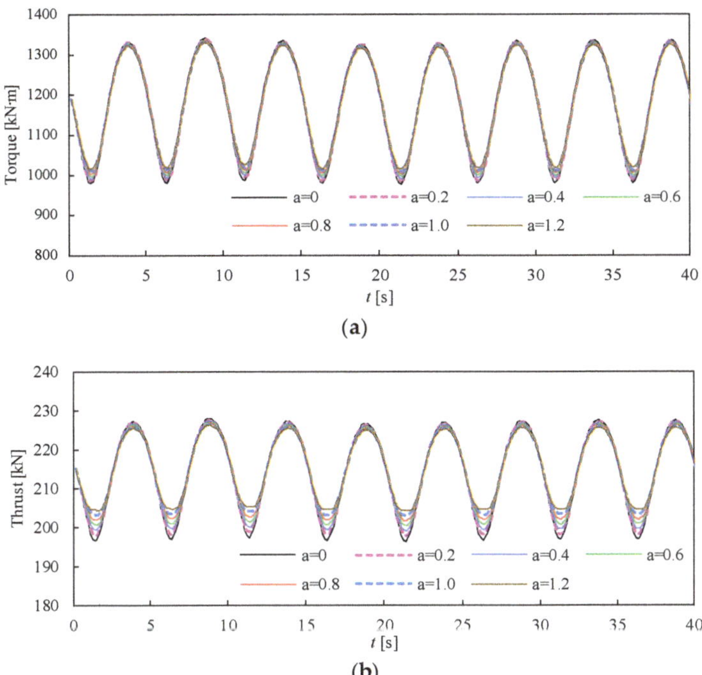

Figure 15. Torque and thrust response of a single blade under wind shear conditions for $U_{hub} = 11.4$ m/s. (**a**) Torque response; (**b**) thrust response.

Different values of the control factors also have different effects. As shown in Figure 14, a change in the value of the flap control factor *a* significantly changes the amplitude of the torque and power fluctuations. It is noteworthy that the torque curve of the wind turbine has the smallest fluctuation range at $a = 0.8$ but the adjustment of the thrust has overshoot under this condition; the thrust curve has the smallest fluctuation range at $a = 0.4$. Whereas, the torque fluctuation is large at this value. Figure 15 shows that under this wind condition, the effect of the flaps on the torque fluctuation of a single blade is considerably less than that on the thrust fluctuation and the effect on the curve's crest is less than that on the trough. This is also the reason why the optimal control factors (*a*) of the torque and thrust of the wind turbine are different under this condition.

When the wind speed is higher than the rated wind speed, the control system limits an additional increase in the power of the wind turbine by increasing the pitch angle. Increasing the pitch angle will reduce the blade's aerodynamic angle of attack and the aerodynamic characteristics of the flaps (Section 4) show that the flaps with a smaller angle of attack have a larger and more stable adjustment range, which improves the ability of the trailing-edge flaps to control the load.

Because the origin of the flap's deflection angle is −5°, the torque and thrust of the wind turbine will be significantly less than the original thrust and torque of the NREL 5-MW wind turbine at high wind speeds. When the wind speed is greater than 11.4 m/s, the original pitch angle should be reduced by one unit in order to maintain the wind turbine power stable at around 5 MW, which is a necessary operation in the pitch control system after the trailing-edge flaps are installed. The results at U_{hub} = 16 m/s are calculated to observe the effect of the controller at a small angle of attack of the flap. A number of tests and data analyses indicate that modifying the original pitch angle of 12.06° to 10.5° ensures that the power of the wind turbine can be maintained at around 5 MW. The result is shown in Figure 16. It is observed that the torque and thrust responses are synchronous. When $a = 0.6$, the fluctuations of the thrust curve and torque curve are very small. Compared with Figure 15, high wind speed and small angle of attack are suitable when using trailing-edge flaps.

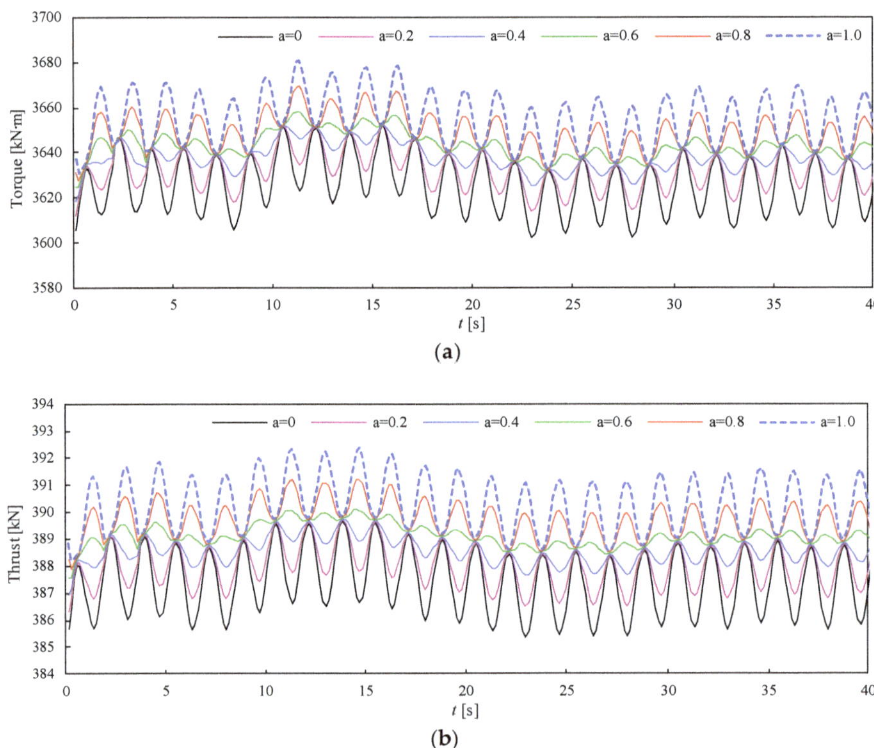

Figure 16. Torque and thrust response of the wind turbine rotor under wind shear conditions for U_{hub} = 16 m/s and θ_p = 10.5. (**a**) Torque response; (**b**) Thrust response.

7.2. Calculation Results of Turbulent Wind

The performance of the flaps under turbulent wind conditions is an important index to test their load mitigation ability. Figure 17 shows the thrust and torque responses for U_{mean} = 11.4 m/s and Figure 18 shows the thrust and torque responses for U_{mean} = 16 m/s. The simulation time is 200 s. For the convenience of observation, the figures show the first 100 s. The value range of b is 0–15. The results of $b = 0$ are these which were conducted without using control strategies but at a constant flap angle of 0°. It is evident that an appropriate control factor can mitigate the load fluctuations of the wind turbine, especially for low-frequency fluctuations. For high-frequency fluctuations near the average value, the effect is very small. As shown in Figure 17, when b equals the maximum value of 15, the thrust curve fluctuates in a small range above and below the constant wind curve, while there is a small deviation between the torque curve and the constant curve. At the same turbulence intensity, the turbulence fluctuation amplitude at the average wind speed of 16 m/s is larger than that at the average wind speed of 11.4 m/s. Therefore, the amplitudes of the torque and thrust are larger at 16 m/s than at 11.4 m/s and the performance of the flap is better for the same control factor (see Figure 18). When b equals the maximum value of 15, the torque curves and thrust curves remain near the average value and their amplitudes remain low. This is in agreement with our finding that the trailing-edge flap performs better at high wind speeds and a small angle of attack.

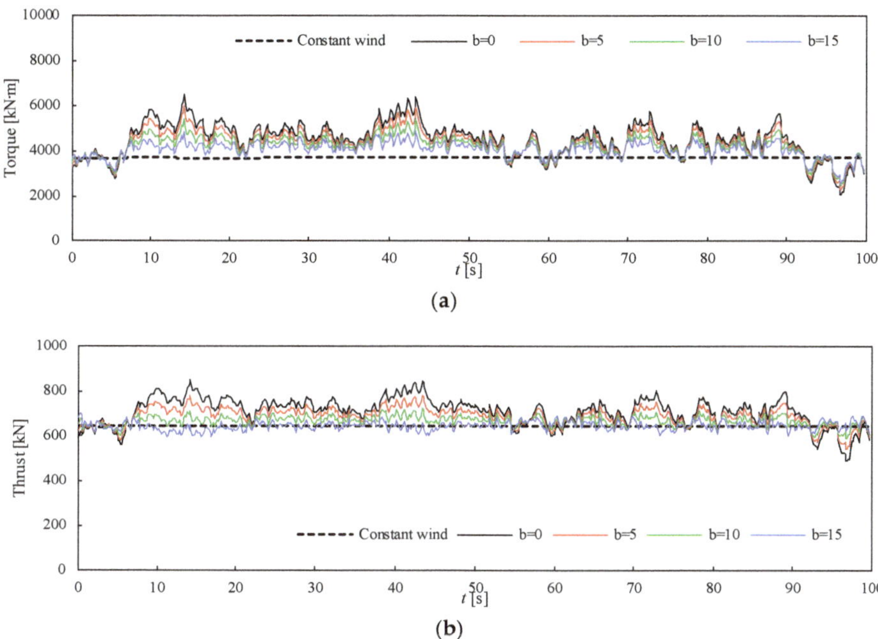

Figure 17. Torque and thrust response of the wind turbine rotor under turbulent wind conditions for $U_{\text{mean}} = 11.4$ m/s. (**a**) Torque response; (**b**) thrust response.

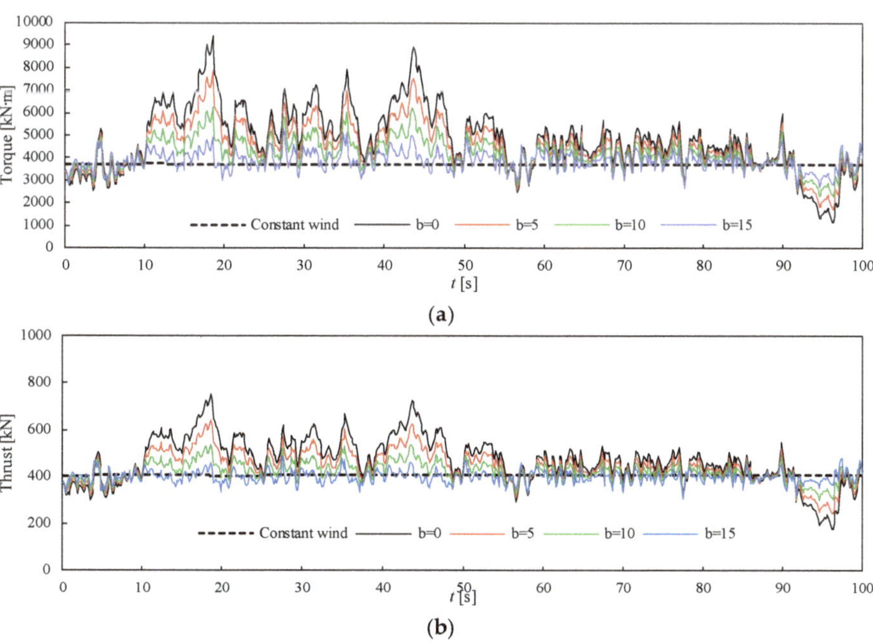

Figure 18. Torque and thrust response of the wind turbine rotor under turbulent wind conditions for $U_{\text{mean}} = 16$ m/s and $\theta_{\text{p}} = 10.5°$. (**a**) Torque response; (**b**) thrust response.

8. Conclusions

In this study, the effect of a trailing-edge flap on load mitigation in a large-scale offshore NREL 5-MW wind turbine was analyzed by using the FVW model.

Firstly, the variation of airfoil aerodynamic performance due to the trailing edge flap deflection is obvious. The deflection of flap has a significant effect especially on the loads that are generated in the radial distribution of the flaps. The smaller the angle of attack of the profile, the greater the change in the blade's aerodynamic force is when the flap deflection angle changes. Besides, the deflection of flap has a significant effect on the axial velocity of wake especially in the near wake of the outer part of the blade.

Secondly, control strategies of the trailing-edge flap for the shear wind and turbulent wind conditions were developed. The application of the trailing-edge flap control strategy can mitigate the fluctuation of load (torque and thrust) well in above unsteady conditions. The proposed control factor values of $a = 0.6$ and $b = 15$ were obtained for the NREL 5-MW wind turbine.

In a word, the control factors a and b are essential to the effect of load mitigation and are dependent on the design of the wind turbine. Some practical applications will be conducted in the future.

Author Contributions: Conceptualization, X.C. and Y.W.; methodology, B.X.; software, Y.W.; validation, X.C., Y.W., B.X. and J.F.; formal analysis, B.X.; investigation, X.C. and Y.W.; resources, X.C.; data curation, J.F.; writing—original draft preparation, Y.W. and J.F.; writing—review and editing, B.X.; visualization, Y.W.; supervision, X.C.; project administration, X.C.; funding acquisition, X.C. and B.X. All authors have read and agreed to the published version of the manuscript.

Funding: This research was funded by the Fundamental Research Funds for the Central Universities (grant number 2018B48614, 2019B14614); and the National Natural Science Foundation of China (grant number 51607058).

Conflicts of Interest: The authors declare no conflict of interest.

References

1. Li, Y.; Le, C.; Ding, H.; Zhang, P.; Zhang, J. Dynamic Response for a Submerged Floating Offshore Wind Turbine with Different Mooring Configurations. *J. Mar. Sci. Eng.* **2019**, *7*, 115. [CrossRef]
2. Loth, E.; Steele, A.; Qin, C.; Ichter, B.; Selig, M.S.; Moriarty, P. Downwind pre-aligned rotors for extreme-scale wind turbines. *Wind Energy* **2017**, *20*, 1241–1259. [CrossRef]
3. Vamsi, I.; Sabareesh, G.R.; Penumakala, P.K. Comparison of condition monitoring techniques is assessing fault severity for a wind turbine gearbox under non-stationary loading. *Mech. Syst. Signal Process.* **2019**, *124*, 1–20. [CrossRef]
4. Barlas, T.K.; van Kuik, G.A.M. Review of state of the art in smart rotor control research for wind turbines. *Prog. Aerosp. Sci.* **2010**, *46*, 1–27. [CrossRef]
5. Li, M.; He, Y.L.; Li, C.W. Study and analysis on collaborated control simulation for the new typed transmission system of wind turbine set. *J. Mach. Des.* **2008**, *25*, 36–40.
6. Cheng, Y.; Xue, Z.; Jiang, T.; Wang, W.; Wang, Y. Numerical simulation on dynamic response of flexible multi-body tower blade coupling in large wind turbine. *Energy* **2018**, *152*, 601–612. [CrossRef]
7. Xia, Y.; Sun, D.Y. Characteristic analysis on a new hydro-mechanical continuously variable transmission system. *Mech. Mach. Theory* **2018**, *126*, 457–467. [CrossRef]
8. Jang, M.H.; Lee, J.H.; Chung, Y.D. A study of a hydraulic torque converter for the offshore wind turbine system. In Proceedings of the International Conference on Electrical Machines and Systems, Busan, Korea, 26–29 October 2013; pp. 2305–2308.
9. Chopra, I. Review of state of art of smart structures and integrated systems. *AIAA J.* **2002**, *40*, 2145–2187. [CrossRef]
10. Hansen, M.O.; Madsen, H.A. Review paper on wind turbine aerodynamics. *J. Fluids Eng.* **2011**, *133*, 114001. [CrossRef]
11. Chen, H.; Qin, N. Trailing-edge flow control for wind turbine performance and load control. *Renew. Energy* **2017**, *105*, 419–435. [CrossRef]
12. Roget, B.; Chopra, I. Wind-tunnel testing of rotor with individually controlled trailing-edge flaps for vibration reduction. *J. Aircr.* **2008**, *45*, 868–879. [CrossRef]

13. Bak, C.; Gaunaa, M.; Andersen, P.B.; Buhl, T.; Hansen, P.; Clemmensen, K. Wind tunnel test on airfoil Risø-B1-18 with an Active Trailing Edge Flap. *Wind Energy* **2010**, *13*, 207–219. [CrossRef]

14. Lee, T.; Su, Y.Y. Lift enhancement and flow structure of airfoil with joint trailing-edge flap and Gurney flap. *Exp. Fluids* **2011**, *50*, 1671–1684. [CrossRef]

15. Lu, W.S.; Tian, Y.; Liu, P.Q. Aerodynamic optimization and mechanism design of flexible variable camber trailing-edge flap. *Chin. J. Aeronaut.* **2017**, *30*, 988–1003. [CrossRef]

16. Lackner, M.A.; Kuik, G.V. A comparison of smart rotor control approaches using trailing edge flaps and individual pitch control. *Wind Energy* **2010**, *13*, 117–134. [CrossRef]

17. Xu, B.F.; Feng, J.H.; Wang, T.G.; Yuan, Y.; Zhao, Z.Z.; Zhong, W. Trailing-edge flap control for mitigating rotor power fluctuations of a large-scale offshore floating wind turbine under the turbulent wind condition. *Entropy* **2018**, *20*, 676. [CrossRef]

18. Xu, B.F.; Wang, T.G.; Yuan, Y.; Cao, J.F. Unsteady aerodynamic analysis for offshore floating wind turbines under different wind conditions. *Philos. Trans. R. Soc. A Math. Phys. Eng. Sci.* **2015**, *373*. [CrossRef]

19. Sabale, A.; Gopal, N. Nonlinear aeroelastic analysis of large wind turbines under turbulent wind conditions. *AIAA J.* **2019**, *57*, 4416–4432. [CrossRef]

20. Weissinger, J. The lift distribution of swept-back wings. In *Technical Report NACA-TM-1120*; National Advisory Committee for Aeronautics: Washington, DC, USA, 1947.

21. Xu, B.F.; Wang, T.G.; Yuan, Y.; Zhao, Z.Z.; Liu, H.M. A simplified free vortex wake model of wind turbines for axial steady conditions. *Appl. Sci.* **2018**, *8*, 866. [CrossRef]

22. Jonkman, J.; Butterfield, S.; Musial, W.; Scott, G. *Definition of a 5 mw Reference Wind Turbine for Offshore System Development*; Office of Scientific & Technical Information Technical Reports; National Renewable Energy Lab. (NREL): Golden, CO, USA, 2009.

23. Xu, B.F.; Feng, J.H.; Xu, C.; Zhao, Z.Z.; Yuan, Y. Aerodynamic performance analysis of a trailing-edge flap for wind turbines. *J. Phys. Conf. Ser.* **2018**, *1037*, 022020.

24. Zhang, W.G.; Wang, Y.F.; Liu, R.J. Unsteady aerodynamic modeling and control of the wind turbine with trailing edge flap. *J. Renew. Sustain. Energy* **2018**, *20*, 063304. [CrossRef]

25. Large, W.G.; McWilliams, J.C.; Doney, S.C. Oceanic vertical mixing: A review and a model with a nonlocal boundary layer parameterization. *Rev. Geophys.* **1994**, *32*, 363. [CrossRef]

26. Khosravi, M.; Sarkar, P.; Hu, H. An experimental investigation on the performance and the wake characteristics of a wind turbine subjected to surge motion. *AIAA J.* **2016**, *138*, 042602.

27. Bossanyi, E.A. *GH Bladed Theory Manual*; Garrad Hassan and Partners Ltd.: Bristol, UK, 2009.

28. Kitagawa, T.; Nomura, T. A wavelet-based method to generate artificial wind fluctuation data. *J. Wind Eng.* **2003**, *91*, 943–964. [CrossRef]

Journal of
*Marine Science
and Engineering*

Article

Loads and Response of a Tension Leg Platform Wind Turbine with Non-Rotating Blades: An Experimental Study

Timothy Murfet and Nagi Abdussamie *

National Centre for Maritime Engineering and Hydrodynamics, Australian Maritime College, University of Tasmania, Launceston, TAS 7250, Australia; tdmurfet@utas.edu.au
* Correspondence: nagia@utas.edu.au; Tel.: +61-3-6324-9732

Received: 31 December 2018; Accepted: 22 February 2019; Published: 27 February 2019

Abstract: This paper describes model testing of a Tension Leg Platform Wind Turbine (TLPWT) with non-rotating blades to better understand its motion and tendon responses when subjected to combined wind and unidirectional regular wave conditions. The TLPWT structure is closely based on the National Renewable Energy Laboratory (NREL) 5 MW concept. Multiple free decay tests were performed to evaluate the natural periods of the model in the key degrees of freedom, whilst Response Amplitude Operators (RAOs) were derived to show the motion and tendon characteristics. The natural periods in surge and pitch motions evaluated from the decay tests had a relatively close agreement to the theoretical values. Overall, the tested TLPWT model exhibited typical motion responses to that of a generalised TLP with significant surge offsets along with stiff heave and pitch motions. The maximum magnitudes for the RAOs of surge motion and all tendons occurred at the longest wave period of 1.23 s (~13.0 s at full-scale) tested in this study. From the attained results, there was evidence that static wind loading on the turbine structure had some impact on the motions and tendon response, particularly in the heave direction, with an average increase of 13.1% in motion amplitude for the tested wind conditions. The wind had a negligible effect on the surge motion and slightly decreased the tendon tensions in all tendons. The results also showed the set-down magnitudes amounting to approximately 2–5% of the offset. Furthermore, the waves are the dominant factor contributing to the set-down of the TLPWT, with a minimal contribution from the static wind loading. The results of this study could be used for calibrating numerical tools such as CFD codes.

Keywords: offshore wind; tension leg platforms; loads and response; model testing

---

1. Introduction

There is an increasing demand worldwide for renewable energy generation, largely due to the increasing awareness of climate change and limited fossil fuel resources [1]. The total capacity of offshore wind has increased considerably in the last decade, with global capacity reaching a recorded 19.27 gigawatts (GW) in 2017, up from only 1.44 GW in 2008 [2]. Many major countries are continuing to develop offshore wind technology. The current rate of development is only set to increase, with predictions of up to 120 GW to be installed by 2030 [3].

Wind energy is considered a potential solution to cope with increasing energy demand, but development has largely been limited to onshore applications. This is particularly evident with 88% of the global offshore wind energy generation capacity located in European shallow waters as of the end of 2016 [4]. A major reason for this is the increased complexity of offshore turbine support structures, combined with additional factors from the maritime environment [5]. Typically, support structures include gravity bases, monopole and jacket structures, with monopoles being the most common, based on competitive fabrication and installation costs [6–8]. However, with space fast becoming a limiting factor for land-based wind power generation, significant research and development has been directed towards alternatives better suited for deeper waters [9]. In some cases, nearshore developments are undesirable due to their visual impact, further supporting these developments. For deeper water, floating structures appear to be a viable alternative to the restrictions of piled and jacket-based designs. However, floating structures are more challenging to design, based upon considerations of coupling between the turbine and support structure. Other factors such as mooring configurations and sea state conditions are likely to have a greater effect on the performance of the structure [10].

Currently, there has only been one full-scale Floating Offshore Wind Turbine (FOWT) commercial project commissioned. The Hywind project off the coast of Peterhead, Scotland began commercial generation in October 2017 and consists of five 6 MW turbines supported by spar-buoy floating structures. With the implementation of the first FOWTs, there is potential for FOWTs to play a more prominent role in the offshore wind industry [11]. This is most likely to be evident for larger countries such as the United States, China, Japan and Norway, which are limited in the amount of shallow water areas to place turbines [9].

One proposal for FOWT developments is the concept of Tension Leg Platforms (TLPs). TLPs have long been utilised in the offshore oil and gas industry, with the potential for this expertise to be applied to offshore renewable energy technology [1,12,13]. TLPs are a promising option for intermediate water depths due to the limited motions of the platform, allowing for the reduction of turbine motions and loads [14]. TLPs may also prove more effective for the relatively light topside conditions.

TLPs consist of a floating structure that uses a vertical tether system connected to the seafloor to achieve its required stability [15]. There are a wide range of TLP structure arrangements that have been developed for the different purposes they serve. These different types can be categorised into mono-column and conventional multi-column TLPs [16]. Up until the late 1990s, most oil and gas production platforms consisted of square four-column configurations. However, as time has progressed, more unique designs have been developed such as the single-column SeaStar TLP and the extended pontoon TLP [17].

The intact tendon system provides sufficient righting moments in response to small deformations due to the high vertical tension. This is unique compared to ships or other offshore structures that use conventional mooring systems [16]. This limits the structural loadings on the topside without the need of a deep draft or spread mooring system [14]. The design of the tendons has significant influence over the motion response of TLP structures. The stiff mooring system significantly limits the motions in the heave, pitch, and roll directions when subjected to environmental forces [18]. The tendons also assist in ensuring the natural periods of the structure are outside the typical range of appreciable ocean waves of 6–20 s [19]. However, because of the high axial tension, higher order resonant responses from second order waves can occur in low and high frequency regimes due to the random nature of the sea state [20]. An investigation by Srinivasan et al. [21] has analysed non-linear phenomena such as ringing and springing responses [12,22] that have been observed in TLPs under impact and non-impact wave conditions. These phenomena can pose a threat to platform stability and can result in the eventual fatigue failure of the tendons [21,23].

According to Nihei et al. [24], typical turbine structures of around 450 tonnes total weight could allow for a reduction of a total water plane area and overall hull displacement. These alterations could lead to a reduction in cost and spatial requirements whilst also potentially leading to less tendon tension requirements. There can be major differences in the requirements of the support structure based on the size and rated output of the turbine. Over recent decades, the rated output of wind turbines has substantially increased from 75 kW to the largest current concepts ranging from 5–10 MW [4]. As a way of supporting research and development into TLPWTs, the NREL concept is based on a 5 MW turbine to represent the current technology for typical three-blade designs [25]. This turbine has been used in model experiments and numerical simulations such as Kimball et al. [26] and Koo et al. [10]. This has been applied in a conceptual NREL-MIT TLPWT design developed by Tracy [27].

There are variants of mono-column TLPWTs that have been examined in a parametric study by Bachynski and Moan [14], with five different structures having been investigated that include different hull arrangements and different sizes of submerged pontoons. Up until now, numerical simulations and codes such as FAST, Bladed or FLEX have been used to perform dynamic analysis [9,11,23,28–30]. Nevertheless, any numerical simulation techniques can only be trusted by the industry if their results have been thoroughly validated against experimental data first. To date there have been limited scaled model tests performed for FOWTs, particularly with TLPWTs [24,28,31–33]. The main purpose of this study is to fill this gap by conducting an experimental study into the hydrodynamic performance of a generic TLPWT model. The outcome of this study can serve as preliminary work to better understanding the motion and tendon responses under the influence of waves and wind forcing. Furthermore, the study aims at providing valuable data to verify/validate the results of numerical simulations to be conducted in future. In order to easily identify the effect of the wind loading on the global loads and responses of the TLPWT model, the wind turbine structure was modelled without considering the turbine thrust generated by spinning blades.

The main scope of this study is to investigate a conceptual TLPWT with a static rotor (i.e., non-rotating blades) using experimental tests at a scale of 1:112 with emphasis on the global hydrodynamic performance under combined wave and wind conditions. The scaled TLP model was based on a generic TLPWT derived from concept designs developed by Matha [9] and Bachynski and Moan [14]. Whilst the wind turbine structure was closely based on the NREL 5 MW turbine to represent the current technology used in the offshore wind industry [4]. To better understand the motion and tendon responses, the model was subjected to several wind and unidirectional regular wave conditions derived from Bachynski and Moan [14]. The study presents the differences in the platform wave-induced motions and tendon response with and without wind acting on the structure. An analysis of how offset and set-down correlate under changing wind and wave conditions was also performed. The materials of this paper are set out as follows: Section 2 describes the TLPWT model, instrumentations and the experimental setup. Furthermore, the wave and wind conditions selected for this study are included in Section 2. Section 3 introduces the results of free decay tests in different degrees of freedom and introduces the results of the uncertainty analyses. Section 4 discusses the obtained results of the model's dynamic response and tendon tensions and response amplitude operators.

2. Materials and Methods

2.1. TLP Model Description

The TLPWT model used for the testing has been closely based in the NREL concept developed by Matha [9], from an initial investigation from Tracy [27]. This TLPWT concept has been used as a basis for experimental testing, including investigations by Nihei et al. [24] and Zamora-Rodriguez et al. [34]. The primary motivation behind using a similar hull and turbine arrangement allows for the close comparison of similarities and differences in results from this set of testing to previous experiments. This arrangement also represents a potential concept that could be used for full scale commercial development in the future.

Froude scaling law was applied to the TLP structure and turbine model to achieve the best possible scaled geometrical and mass properties for the TLPWT model. This methodology is commonly used for offshore structures for experimental testing in wave tanks [19,35,36]. To capture and measure the hydrodynamic behaviour of the model, an appropriate and practical scale of 1:112 was chosen. Based on the platform column diameter and the tested wave conditions discussed thereafter, the Reynolds number was estimated to be in the range of 1.69×10^4–2.90×10^4 (2.00×10^7–3.43×10^7 at full-scale). This scale was chosen due to laboratory and wind/wave generation constraints. However, such a selection does have implications for the scaling of the water depth, as the maximum achievable depth in the testing facility was limited to 900 mm which corresponded to a full-scale water depth of 100.8 m.

This approach is considered acceptable since this study aims at investigating the hydrodynamic and wind loads of a generic TLPWT platform rather than the response of a specific TLPWT to be installed at a specific water depth. It is worth mentioning that the 1:112 model scale of the experiment is outside the typically chosen scale range (1:30–1:100) for hydrodynamic model testing [36] which might affect the quality of obtained data. Not only are smaller scales rarely used due to increased uncertainties and less repeatability in the modelling, but also due to scaling effects [36]. However, Hansen et al. [35] constructed and tested a floating TLP wind turbine at 1:200 scale to analyse its dynamic response experimentally in co-directional wind and waves. The authors concluded that their experiments have demonstrated the potential of the model scale floating wind turbine and the measurement set-up to provide data and insight into the dynamic response of a floating wind turbine in different mooring and weather conditions.

The scaled and 'as-constructed' parameters for the TLPWT hull are shown in Table 1, with reference to the literature and concept designs to which it is based. A combination of interpretations of this concept design from Matha [9] and Bachynski and Moan [14] were used as a basis for the full-scale parameters. The TLP model was constructed by Chia [37], with the structural geometry remaining the same for this experiment. Some changes were made based on construction and facility limitations, most notably the column height, freeboard and draft. The mass of the structure is greater which resulted in a higher pre-tension, however this proved beneficial for obtaining more reliable tendon tension data during the model testing. Although these changes increase the full-scale footprint and mass of the structure, the general hydrodynamic behaviour will still provide meaningful relationships and trends for the hydrodynamic performance of the model.

Table 1. 'As constructed' parameters of TLP hull.

Parameter	Planned		Tested Model-Scale	Reference for Planned Full-Scale Values
	Full-Scale	Model-Scale		
Hull column diameter	18.00 m	160.71 mm	160.00 mm	Bachynski and Moan [14]
Hull column height	52.60 m	469.64 mm	600.00 mm	Bachynski and Moan [14]
Draft	47.89 m	427.59 mm	500.00 mm	Matha [9]
Freeboard	5.00 m	44.64 mm	100.00 mm	Matha [9]
Pontoon length	18.00 m	160.71 mm	160.00 mm	Bachynski and Moan [14]
Pontoon width	2.40 m	21.42 mm	21.70 mm	Bachynski and Moan [14]
Pontoon height	2.40 m	21.42 mm	21.70 mm	Bachynski and Moan [14]
Hull structural mass	-	-	2.63 kg	-
Ballast mass	-	-	1.82 kg	-
Total pre-tension	-	-	5.11 kg	-
Volumetric displacement	11.80×10^7 m^3	8.44×10^6 mm^3	1.04×10^7 mm^3	Bachynski and Moan [14]
Water depth	150.00 m	1.34 m	0.90m	Bachynski and Moan [14]

2.2. Turbine Model Description

The turbine model was scaled using the same factor as that used for the TLP hull, to ensure similarity for both components. The turbine was constructed as a fixed structure so that only static wind loading would be experienced on the structure. PLA plastic was used for the turbine assembly, which was created using a 3D printer. As can be seen in Table 2, all the geometrical parameters were able to be represented with a high accuracy. However, due to limitations with 3D printing and the scaling of the turbine blades, the weight of the blades was slightly higher compared to the intended scaled value. The turbine blade twist was also neglected, whilst the nacelle dimensions were chosen such that the nacelle mass was scaled appropriately.

Table 2. 'As constructed' parameters of the scaled NREL 5 MW turbine.

Parameter	Planned		Tested	Reference for Planned
	Full-Scale	Model-Scale	Model-Scale	Full-Scale Values
Tower height	90.00 m	803.00 mm	800.00 mm	
Tower bottom diameter	6.00 m	53.50 mm	53.30 mm	
Tower top diameter	3.87 m	34.50 mm	34.50 mm	
Hub height	90.00 m	803.00 mm	800.00 mm	
Rotor hub diameter	3.00 m	26.70 mm	35.00 mm	
Rotor diameter	126.00 m	1.12 m	1.13 m	Matha [9]
Overhang	5.00 m	44.60 mm	45.00 mm	
Blade length	61.50 m	549.00 mm	550.00 mm	
Tower mass	3.47×10^5 kg	247.00 g	246.00 g	
Nacelle mass	2.40×10^5 kg	171.00 g	170.00 g	
Rotor mass	1.10×10^3 kg	78.20 g	160.00 g	

2.3. TLPWT Model Construction

The constructed TLPWT model is shown in Figure 1A. The materials that were used for each component of the model are also shown in Table 3, with their respective weights. To ensure the model was water-tight for the testing, silicon sealant and waterproof tape were applied to the model. These have not been explicitly stated but have been accounted for in the total weight of the model.

Figure 2 shows the arrangement of the tendons and equipment within them. Each of the tendon lines consisted of 7-strand, coated steel wire, each with a spring (spring constant of 16.8 N/mm) and load cell attached to record the tendon tension data. Two different types of load cells (LCs) were used during the experiment, with the forward and aft pontoons using Futek LCs, whilst the port and starboard pontoons consisted of X-Trans LCs. The induced tension on the forward and aft pontoons was increased to account for the variance in weight between the two LC models. To achieve the required draft, a tension load of 1.28 kg was induced into each line, then threaded through eyelets on the base-plate sitting on the floor of the basin.

(A)　　　　　　　　　　　　　　　　(B)

Figure 1. Constructed TLPWT model (**A**), experimental setup (**B**).

Table 3. Summary of component materials and masses.

Model Component	Material	Mass (g)
Hull cylinder	Polyvinyl Chloride (PVC)	1890
External pontoons	Timber	72.00 (each)
Bottom and top caps	Polyvinyl Chloride (PVC)	173.00 (each)
Ballast	Lead	1866.00
Turbine	PLA plastic	576.00
Qualisys probes	-	50.00
Tendon lines	7-Strand, coated steel wire	-

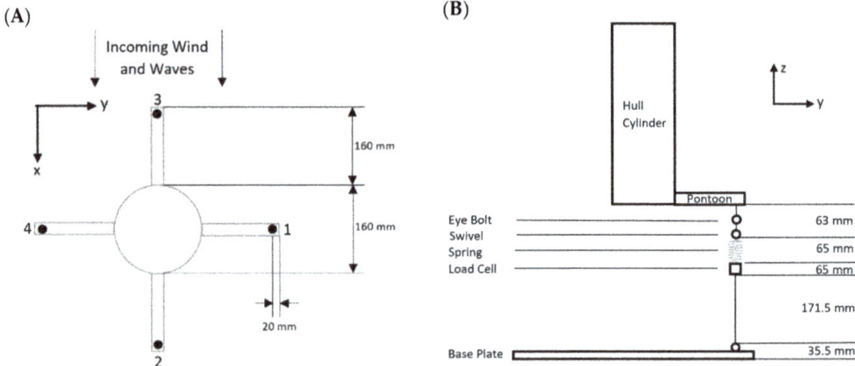

Figure 2. Plan view of external pontoon configuration (**A**) and profile view of tendon arrangement (**B**).

Before the experiment was conducted, the vertical centre of gravity was verified experimentally. The TLP model and turbine topside were placed on a metal bar to measure the point of equilibrium, with the distance from the keel of the model to the point of equilibrium forming the *KG*. A computerised model of the physical TLPWT model was developed to determine the mass moments of inertia for the model, and the obtained results are presented in Table 4.

Table 4. Summary of mass and inertia parameters of the TLPWT model.

Parameter	Value	Unit
Vertical Centre of Gravity (KG)	0.29	m
Vertical Centre of Buoyancy (KB)	0.24	m
Mass Moment of Inertia about x-axis, I_{xx}	9.71×10^5	kg·mm^2
Mass Moment of Inertia about y-axis, I_{yy}	9.79×10^5	kg·mm^2
Mass Moment of Inertia about z-axis, I_{zz}	3.26×10^4	kg·mm^2

2.4. Experimental Setup and Test Matrix

The model testing was carried out in the Model Test Basin (MTB) at the Australian Maritime College (AMC). The MTB is 35 m long × 12 m wide (Figure 3), with a water depth of 0.9 m being consistent across all tests. This allowed for the best possible scaling of the model parameters and tendons considering the facility limitations. The model was placed at a distance of 5.8 m from the wave-maker such that the fan and motion capture system could be appropriately placed. Wave tank model tests may exhibit large experimental scatter due to wall interaction [19]. As such the centreline of the model's column was positioned at approximately 3 m from the side of the basin which yields 18.75 times the column diameter. The fans were placed 2 m in front of the model to obtain the best possible air-flow, however this was not optimised for this experiment. While the TLPWT was in the basin, two wave probes (WP) placed along the side of the tank approximately 0.6 m from the wall, 2.8 m (WP1) and 5.8 m (WP2) away from the wavemaker, and 0.6 m off the basin wall were used to measure the wave elevations along the basin. As the wave height depends on the location of the wave probes, WP1 was employed for the wavemaker calibration whereas WP2 (in-line with the model) was employed to derive response amplitude operators. It is worth mentioning that the aspect of wave evolution along the physical wave tank is beyond the scope of this study. Several studies have recently been conducted on experimental waves generated in the model test basin of AMC and the quality of such waves along the basin has been documented in [13,38,39]. A digital *Qualisys* motion tracking system was used to measure all model motion response in six degrees of freedom [40]. The system consists of 8 cameras located around the test basin which provide the reference coordinates of the *Qualisys* markers placed on the model's tower by picking up reflections of the markers from an infra-red signal. The coordinates are plotted in relation to the model's VCG and accurately capture all motions during wave testing. A sampling frequency of 200 Hz was used in the data acquisition system.

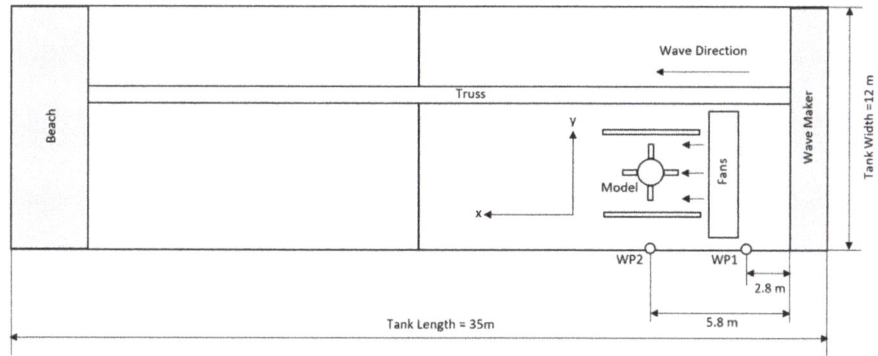

Figure 3. Plan view of TLPWT experimental setup (not to scale).

The basis for the testing program was derived from the Bachynski and Moan [14] environmental condition 3 (EC3); representative of operational wind and wave conditions, with a significant wave height of 4.4 m, peak period of 10.6 s, and mean wind speed of 18.0 m/s at the turbine hub (Table 5). According to Bachynski and Moan [14], the EC3 represents an above-rated condition where the generator is operational. At model scale, this translated to a wave height of 0.039 m, peak period of 1 s and wind speed of 1.70 m/s. The test matrix for the model testing involved a series of regular wave runs, both with varying wave height and frequency conditions, to be repeated for conditions with and without wind present. The use of regular waves in model testing has been common in practice, as it provides a practical starting point towards more complex conditions such as testing in irregular waves. As seen in Table 6, test conditions 1–10 of the experiment involved changing the wave period (for T/T_p from 0.777 to 1.23) for 'wind' and 'no-wind' conditions, whilst conditions 11–20 analysed increasing wave height at the peak period of the EC3 wave spectrum (for H/H_s form 0.0.641 to 1.795). The corresponding values of wave length (L) estimated based on the dispersion relationship [36] yields a d/L range of 0.39 to 0.95 (i.e., intermediate to deep water conditions) and a D/L range of 0.07 to 0.17. As the D/L ratio (column diameter to wave length ratio) is below 0.2 (i.e., the limit of small structures), wave diffraction and reflection effects due to the model presence can be neglected [36,41]. As such the model was placed at a closer distance (5.8 m) from the wave-maker which was controlled by the wind quality that could be produced at the MTB.

Table 5. Environmental condition to be tested.

Parameter	Full-Scale	Model-Scale
Significant wave height, H_s (m)	4.40	0.039
Peak wave period, T_p (s)	10.60	1.00
Mean wind speed at hub, U (m/s)	18.00	1.70

Table 6. Experimental test matrix for regular waves.

Condition	*H* (m)	*T* (s)	Condition	*H* (m)	*H*/*H*$_s$ (-)	*T* (s)
1		1.230	11	0.025	0.641	
2		1.147	12	0.030	0.769	
3		1.115	13	0.035	0.897	
4		1.078	14	0.040	1.026	
5	0.039	1.043	15	0.045	1.154	1.000
6		1.000	16	0.050	1.282	
7		0.963	17	0.055	1.410	
8		0.935	18	0.060	1.539	
9		0.905	19	0.065	1.667	
10		0.777	20	0.070	1.795	

3. Model Calibrations

3.1. Wave Calibration

The wave probes used in the experiment were calibrated on daily basis by positioning them at identified heights in a still water condition and fitting a linear relationship to the corresponding measured voltage. Without the TLPWT being in the basin, the change in wave height across the basin was investigated using two wave probes, WP2 (0.6 m from the basin side wall) and the other one at the model's virtual location (3.0 m from the basin side wall). As seen in Figure 4, the wave height measured across the tank at these two points was quite consistent; the deviation in data started at $H > 60$ mm. This reveals that the effects of the tank walls have negligible implications on the estimation of RAOs which were obtained using $H \sim 40$ mm (refer to Table 6). In comparison with wave theory, Figure 5 shows an example of a time-history of the waves for condition 5 as recorded by WP2 (0.6 m from the basin side wall). The Power Spectral Density (PSD) graph was generated from this time-series which shows a peak wave frequency of 0.959 Hz ($T = 1.043$ s). Upon comparing the obtained wave elevation to the Airy and Stokes 2nd order wave theories, as expected the measured waves exhibited more and less as Stokes 2nd order, with slight differences in the magnitudes of the peaks and troughs as shown in Figure 5C.

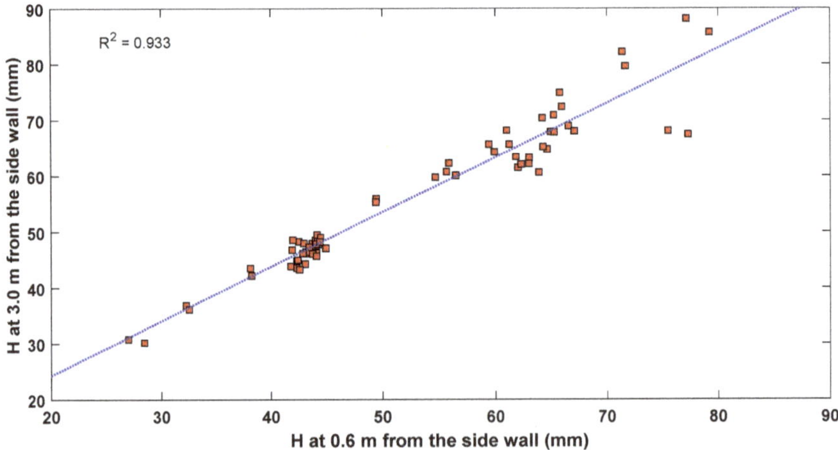

Figure 4. Wave height data measured at 5.8 m away from the wavemaker and at different distances from the basin side wall.

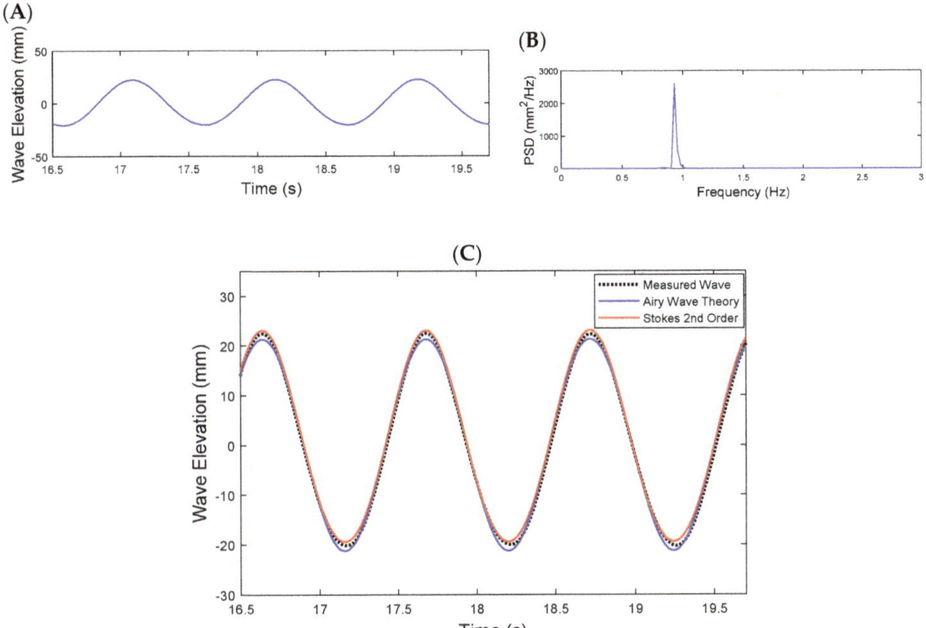

Figure 5. Measured wave elevation time history at WP2 (**A**), PSD for wave elevation time history (**B**) and comparisons with wave theory (**C**).

3.2. Wind Calibration

The wind generation was performed by using an array of fans placed 2 m in front of the model. For this experiment, a single wind speed was tested to understand the differences in model motion compared to a 'no-wind' condition. As already mentioned, the tested wind speed was based on the environmental condition 3 (EC3) from Bachynski and Moan [14] as a representative operational wind condition. The wind speed was scaled in line with the model scaling, resulting in an input wind speed of 1.70 m/s. To measure the mean wind speed, a hand-held anemometer was used to assist in calibrating the fans to achieve the best possible representation of the scaled wind speed. Figure 6 shows the different recording locations for wind speed, whilst Figure 7 shows the recorded wind speeds at each location, after a three-minute period of monitoring. It should be noted that the anemometer used for the testing did not have time-series data collection capabilities, thus the recorded range is based on visual observations of the wind speed reading. From Figure 7 the calibrated wind speed was closest to the target wind speed (U) of 1.70 m/s at the recording points located near the turbine hub (locations #4–9). Further away from the hub, the wind speed was not as accurate, however this had less bearing on the results due to less impact on the structure.

3.3. Free Decay Tests

Decay tests were performed on the TLPWT model in the heave, pitch, surge, and yaw directions to estimate the natural periods of the structure. The surge and yaw directions (Figure 8A,B) were obtained from the motion capture, whilst the heave and pitch natural periods (Figure 8C,D) were found using the dynamic tension in the tendon load cells. A single impact load on the model allowed for the most accurate measurement in each direction. The logarithmic decrement method was used for the most consistent signal for each test, checked against the spectral method using a Fast Fourier Transform (FFT). From these tests, it was found that the heave and pitch natural periods of 0.256 s and 0.260 s were very similar. The full-scale surge and pitch natural periods were consistent with the results of Oguz et al. [28] despite the difference in the tested water depths. The full-scale values obtained for surge, heave and pitch were all outside the typically experienced wave periods of 6–20 s [19], whilst yaw fell inside this range. The evaluation for yaw is significant in the selection process for TLPWTs. For full-scale turbines, yaw control of the nacelle is typically used to orient the turbine towards the predominant wind direction, which could lead to an increase in yaw motion if excited by wave motion.

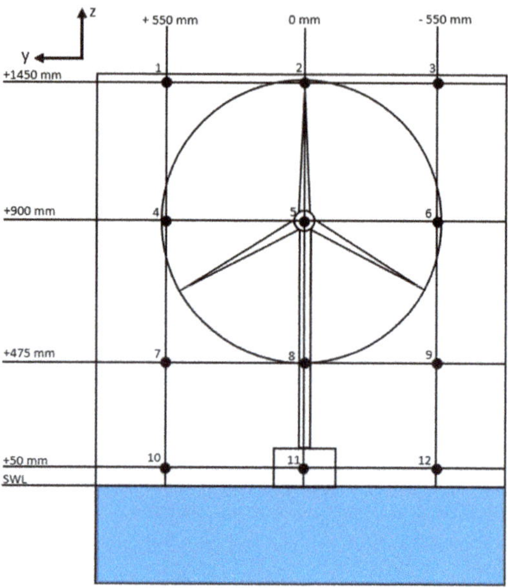

Figure 6. Anemometer recording locations for wind speed calibration.

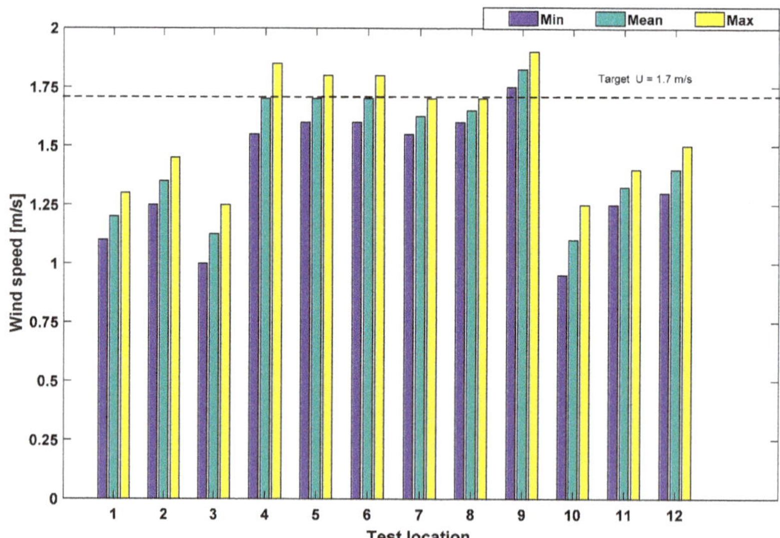

Figure 7. Recorded wind speeds for each test location.

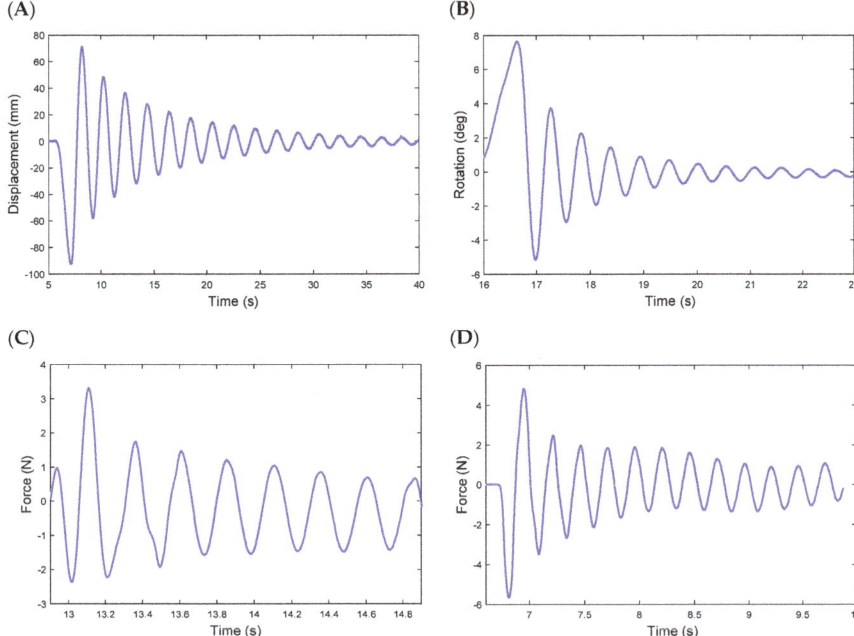

Figure 8. Free decay test results for surge motion (**A**), yaw motion (**B**), heave motion (**C**) and pitch motion (**D**).

The natural periods evaluated from the decay tests have a relatively close agreement to the theoretical values using equations from DNV-RP-C205 [36] and Naess and Moan [42]. It was found that the heave and yaw errors were the most significant at 20.2% and 15.2% exactly, whilst surge and pitch were more accurate with 1.68% and 6.99% error, respectively (Table 7).

Table 7. Comparison between theoretical and obtained natural periods, with full-scale equivalent values.

Degree of Freedom	Predicted T_n (s)	Measured T_n (s)	Full-Scale Equivalent T_n (s)
Surge	2.113	2.078	21.99
Heave	0.213	0.256	2.71
Pitch	0.243	0.260	2.75
Yaw	0.675	0.586	6.20

3.4. Data Analysis

The time series of the data was collected with a sample frequency of 200 Hz for a collection period of 40 s. Due to the disturbance caused by the start-up condition of the wavemaker (the initial transient periods in Figure 9) and reflected waves travelling back up the tank, a steady state period of fully developed waves was selected for analysis in each run. The steady state period to be analysed was first determined by examining and trimming the phase wave probe data (WP2), and subsequently trimming the motion data relative to the trimmed WP2 data. An example of this process in presented in Figure 9 where the vertical lines represent the range of data to be trimmed.

According to DNV-RP-C205 [36], the repeatability analysis of tank measurements should be documented. A repeatability analysis was performed across all data recording platforms to understand the variability of results from run to run. The generated wave data, load cell data, and motion response data for condition 5 were compared for five repeated runs with identical input wave parameters with and without wind. For the motion response data, the surge motion (denoted by X) was analysed, as it experienced the most significant motions of any direction. Figure 10A,C,E show the time-series data for wave elevation, dynamic tendon tension (i.e., the pretension was subtracted from the total tension), and surge motion for the wave only condition, whilst Figure 10B,D,F show the runs inclusive of wind.

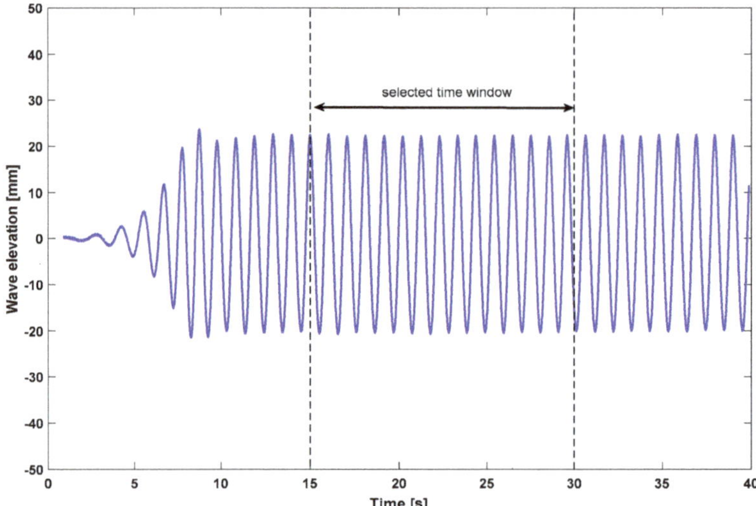

Figure 9. Time history of wave elevation at WP2 and the selected time window for data analysis.

By referring to the details of data variances shown in Tables 8 and 9, it can be appreciated that the wave elevation and surge motion variation between each run was within expected limits, with a maximum coefficient of variance (CV) and a maximum Normalised Root Mean Square Error (NRMSE) of 0.99 and 0.91%, respectively. Furthermore, a cross-correlation R^2 was obtained for the collected time series data in which the minimum R^2 was 0.9745 for the tendon tensions of LC3. The load cell data encountered slightly more noise, particularly for the wind assisted conditions with a maximum CV of 5.23% (4.97% NRMSE) for the minimum tension values. The minimum tension demonstrated a large variability, which could be caused by the dynamic response in the tendons [43]. Overall, the maximum and minimum tensions were consistent between wave peaks and runs, resulting in usable data for analysis. These findings were also supported by a previous work [12,38,43,44] conducted at a smaller scale (1:125) which has indicated that good qualitative repeatability can be achieved among multiple repeated runs for all wave probes, *Qualisys* system and load cells used in their model tests such that lower values of CV were obtained during model calibrations.

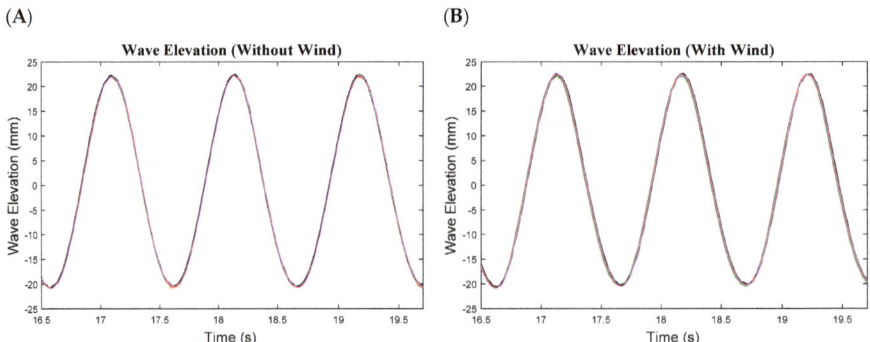

Figure 10. *Cont.*

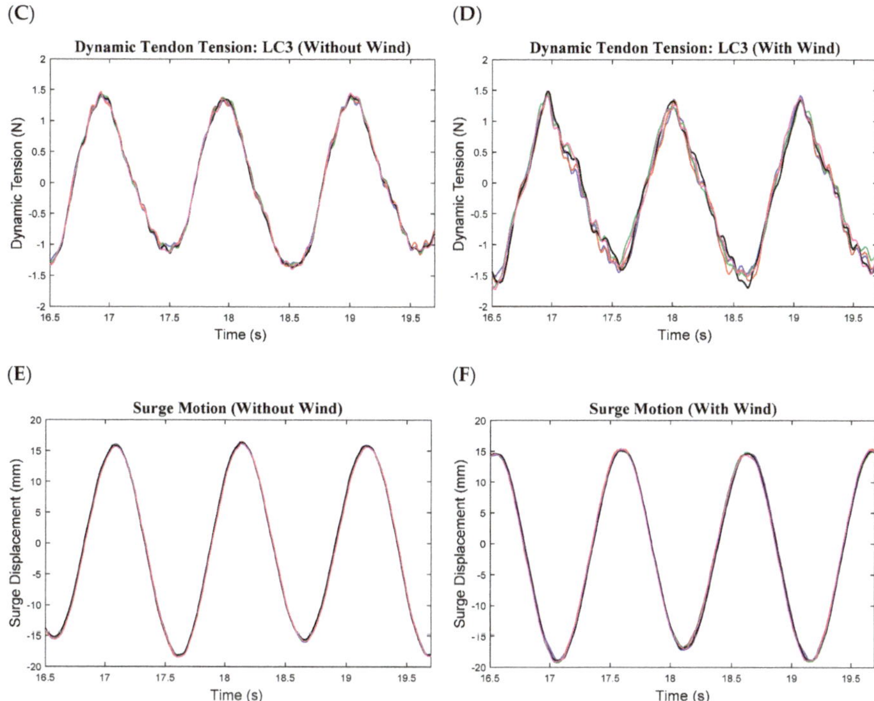

Figure 10. Time history of wave elevations, dynamic tensions and surge motions for five repeated runs.

Table 8. Results of repeatability analysis for wave only runs.

Run	WP2 (mm)		LC3 (N)		*Qualisys, X* (mm)	
	Max	**Min**	**Max**	**Min**	**Max**	**Min**
Run 1	22.24	−20.73	1.41	−1.33	16.14	−18.19
Run 2	22.26	−20.77	1.44	−1.36	16.46	−18.23
Run 3	22.47	−20.75	1.42	−1.33	16.34	−18.23
Run 4	22.50	−20.75	1.42	−1.37	16.39	−18.14
Run 5	22.36	−20.86	1.44	−1.38	16.18	−18.48
Mean, m	22.37	−20.77	1.43	−1.35	16.30	−18.25
Standard deviation, σ	0.12	0.05	0.01	0.02	0.14	0.13
CV (%)	0.54	0.24	0.70	1.48	0.86	0.71
NRMSE (%)	0.47	0.22	0.88	1.55	0.75	0.65
R^2 (-)	0.9926		0.9745		0.9978	

Table 9. Results of repeatability analysis for wave with wind runs.

Run	WP2 (mm)		LC3 (N)		*Qualisys, X* (mm)	
	Max	**Min**	**Max**	**Min**	**Max**	**Min**
Run 1	22.61	−20.64	1.36	−1.46	15.38	−18.92
Run 2	22.52	−20.55	1.39	−1.56	15.29	−19.29
Run 3	22.33	−20.75	1.38	−1.48	15.33	−19.29
Run 4	22.63	−20.57	1.43	−1.66	15.11	−18.90
Run 5	22.48	−20.50	1.36	−1.49	15.33	−19.18
Mean, m	22.55	−20.60	1.38	−1.53	15.28	−19.12
Standard deviation, σ	0.12	0.10	0.03	0.08	0.10	0.19
CV (%)	0.53	0.49	2.17	5.23	0.65	0.99
NRMSE (%)	0.50	0.42	1.89	4.79	0.61	0.91
R^2 (-)	0.9997		0.9993		0.9994	

4. Results and Discussion

4.1. Time Series of Motion and Tendon Responses

Figure 11 shows the steady state motions of the model observed in condition 3 (no wind action). Overall, the tested TLPWT model exhibited typical motion responses to that of a generalised TLP with significant surge offsets along with stiff heave and pitch motions [41,43]. Moreover, there was an evidence of a potential coupling between multiple DOFs, particularly for heave and surge motions. This could be attributed to the relationship of offset and set-down for TLPs, based on their tethered nature [45]. Such a relationship will be discussed in detail thereafter in Section 4.3. There are larger surge motions in the positive direction as the model was constantly subjected to waves, which did not allow for even amplitude in both directions. This observation was also noticeable for the heave motion, with the magnitude of the downward motion proving to be more significant compared to the upward motion of the model. The pitch angles experienced were consistent, with the amplitude in the positive direction only slightly greater than that in the negative direction. Such observations in motion nonlinearities were anticipated as the nature of the physical wave was nonlinear as well. As expected, the magnitude of sway, roll and yaw motions was negligible in this case, as tests were only performed in head seas [35].

Figure 12 shows the time-series data of the dynamic tendon tension from each of the load cells using an example from condition 3 (no wind action). These graphs clearly show the excitation in the tendons from the model motions, with the forward (up-wave) tendon (LC3) experiencing the highest magnitudes for dynamic tension. The port and starboard tendons (LC1 and LC4) experienced almost similar magnitudes of tension; the variation among them can be attributed to the yaw motion of the model (Figure 11D). The aft (down-wave) tendon (LC2) experienced higher tension when compared to the port and starboard tendons, which showed relatively consistent tension fluctuations.

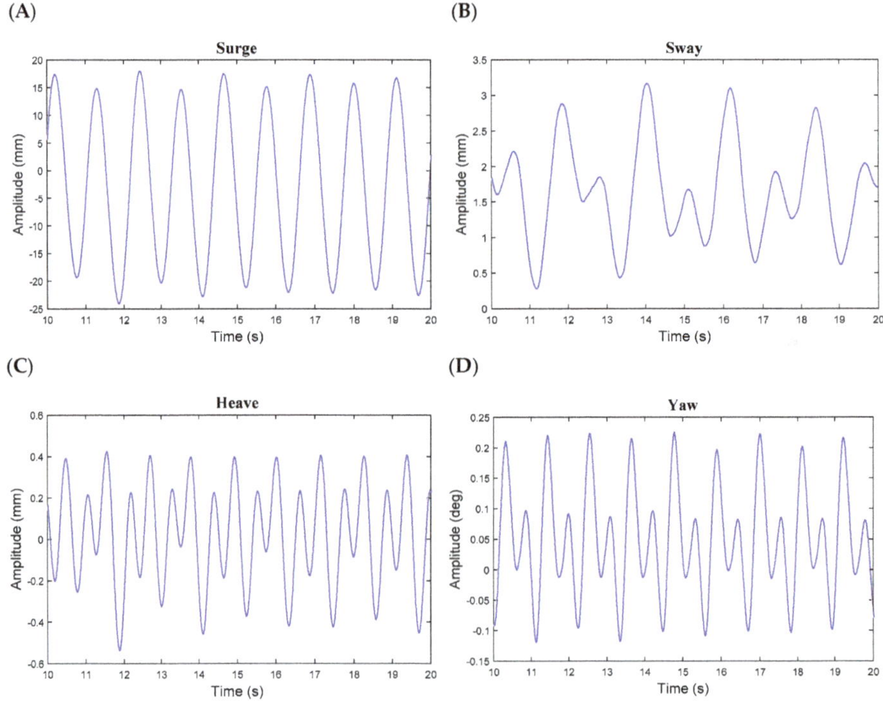

Figure 11. *Cont.*

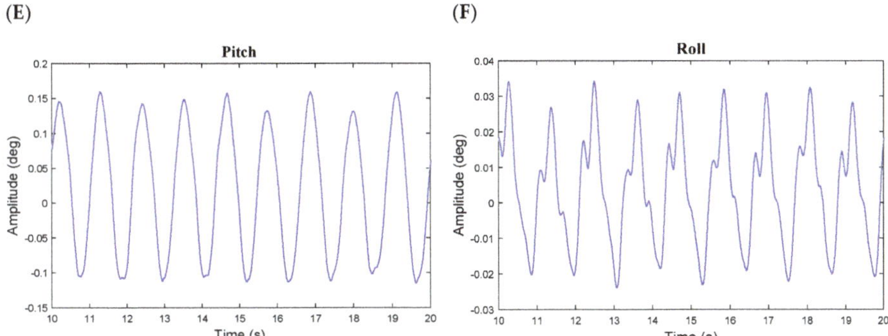

Figure 11. Time history of model motions recorded for condition 3: surge motion (**A**), sway motion (**B**), heave motion (**C**), yaw motion (**D**), pitch motion (**E**) and roll motion (**F**).

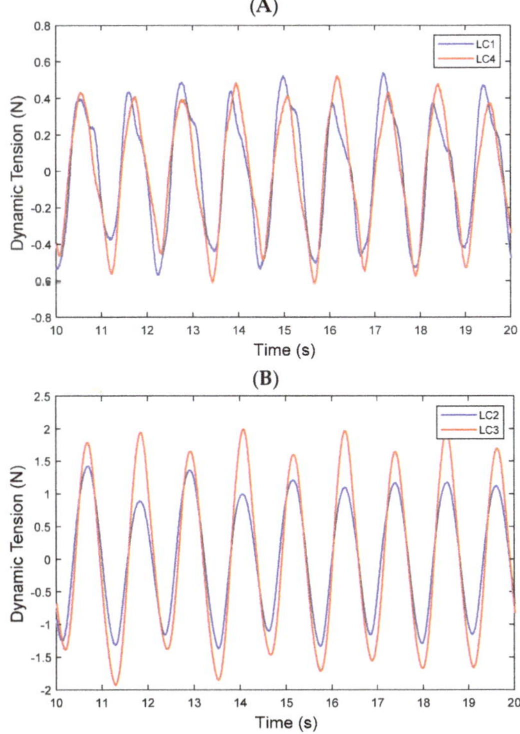

Figure 12. Time history of dynamic tendon tension measured by load cells for condition 3: LC1 and LC4 (**A**) and LC2 and LC3 (**B**).

4.2. Response Amplitude Operators (RAOs)

Wave frequency Response Amplitude Operators (RAOs) of motion and tendon responses are discussed in this section. Figure 13 A–C show the translational RAOs for surge, sway and heave, respectively, whilst Figure 13 D–F show the rotational RAOs for roll, pitch, and yaw. FFT analysis was used for each run to find the RAOs at each wave frequency. As already mentioned, the wave and wind conditions tested for the analysis of the RAOs are based on EC3 [14] to represent an operational case. For the key motions of surge, heave, and pitch, there were some clear trends apparent from the 'wave only' and wind assisted runs. By studying the effect of wind, the magnitude of heave motion was found to increase due to wind forcing with an average increase of 13.1% for the tested conditions. The effects of natural periods on the motion amplitudes cannot be identified as the results obtained in Table 7 fall out of the tested range of 0.777–1.230. Furthermore, it should be stressed that it is unlikely that the maximum motions of the model were captured during these tests conducted in this study. A clearer picture would likely be obtained with further testing in a survivability sea state ($H_{max} \sim 1.86\ H_s$ corresponding to the design return period) [36]. Analysing similar tests performed by Zamora–Rodriguez et al. [34], trends in surge results were basically comparable, with only minimal variation between each wind condition. An increase in heave motions with wind was also found to be similar, albeit to a different magnitude. The variations experienced in pitch were minimal, however were in a similar range when compared to Zamora–Rodriguez et al. [34].

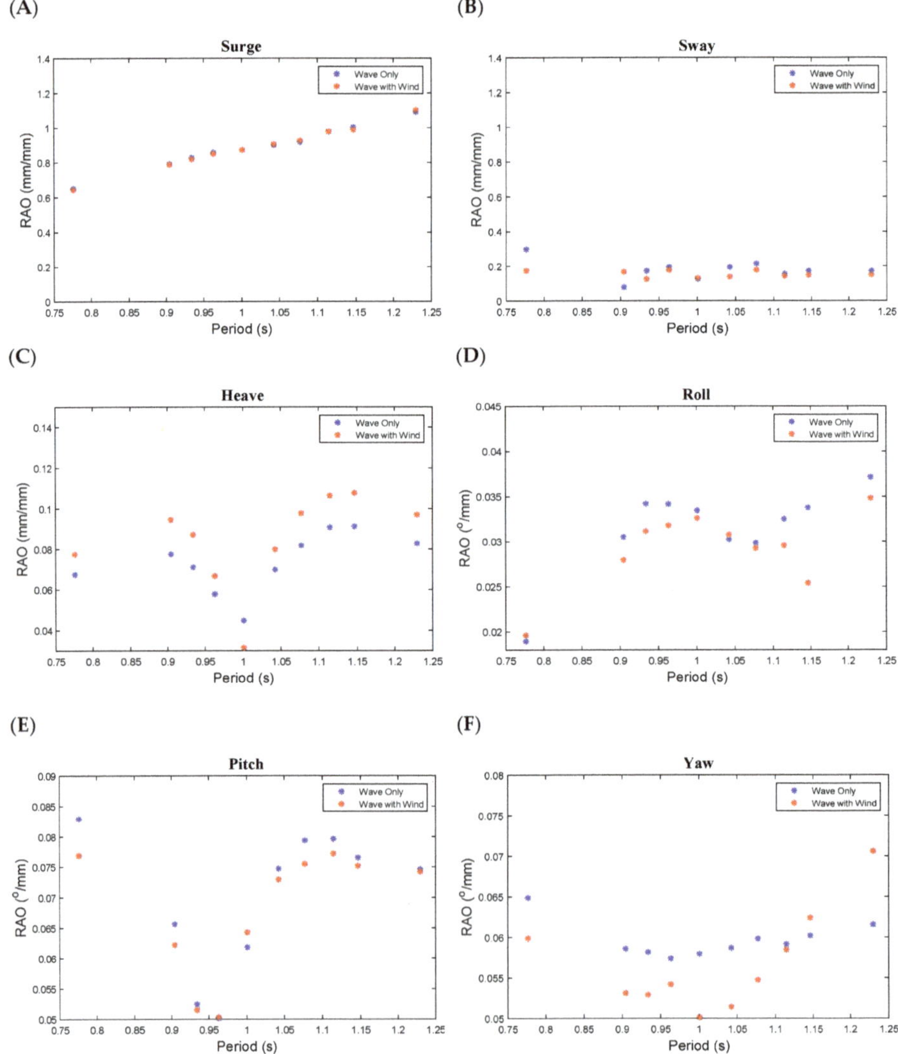

Figure 13. Motion response amplitude operators for each degree of freedom: Surge RAO (**A**), sway RAO (**B**), heave RAO (**C**), roll RAO (**D**), pitch RAO (**E**) and yaw RAO (**F**).

The tendon RAOs were also evaluated for the same wind and wave conditions as the motion RAOs (Figure 14). Each LC was analysed for the maximum dynamic tension for each of the different wave periods, with clear trends being observed. Overall, it is noted that the dynamic tension in all tendons was lower for most of the wind assisted runs, whilst the largest dynamic tensions were experienced with the lower frequency waves. As expected, the RAOs of the port and starboard tendons (LC1 and LC4) were quite similar due to symmetry. Likewise, the maximum dynamic tensions of the model's tendons were captured during these tests for a mild to moderate sea state. A testing in a survivability sea state test would likely provide such information. It should be noted that the RAOs of the surge motion and all tendon tensions follow a similar trend with a larger magnitude response to larger wave periods. However, these responses are also a function of tendon length and stiffness which should be considered when optimising and designing a mooring arrangement. The maximum magnitudes for these RAOs occurred at the longest wave period of 1.23 s tested (~ 13.0 s at full-scale) with Table 10 outlining the values. As can be seen the wind had a negligible effect on the surge motion and slightly decreased the tendon tensions in all tendons except the forward/up-wave tendon (LC3). These findings were consistent with the results of Nihei and Fujioka [31] who stated that the wind effect decreases the dynamic response of the tendons in waves and wind coexisting field.

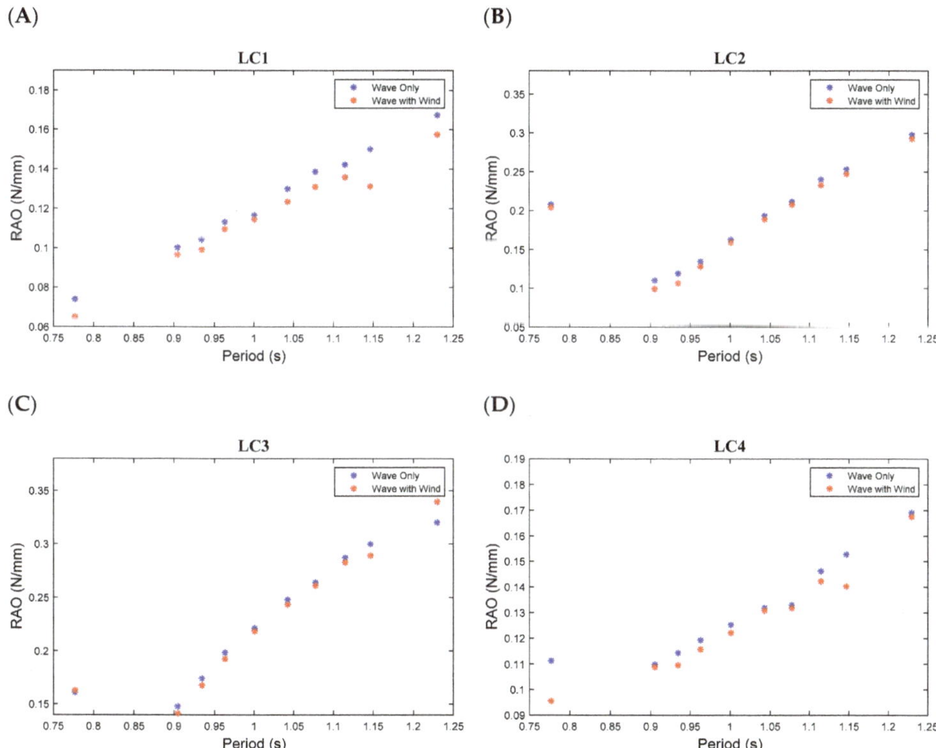

Figure 14. Dynamic tendon tension response amplitude operators for each tendon: LC1 RAO (**A**), LC2 RAO (**B**), LC3 RAO (**C**), LC4 RAO (**D**).

Table 10. Maximum RAO values of the TLPWT model at $T = 1.23$ s.

RAO Parameter	Max RAO (Wave Only)	Max RAO (Wave and Wind)	Wind Effect
Surge motion	1.091 mm/mm	1.102 mm/mm	1.0%
LC1	0.167 N/mm	0.158 N/mm	−6.0%
LC2	0.298 N/mm	0.292 N/mm	−2.0%
LC3	0.320 N/mm	0.340 N/mm	6.0%
LC4	0.169 N/mm	0.168 N/mm	−1.0%

4.3. Model's Offset Versus Set-Down

An analysis of how offset and set-down correlate under changing wind and wave conditions was performed. For a TLP, the kinematic coupling between the horizontal surge/sway motions and the vertical heave motions results in the so-called platform set-down [46]. The set-down refers to the vertical downward movement of the hull when the platform moves in its compliant modes (surge, sway and yaw). The relationship between offset and set-down changes depending on the input wave frequency, whilst wind loading can also have an effect. In this case, a mathematical formula developed by Demirbilek [45] can be used to evaluate the set-down based on the offset experienced by the model and the length of the tendons. By referring to the definition sketch shown in Figure 15, this relationship is given in equation (1) where L_o is the initial tendon length and $X(t)$ is the surge motion time-series:

$$Z(t) = L_o - \sqrt{\left(L_o^2 - X(t)^2\right)} \tag{1}$$

This was done with the assumption of zero pitch rotational motion while the platform is moving in the x-direction and neglecting the elongation in the tendon [43]. The magnitude of the set-down fluctuated during this set of tests as the model was moving back and forth. Overall, the magnitude of the set-down was minimal, likely due to the high stiffness of the tendons as discussed in Section 3.3. Figure 16A shows the set-down motion in the following wave direction i.e., wave travelling direction being greater than that in the opposite direction. This is due to the impact of the waves not allowing the structure to move equally in the opposite direction. Furthermore, the effect of wind was more pronounced in in the following wave direction than in the in the opposite direction. Figure 16B illustrates the differences experienced with and without wind acting on the structure as a function of wave period. The set-down followed the same trend as the surge motion with low frequency waves (long waves) inducing the most offset and contributing to the most set-down experienced by the model. By comparing the magnitudes of the offset and set-down (Figures 13A and 16B), the ratio of Z/X was found to be 2–5%. Such findings can be useful in the calculation of tendon forces and in the calculation of responses in power take-off cables as well as the evaluation of the air gap of an FOWT system in accordance with the DNVGL-ST-0119 standard [46].

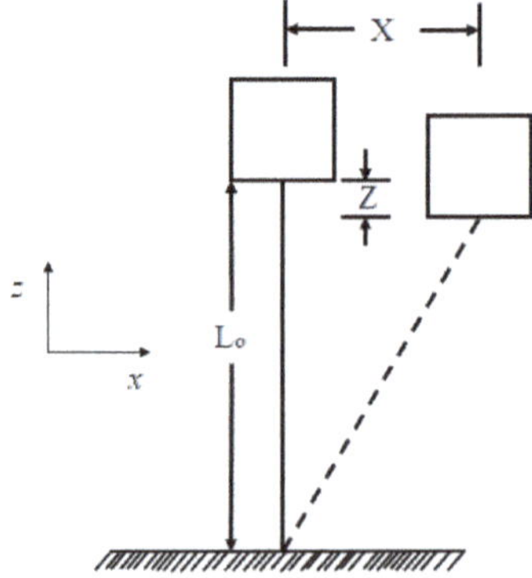

Figure 15. Definition sketch for offset and set-down relationship.

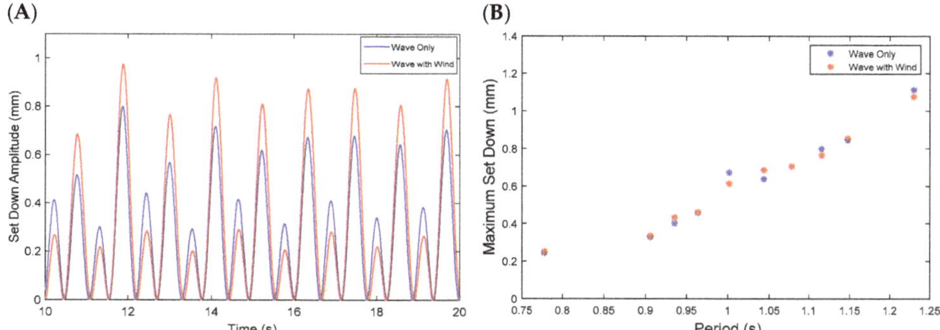

Figure 16. Comparison of set-down time history for condition 3 (**A**) and maximum set-down experienced for all conditions (**B**).

5. Conclusions and Recommendations

This paper describes model testing of a TLPWT model with non-rotating blades to better understand its motion and tendon responses when subjected to combined wind and unidirectional regular wave conditions. The turbine structure of the TLPWT model was closely based on the NREL 5 MW concept. The analysis of the measurements and observations of the model response enabled several general conclusions to be drawn as follows:

1. Several free decay tests were performed to evaluate the natural periods of the model in the key degrees of freedom including surge, heave, pitch, and yaw. The natural periods in the surge and pitch motions evaluated from the decay tests had a relatively close agreement to the theoretical values. Furthermore, the natural periods in the surge, heave, and pitch degrees of freedom were all outside the range of 6–20 s, whilst the yaw natural period fell inside this range. As the yaw control of the nacelle is typically used to orient the turbine towards the predominant wind direction, such a situation could lead to an increase in the yaw motion if excited by waves. Therefore, it is recommended to investigate the effect of yaw motion on the performance of a TLPWT.

2. The tested TLPWT model showed typical motion responses to that of generalised TLP systems with significant surge offsets along with stiff heave and pitch motions. The maximum magnitudes for the RAOs of surge motion and all tendons occurred at the longest wave period of 1.23 s (~13.0 s at full-scale) tested in this study.

3. There was evidence that static wind loading on the turbine structure had some impact on the motions and tendon response, particularly in the heave direction, with an average increase of 13.1% in motion magnitude for the tested wind conditions. The wind had a negligible effect on the surge motion and slightly decreased the tendon tensions in all tendons.

4. The results also showed the set-down magnitudes amounting to approximately 2–5% of the offset. Furthermore, the waves are the dominant factor contributing to the set-down of the TLPWT, with a minimal contribution from the static wind forcing.

5. As the environmental condition tested in this study is considered a mild to moderate sea state, it should be stressed that it is unlikely that the maximum motions and loads of the model were captured during these tests. A testing in a survivability sea state condition would likely provide such information. It is therefore recommended that further testing into the survivability of the TLPWT should be performed. Furthermore, the use of a drag plate instead of the static rotor tested in this study can be investigated in future studies, as the thrust distribution would be more uniform. The results of this study could be used for calibrating numerical tools such as CFD codes which can then be used for further investigations.

J. Mar. Sci. Eng. **2019**, *7*, 56

Author Contributions: Formal analysis, T.M.; investigation, T.M. and N.A.; supervision, N.A.; writing—original draft, T.M.; writing—review & editing, N.A.

Funding: This research received no external funding.

Acknowledgments: The authors would like to thank William Chia for his preliminary work conducted on the topics covered in this study along with thanks to Tim Lilienthal, Christopher Dunn, Uddish Singh, Darren Lee, and Thiban Subramaniam for assistance during the experimental setup, and Jock Ferguson for assistance in 3D printing the turbine model.

Conflicts of Interest: The authors declare no conflict of interest.

Abbreviations and Notations

AMC	Australian Maritime College
CFD	Computational Fluid Dynamics
CV	Coefficient of Variation (%)
D	Column diameter (m)
d	Water depth (m)
EC	Environmental Condition
FAST	Fatigue, Aerodynamics, Structures, and Turbulence
FFT	Fast Fourier Transform
FOWT	Floating Offshore Wind Turbine
H	Wave height (m)
H_{max}	Maximum wave height (m)
H_s	Significant wave height (m)
I_{xx}	Mass Moment of Inertia about x-axis (kg·mm^2)
I_{yy}	Mass Moment of Inertia about y-axis (kg·mm^2)
KB	Vertical distance measured from the model's keel to the centre of buoyancy (m)
KG	Vertical distance measured from the model's keel to the VCG (m)
L	Wave length (m)
LC	Load Cell
L_o	Original tendon length (m)
m	Mean value (vary)
MIT	Massachusetts Institute of Technology
MTB	Model Test Basin
MW	Mega Watt
NREL	National Renewable Energy Laboratory
NRMSE	Normalised Root Mean Square Error
PSD	Power Spectral Density (m^2/Hz)
R^2	Correlation coefficient (-)
RAO	Response Amplitude Operator
t	Time (s)
T	Wave period (s)
TLP	Tension Leg Platform
T_n	Natural period (s)
T_p	Peak period (s)
U	Mean wind speed (m/s)
VCG	Vertical Centre of Gravity
WP	Wave Probe
X	Horizontal offset (m)
Z	Set-down (m)
σ	Standard deviation (vary)

References

1. Breton, S.-P.; Moe, G. Status, plans and technologies for offshore wind turbines in Europe and North America. *Renew. Energy* **2009**, *34*, 646–654. [CrossRef]

2. IRENA. *Renewable Capacity Statistics 2018*; The International Renewable Energy Agency (IRENA): Abu Dhabi, UAE, 2018.

3. Global Wind Energy Council (GWEC). Global Wind Report: Annual Market Update 2017. Available online: https://www.tuulivoimayhdistys.fi/filebank/1191-GWEC_Global_Wind_Report_April_2018.pdf (accessed on 31 December 2018).

4. Wang, X.; Zeng, X.; Li, J.; Yang, X.; Wang, H. A review on recent advancements of substructures for offshore wind turbines. *Energy Convers. Manag.* **2018**, *158*, 103–119. [CrossRef]

5. Arshad, M.; O'Kelly, B.C. Offshore wind-turbine structures: A review. *Proc. Inst. Civ. Eng. Energy* **2013**, *166*, 139–152. [CrossRef]

6. Doherty, P.; Gavin, K. Laterally loaded monopile design for offshore wind farms. *Proc. ICE-Energy* **2011**, *165*, 7–17. [CrossRef]

7. Sclavounos, P.; Tracy, C.; Lee, S. Floating offshore wind turbines: Responses in a seastate pareto optimal designs and economic assessment. In Proceedings of the ASME 2008 27th International Conference on Offshore Mechanics and Arctic Engineering, Estoril, Portugal, 15–20 June 2008; American Society of Mechanical Engineers: New York, NY, USA, 2008.

8. Wang, C.; Utsunomiya, T.; Wee, S.; Choo, Y. Research on floating wind turbines: A literature survey. *IES J. Part A Civ. Struct. Eng.* **2010**, *3*, 267–277. [CrossRef]

9. Matha, D. *Model Development and Loads Analysis of an Offshore Wind Turbine on a Tension Leg Platform with a Comparison to Other Floating Turbine Concepts: April 2009*; National Renewable Energy Lab. (NREL): Golden, CO, USA, 2009.

10. Koo, B.J.; Goupee, A.J.; Kimball, R.W.; Lambrakos, K.F. Model tests for a floating wind turbine on three different floaters. *J. Offshore Mech. Arct. Eng.* **2014**, *136*, 020907. [CrossRef]

11. Skaare, B.; Hanson, T.D.; Nielsen, F.G.; Yttervik, R.; Hansen, A.M.; Thomsen, K.; Larsen, T.J. Integrated Dynamic Analysis of Floating Offshore Wind Turbines. In Proceedings of the 2007 European Wind Energy Conference and Exhibition, Milan, Italy, 7–10 May 2007.

12. Abdussamie, N.; Drobyshevski, Y.; Ojeda, R.; Thomas, G.; Amin, W. Experimental investigation of wave-in-deck impact events on a TLP model. *Ocean Eng.* **2017**, *142*, 541–562. [CrossRef]

13. Hore, J.; Abdussamie, N.; Ojeda, R.; Penesis, I. Experimental investigation into the hydrodynamic performance of a TLP-OWC device. In Proceedings of the 28th International Ocean and Polar Engineering Conference, Sapporo, Japan, 10–15 June 2018.

14. Bachynski, E.E.; Moan, T. Design considerations for tension leg platform wind turbines. *Mar. Struct.* **2012**, *29*, 89–114. [CrossRef]

15. Sclavounos, P.; Lee, S.; DiPietro, J.; Potenza, G.; Caramuscio, P.; De Michele, G. Floating offshore wind turbines: tension leg platform and taught leg buoy concepts supporting 3-5 MW wind turbines. In Proceedings of the European Wind Energy Conference, Warsaw, Poland, 20–23 April 2010.

16. Xu, N.; Zhang, J. Pitch/Roll Static Stability of Tension Leg Platforms. In Proceedings of the ASME 29th International Conference on Ocean, Offshore and Arctic Engineering, Shanghai, China, 6–11 June 2010; American Society of Mechanical Engineers: New York, NY, USA, 2010.

17. Halkyard, J. *Floating Offshore Platform Design, in Handbook of Offshore Engineering*; Elsevier: Amsterdam, The Netherlands, 2005; pp. 419–661.

18. Nihei, Y.; Iijima, K.; Murai, M.; Ikoma, T. A comparative study of motion performance of four different FOWT designs in combined wind and wave loads. In Proceedings of the ASME 2014 33rd International Conference on Ocean, Offshore and Arctic Engineering, San Francisco, CA, USA, 8–13 June 2014; American Society of Mechanical Engineers: New York, NY, USA, 2014.

19. Chakrabarti, S.K. *Offshore Structure Modeling*; World Scientific: Singapore, 1994; Volume 9.

20. Low, Y. Frequency domain analysis of a tension leg platform with statistical linearization of the tendon restoring forces. *Mar. Struct.* **2009**, *22*, 480–503. [CrossRef]

21. Srinivasan, C.; Gaurav, G.; Serino, G.; Miranda, S. Ringing and springing response of triangular TLPs. *Int. Shipbuild. Prog.* **2011**, *58*, 141–163.

22. Matsui, T.; Sakoh, Y.; Nozu, T. Second-order sum-frequency oscillations of tension-leg platforms: Prediction and measurement. *Appl. Ocean Res.* **1993**, *15*, 107–118. [CrossRef]

23. Feng, W.H.; Hua, F.Y.; Yang, L. Dynamic analysis of a tension leg platform for offshore wind turbines. *J. Power Technol.* **2014**, *94*, 42–49.

24. Nihei, Y.; Matsuura, M.; Murai, M.; Iijima, K.; Ikoma, T. New Design Proposal for the TLP Type Offshore Wind Turbines. In Proceedings of the ASME 2013 32nd International Conference on Ocean, Offshore and Arctic Engineering, Nantes, France, 9–14 June 2013; The American Society of Mechanical Engineers (ASME): Nantes, France, 2013.

25. Jonkman, J.; Butterfield, S.; Musial, W.; Scott, G. *Definition of a 5-MW Reference Wind Turbine for Offshore System Development*; Technical Report No. NREL/TP-500-38060; National Renewable Energy Laboratory: Golden, CO, USA, 2009.

26. Kimball, R.; Goupee, A.J.; Fowler, M.J.; de Ridder, E.-J.; Helder, J. Wind/wave basin verification of a performance-matched scale-model wind turbine on a floating offshore wind turbine platform. In Proceedings of the ASME 2014 33rd International Conference on Ocean, Offshore and Arctic Engineering, San Francisco, CA, USA, 8–13 June 2014; American Society of Mechanical Engineers: New York, NY, USA, 2014.

27. Tracy, C.C.H. *Parametric Design of Floating Wind Turbines*; Massachusetts Institute of Technology: Cambridge, MA, USA, 2007.

28. Oguz, E.; Clelland, D.; Day, A.H.; Incecik, A.; López, J.A.; Sánchez, G.; Almeria, G.G. Experimental and numerical analysis of a TLP floating offshore wind turbine. *Ocean Eng.* **2018**, *147*, 591–605. [CrossRef]

29. Jonkman, J.M. *Dynamics Modeling and Loads Analysis of an Offshore Floating Wind Turbine*; Technical Report NREL/TP-500-41958; National Renewable Energy Laboratory: Golden, CO, USA, 2008.

30. Bachynski, E.E. Design and Dynamic Analysis of Tension Leg Platform Wind Turbines. Ph.D. Thesis, Norwegian University of Science and Technology, Trondheim, Norway, March 2014.

31. Nihei, Y.; Fujioka, H. Motion characteristics of TLP type offshore wind turbine in waves and wind. In Proceedings of the ASME 2010 29th International Conference on Ocean, Offshore and Arctic Engineering, Shanghai, China, 6–11 June 2010; American Society of Mechanical Engineers: New York, NY, USA, 2010.

32. Martin, H.R.; Kimball, R.W.; Viselli, A.M.; Goupee, A.J. Methodology for wind/wave basin testing of floating offshore wind turbines. *J. Offshore Mech. Arct. Eng.* **2014**, *136*, 020905. [CrossRef]

33. Zhao, Y.; Yang, J.; He, Y. Preliminary design of a multi-column TLP foundation for a 5-MW offshore wind turbine. *Energies* **2012**, *5*, 3874–3891. [CrossRef]

34. Zamora-Rodriguez, R.; Gomez-Alonso, P.; Amate-Lopez, J.; De-Diego-Martin, V.; Dinoi, P.; Simos, A.N.; Souto-Iglesias, A. Model scale analysis of a TLP floating offshore wind turbine. In Proceedings of the ASME 2014 33rd International Conference on Ocean, Offshore and Arctic Engineering, San Francisco, CA, USA, 8–13 June 2014; American Society of Mechanical Engineers: San Francisco, CA, USA, 2014.

35. Hansen, A.M.; Laugesen, R.; Bredmose, H.; Mikkelsen, R.; Psichogios, N. Small scale experimental study of the dynamic response of a tension leg platform wind turbine. *J. Renew. Sustain. Energy* **2014**, *6*, 053108. [CrossRef]

36. DNV. *Recommended Practice DNV-RP-C205: Environmental Conditions and Environmental Loads*; DNV: Oslo, Norway, 2010.

37. Chia, C.K.W. Experimental Investigation into Extreme Wave Impact on a TLP Offshore Wind Turbine. In *National Centre for Maritime Engineering & Hydrodynamics*; University of Tasmania: Launceston, Australia, 2017.

38. Banks, M.; Abdussamie, N. The response of a semisubmersible model under focused wave groups: Experimental investigation. *J. Ocean Eng. Sci.* **2017**, *2*, 161–171. [CrossRef]

39. Groves, B.; Abdussamie, N. Generation of rogue waves at model scale. *J. Ocean Eng. Sci.* **2019**, in press. [CrossRef]

40. Qualisys. Qualisys Motion Tracking System. 2018. Available online: https://www.qualisys.com/ (accessed on 31 December 2018).

41. Chakrabarti, S.K. *Hydrodynamics of Offshore Structures*; WIT Press: Southampton, UK, 1987.

42. Naess, A.; Moan, T. *Stochastic Dynamics of Marine Structures*; Cambridge University Press: Cambridge, UK, 2012.

43. Abdussamie, N.; Ojeda, R.; Drobyshevski, Y.; Thomas, G.; Amin, W. Dynamic behaviour of a TLP in waves: CFD versus model tests. In Proceedings of the 28th International Ocean and Polar Engineering Conference, Sapporo, Japan, 10–15 June 2018.

44. Abdussamie, N. Towards Reliable Prediction of Wave-in-Deck Loads and Response of Offshore Structures. Ph.D. Thesis, University of Tasmania, Launceston, Australia, June 2016.

45. Demirbilek, Z. Design formulae for offset, set down and tether loads of a tension leg platform (TLP). *Ocean Eng.* **1990**, *17*, 517–523. [CrossRef]

46. DNVGL. *Floating Wind Turbine Structures—Standard—DNVGL-ST-0119*; DNV.GL: Høvik, Norway, 2018.

MDPI

St. Alban-Anlage 66

4052 Basel

Switzerland

Tel. +41 61 683 77 34

Fax +41 61 302 89 18

www.mdpi.com

Journal of Marine Science and Engineering Editorial Office

E-mail: jmse@mdpi.com

www.mdpi.com/journal/jmse

9 783039 285624